Engineering Tools and Solutions for Sustainable Transportation Planning

Hermann Knoflacher
Vienna University of Technology, Austria

Ebru V. Ocalir-Akunal
Gazi University, Turkey

A volume in the Advances in Civil and Industrial Engineering (ACIE) Book Series

IGI GLOBAL
DISSEMINATOR OF KNOWLEDGE

www.igi-global.com

Published in the United States of America by
IGI Global
Engineering Science Reference (an imprint of IGI Global)
701 E. Chocolate Avenue
Hershey PA, USA 17033
Tel: 717-533-8845
Fax: 717-533-8661
E-mail: cust@igi-global.com
Web site: http://www.igi-global.com

Library of Congress Cataloging-in-Publication Data

Names: Knoflacher, Hermann., editor. | Ocalir-Akunal, Ebru, 1972- editor.
Title: Engineering tools and solutions for sustainable transportation planning / Hermann Knoflacher and Ebru V. Ocalir-Akunal, editors.
Description: Hershey PA : Engineering Science Reference, [2017] | Includes bibliographical references.
Identifiers: LCCN 2016054451| ISBN 9781522521167 (hardcover) | ISBN 9781522521174 (ebook)
Subjects: LCSH: Transportation engineering. | Transportation--Planning.
Classification: LCC TA1145 .E565 2017 | DDC 388.01/1--dc23 LC record available at https://lccn.loc.gov/2016054451

This book is published in the IGI Global book series Advances in Civil and Industrial Engineering (ACIE) (ISSN: 2326-6139; eISSN: 2326-6155)

British Cataloguing in Publication Data
A Cataloguing in Publication record for this book is available from the British Library.

For electronic access to this publication, please contact: eresources@igi-global.com.

Advances in Civil and Industrial Engineering (ACIE) Book Series

Ioan Constantin Dima
University Valahia of Târgovişte, Romania

ISSN:2326-6139
EISSN:2326-6155

Mission

Private and public sector infrastructures begin to age, or require change in the face of developing technologies, the fields of civil and industrial engineering have become increasingly important as a method to mitigate and manage these changes. As governments and the public at large begin to grapple with climate change and growing populations, civil engineering has become more interdisciplinary and the need for publications that discuss the rapid changes and advancements in the field have become more in-demand. Additionally, private corporations and companies are facing similar changes and challenges, with the pressure for new and innovative methods being placed on those involved in industrial engineering.

The **Advances in Civil and Industrial Engineering (ACIE) Book Series** aims to present research and methodology that will provide solutions and discussions to meet such needs. The latest methodologies, applications, tools, and analysis will be published through the books included in **ACIE** in order to keep the available research in civil and industrial engineering as current and timely as possible.

Coverage

- Construction Engineering
- Materials Management
- Production Planning and Control
- Structural Engineering
- Coastal Engineering
- Optimization Techniques
- Transportation Engineering
- Quality Engineering
- Operations Research
- Ergonomics

IGI Global is currently accepting manuscripts for publication within this series. To submit a proposal for a volume in this series, please contact our Acquisition Editors at Acquisitions@igi-global.com or visit: http://www.igi-global.com/publish/.

The Advances in Civil and Industrial Engineering (ACIE) Book Series (ISSN 2326-6139) is published by IGI Global, 701 E. Chocolate Avenue, Hershey, PA 17033-1240, USA, www.igi-global.com. This series is composed of titles available for purchase individually; each title is edited to be contextually exclusive from any other title within the series. For pricing and ordering information please visit http://www.igi-global.com/book-series/advances-civil-industrial-engineering/73673. Postmaster: Send all address changes to above address.

Titles in this Series

For a list of additional titles in this series, please visit: www.igi-global.com

Modeling and Simulation Techniques in Structural Engineering
Pijush Samui (National Institute of Technology Patna, India) Subrata Chakraborty (Indian Institute of Engineering Science and Technology (IIEST), Shibpur, India) and Dookie Kim (Kunsan National University, South Korea)
Engineering Science Reference • copyright 2017 • 524pp • H/C (ISBN: 9781522505884) • US $220.00 (our price)

Computational Modeling of Masonry Structures Using the Discrete Element Method
Vasilis Sarhosis (Newcastle University, UK) Katalin Bagi (Budapest University of Technology and Economics, Hungary) José V. Lemos (National Laboratory for Civil Engineering, Portugal) and Gabriele Milani (Technical University in Milan, Italy)
Engineering Science Reference • copyright 2016 • 505pp • H/C (ISBN: 9781522502319) • US $210.00 (our price)

Handbook of Research on Emerging Innovations in Rail Transportation Engineering
B. Umesh Rai (Chennai Metro Rail Limited, India)
Engineering Science Reference • copyright 2016 • 664pp • H/C (ISBN: 9781522500841) • US $235.00 (our price)

Advanced Research on Nanotechnology for Civil Engineering Applications
Anwar Khitab (Mirpur University of Science and Technology, Pakistan) and Waqas Anwar (Mirpur University of Science and Technology, Pakistan)
Engineering Science Reference • copyright 2016 • 339pp • H/C (ISBN: 9781522503446) • US $195.00 (our price)

Emerging Challenges and Opportunities of High Speed Rail Development on Business and Society
Raj Selladurai (Indiana University Northwest, USA) Peggy Daniels Lee (Kelley School of Business, Indiana University, USA) and George VandeWerken (Providence Bank, USA)
Engineering Science Reference • copyright 2016 • 289pp • H/C (ISBN: 9781522501022) • US $210.00 (our price)

Advanced Manufacturing Techniques Using Laser Material Processing
Esther Titilayo Akinlabi (Univeristy of Johannesburg, South Africa) Rasheedat Modupe Mahamood (University of Johannesburg, South Africa & University of Ilorin, Nigeria) and Stephen Akinwale Akinlabi (University of Johannesburg, South Africa)
Engineering Science Reference • copyright 2016 • 288pp • H/C (ISBN: 9781522503293) • US $175.00 (our price)

Handbook of Research on Applied E-Learning in Engineering and Architecture Education
David Fonseca (La Salle Campus Barcelona, Universitat Ramon Llull, Spain) and Ernest Redondo (Universitat Politècnica de Catalunya, BarcelonaTech, Spain)
Engineering Science Reference • copyright 2016 • 569pp • H/C (ISBN: 9781466688032) • US $310.00 (our price)

IGI GLOBAL
DISSEMINATOR OF KNOWLEDGE

www.igi-global.com

701 E. Chocolate Ave., Hershey, PA 17033
Order online at www.igi-global.com or call 717-533-8845 x100
To place a standing order for titles released in this series, contact: cust@igi-global.com
Mon-Fri 8:00 am - 5:00 pm (est) or fax 24 hours a day 717-533-8661

Editorial Advisory Board

Table of Contents

Detailed Table of Contents

On the contrary to other fields of human progress, transport is creating more problems with increasing wealth. Congestion, traffic victims, noise and air pollution are direct unwanted effects, not been solved so far. The cause is the lack of system understanding in the professional world. the existing paradigm in Transport is based on assumptions and extrapolations and not on solid science. The new scientific based paradigm, need also new tools for the sustainable future of transport and cities.

Due to decreasing resources, living in urban regions focus on sustainability in many aspects, including transportation. Sustainable transportation encourages non-motorized modes of walking and cycling as well as public transit (which also relies on walking while accessing a station), as well. However, walking as a mode is still a big mystery itself that needs further attention and research effort especially in the evaluation part. So far, the planners have discussed the concepts of walking and walkability, while engineers have mostly focused on Pedestrian Level of Service (PLOS). The scope of the problem is reflected in the diversity, and consequent inconsistency, in the available PLOS methods, which is one of the problems addressed in this chapter. The second and the bigger problem is the gap between the planning and engineering approaches in evaluating PLOS and walkability producing no consensus or clear relationship between the two, even though they overlap greatly.

Cycling is increasingly being implemented as a viable option for sustainable mobility in transport master plans from the local to the regional and even national level. However, when the implementation of actual measures is nearing the final stage, often mental barriers of decision-makers surface. These mental barriers refer to the resistance to innovation adoption, a phenomenon well documented in the social sciences. In our policy debate we use a list of factors that dissuade people from cycling from the literature to derive a classification of mental barriers in legislation, planning and infrastructure. Mental barriers have an influence on out-turn cycling levels and we provide exemplary evidence from Austria. The nature of these barriers, including the 'vicious cycle of mental barrier construction', is explored, before identifying two possible approaches to reduce, or even eliminate mental barriers. These approaches are: social pressure upon and self-experience by decision-makers.

Public transport is one of the main actors of sustainable transportation. To make an increase in the use of public transport systems and a decrease in private car usage are one of the main purposes of decision makers in many countries. Usage of public transport system has a strong relation with the quality of service. Service quality includes many parameters such as comfort, frequency, information system etc. Fare is another important parameter on demand for public transport systems. All of these parameters are varied by users' perceptions. This study focuses on public transport systems of Denizli, Turkey and investigates users' perceptions. Bus and paratransit (minibus) modes are regarded in the study. The public survey is performed for these modes. In the surveys, different questions have been asked to participants. Answer of each question is evaluated separately and distribution of each question for gender, ages and jobs are analyzed. Results obtained are summarized and discussed.

Having been used extensively since the 1980s, the concept of social exclusion has given a new impetus to the discussions of poverty and disadvantageousness. The concept of social exclusion, which can be defined as the condition in which certain individuals or social groups cannot integrate into the society either socially or economically or politically, leads to social cohesion problems. This triggers segregation of the society, particularly in large cities, in both social and spatial terms. In the context of accessibility, "urban transport" appears as one of the most important factors determining level of social inclusion/exclusion of the individuals or groups. In this study relation between social exclusion and transport has been evaluated as an attempt to identify socio-spatial segregation pattern of Ankara.

The chapter will provide empirical evidence how parking organization (location, number and pricing policies of parking spaces in urban areas) influences the mode choice of transport system users. Following from that, the presently applied practice of parking space modelling within existing transport models will be presented and discussed critically. It will be shown which influence on transport demand the different parking space instruments have in existing 4-stage models (FSM) and where flaws and errors in the transport demand and mode choice estimation can occur. Finally, potential improvements how to overcome these afore mentioned shortcomings are presented and the impacts on future transport planning and transport policy formulation towards sustainability will be introduced.

New development areas impose higher parking supply than needed, and these excessive parking requirements impose the rise of car usage in cities. In this study, it is aimed to analyze the relationship between development patterns of residential areas and supply of parking spaces by focusing gated communities in Ankara. The results of the study revealed that the number of garage parking spaces is positively related to the developments of gated communities, and the ratio of having private open parking space is lesser for the non-estate housing. The results of the analysis also revealed that the ratio of having a garage is higher for new buildings, and house prices are affected significantly by the presence of both garage and open parking spaces. In the study, it is interesting to note the rise of number of road motor vehicles in Turkey after the developments of gated communities started to emerge in the 1990s. Also, the results of the study revealed that there is a positive relationship between the emergence of gated communities and the rise in the number of motor vehicles in Ankara.

Provision of parking space is a constant challenge in urban transport planning because streetscape is a limited resource and therefore highly contested between different uses. In Austria, building regulation competence is subsidiary appointed from the national level to nine provincial parliaments. This leads to nine different parking ordinances. Whereas historically all ordinances have been focusing only on car parking by means of rigid regulations, only recently selected provincial parliaments added detailed bicycle ordinances their building codes. After a detailed overview of the Austrian situation and selected international innovations, the authors identify long-run goals that parking organization in cities should aim at and suggest an improvement of parking ordinances so that city planners and city administrations will have at hand appropriate design tools for mobility regime improvement.

Traditional transport practice and science call motorways as safe roads since the accident rate (accidents per 1 million vehicle kilometres) is lower than on ordinary roads. This is based on the scientific inadmissible comparison. Similar, but not comparable things, are compared: the accident situation of two different kinds of transport systems, which are basically different not only in its construction elements, but also in operation condition, their environments and the composition of their users. All of them are different. To compare, comparability is a necessary precondition. If we take this into consideration we can recognize, that the existing contradiction between accident-research-results and physics can be removed. Systems with a higher speed have a higher risk. But these facts are continuously neglected by traditional educated practitioners: they cling to their myths. One of them is "Since the number of accidents per kilometer driven on motorways is less than on other roads, motorways are safer roads."

This chapter presents a multi-modal method for the assessment of highway performance. It is derived by extending a traditional assessment concept step by step taking into account the capacity and quality of different modes on the road as well as in the corridor. It defines an appropriate performance target and explains why a multi-modal concept has to consider transport demand in persons and goods rather than traffic volumes in vehicle units. It is shown that the concept allows for different options and measures to improve traffic quality and so supports the efficient use of existing infrastructure and the effective allocation of limited funds.

In this study, ambulance location problem, which is a significant research field in transportation in healthcare, is focused on. It is assumed that travelling time of vehicles between two demand points is uncertain and this uncertainty is represented as a crisp and as well as convex set. In order to deal with this uncertainty, robust counterpart approach, which is capable of providing conservatism, is employed. According to results of the simulation process, which is designed for the sake of benchmarking robust and nominal solutions, it can be easily claimed that conservative robustness is able to be achieved with significantly sufferable costs, even with no costs in most of the scenarios, by the proposed model. Ultimately, since robustness is one of the most important keywords for sustainability, on which discussions can be found in several parts of the text, this chapter is concluded with the claim that, by the proposed model, sustainability of transportation in healthcare is enhanced based on the adopted problem.

Logistics in transport considers the physical structure of a city as given and tries to optimize the performance of companies or the transportation system within the given conditions. On the other side, companies choose their location under the given conditions and the expected changes and influence the structure of the city and its economy in a continuous way. City and transport planning methods have not considered these effects in their work so far and are therefore influenced by the driving forces of the economy of scale and demand oriented traffic growth. The introduction of principles of logistics into the early stages of land use and city planning would change the "given conditions" and open the path for a more sustainable development, with more pressure for innovation and fairness in the market.

Foreword

I still remember my first stroll down the Kärntnerstrasse from the central square of Vienna south to the opera building along the ring road built on the site of the former city wall. It was 1962; the street was filled with cars, it was noisy, and the air had a bad smell. One could not safely venture from the narrow sidewalks, because well over 100,000 cars were vying for space on the streets of the first district every day.

I visited Vienna again for several days in September 2016. What a difference! Pedestrians have taken back the Kärntner Strasse. No cars were to be seen on the main street, the only noise was from thousands of happy people visiting the shops and enjoying the clear views. I could wander freely without worrying about getting struck by a car.

This change occurred during a period when personal use of cars globally was growing explosively. Gasoline and diesel consumption today is five times its level when I first was in Vienna. How was that possible?

In the late 1960s a brash young professor of transportation at the Technical University of Austria began to develop a plan for giving the city center back to the people. At first the shop owners were aghast. Certainly prohibiting most personal cars in the first district would drastically reduce their sales. That professor became known as the "enemy of the city center." But the plan was introduced anyway. Kärntner Strasse was converted to pedestrian zone in 1974. It has been a phenomenal success. That pedestrian zone has become one of the most important shopping areas in Europe.

Since then there have been successful efforts in Vienna to extend the car-free zone, move car parking to the periphery, add bicycle lanes, slow down traffic, and strengthen public transportation options. And today Vienna is regularly judged to be among the most livable cities on the planet.

Similar efforts to rethink and revise mobility have produced excellent results in other cities. In 1974 Innsbruck shop owners initially resisted proposals to restrict cars from passing and parking in front of their shops. But their sale receipts increased by 30% as a result of the change, and they were soon paying for studies to extend the car-free zone. In general reducing the cars on a street results in tripling the numbers of people who use it.

But doing this contradicts traditional thinking about the role of a traffic engineer. It requires a new paradigm - one which recognizes that raising the number of motor vehicles, reducing the distance to parking, and increasing the speed of cars do not necessarily bring increased convenience and mobility.

When motorized transport began its phenomenal growth after world War II settlements were smaller, populations lower, land more empty. Perhaps building more roads and parking areas then did increase personal mobility and the quality of life. But now urban areas have spread widely, populations have grown, land is sought for many competing uses, traffic congestion and the constant assault of noise and emissions from vehicles have deteriorated the living environment for many. A new strategy is required.

This edited compendium offers theory, research, and case studies as the bases for developing new approaches for facilitating the movement of goods and people.

Conventional transportation projects will remain the dominant mode; they are not yet fully obsolete. Vienna is a special community and policies that work in its city center are not universally relevant. There still are circumstances that require extensions of the traditional road system. Austria has not fully relinquished the pursuit of new road tunnels, more highways, faster thoroughfares. But now a new paradigm for mobility is emerging and readers of this book will be informed and stimulated to imagine alternatives to the conventional wisdom. The book's many concrete case examples prove that alternatives can be effective.

Paradigm change takes a very long time when it involves massive, long-lived capital investments. Not only those making fortunes off of traditional highway construction must shift their activities. But government bureaucrats have to change their habits; universities need to revise their curricula, parliaments need to design new laws, car manufacturers and oil companies must evolve new business models, professionals require training in the new modes of analysis, society requires new indicators of success to guide its investment in transport system design and construction.

But even a very long journey requires first steps. And this book describes some of those steps.

There are positive reasons for making the change. This book illustrates many win-win policies. When traffic congestion goes down, quality of life goes up, safety improves, costs go down, and commercial enterprises prosper.

But negative reasons are looming that also make it essential for transportation engineers to accelerate the change. Two developments completely outside their control presage major changes in society's reliance on personal cars and trucks for moving people and goods.

First, the negative consequences of climate change will inevitably grow over the coming decades. CO2 from fossil fuel combustion constitutes about two-thirds of the globe's annual greenhouse gas emissions. As concern about climate change rises, there will be growing efforts to limit the use of fossil fuels. Those fuels provide almost 100% of the energy to power vehicles. Climate change will force revisions in traditional transport policies.

Second, every year since 1984 the global use of known petroleum reserves has exceeded global discovery of new oil deposits. Thus fossil fuels production will soon decline as a result of depletion. Some estimates show global oil use falling 50% by 2035. Scarcity will reinforce efforts to restrict the transportation sector's use of these valuable products.

The need to change is urgent. Those two forces - climate disruptions and oil depletion - will begin having major influence on transport systems by 2025 - reinforcing the need for policies that do reduce drastically the demand for fossil fuels. The earlier society can begin adapting its policies and infrastructure to this new reality, the less severe will be the disruptions.

The authors of this book have chosen to impose certain limits on their narrative. The materials they present relate mostly to Austria and to Turkey. Many new transportation strategies in the poorest countries do not appear here. Nor is there any effort to represent directly the thinking and achievements of the many hundreds of innovative transport engineers around the world. Those limits in the scope of the book give space for the text to diversify its coverage of studies and projects relevant to two societies at an intermediate and advanced stage of economic development. Despite the geographical restrictions of the text, the discussion is not parochial.

Dennis Meadows

Preface

Traffic flow problems are only a symptom of deeper underlying causes: Transport and urban structures are optimized not as a system for the benefits of the society in a sustainable way but optimized for individual optimization for car users.

The basis of scientific understanding has traditionally been constructed from the causal relations that exist between the components of a system at a particular time. These relations are used to construct a representation of the system which can be run forward to provide predictions. In economic geography, as well as transportation and land-use, models that are used operationally are still based on equilibrium assumptions, in which the system is supposed to run to equilibrium. The correspondence between the real object and that represented by the model is made through equilibrium relations between the variables. Locations of jobs and residences, land values and traffic flows are assumed to reach equilibrium if some policy or planning action is followed. Such an approach fails to take into account the possibility of processes where growth encourages growth and decline leads to further decline (Allen, 1998, pp. 97-98). In physics and chemistry the predictive models work well. The individual elements that constitute the system obey fixed laws which govern their behaviors. In these disciplines, the mechanisms are fixed and the molecules do not learn. However, for living systems, the situation is completely different. People, for instance, can switch sources of energy on or off and can respond, react, learn and change according to their individual experience and personality (Allen, 1998, p. 101).

The evolution of a city can seldom be considered as a spontaneous self-organization process, since it is artificially driven towards an ideal model. This kind of a guided evolution however, rarely goes as planned. The unpredictable events and learnt behavior result in a development pattern, which is different from that predicted (Butera, 1998, p. 225). The solutions to the problems of cities are in fact hidden within their inner dynamics.

Most problems result from the extension of city beyond the optimal limit determined by the system's organizational rules. That can happen by the emergence of built-up areas on the outskirts of the urban area, either through the growth of small existing settlements or through the new planned developments. The growing separation between constituent parts will have problems of development or even survival (Mossello, 1998, p. 190).

Today's transport system conflicts with the principles of life and evolution. The scale of this system became car, not human any more. The new spatial structure has developed in accordance with car although it is not a sustainable transport mode. It is built in contradiction to the nature and not in agreement with it – creating conflict with everything outside of the transport system – accidents, noise, air pollution, particles, neighborhood separation etc. and also within the system itself – congestion, accidents, cost etc. These are all indicators of an unsustainable system. In order to create sustainable environments, while designing transport structures, system behavior should be taken into consideration. Parking place organization is the key factor to shift to sustainable urban environments.

The editors of this book try to draw attentions to an ironic conflict between traditional methods in traffic engineering and transport planning and real system behavior. The solutions for many complex problems are sometimes hidden in systems' behavior, which we simply have the information inside our evolutionary mechanisms.

The book is organized into fifteen chapters. A brief description of each of the chapters follows:

Chapter 1 identifies the role of engineers and their tools in the transport sector after paradigm change. The chapter sets the scene for a discussion of assumptions and extrapolations to science. In particular the chapter gives an introduction into the issue, the history of the transport system, the problems to cope with the needs of the new system, the dynamic and overlapping of inventions in technology and operation till today. It also identifies the adverse effects, unwanted and unplanned by transport engineers and the struggle to find new supporting hypotheses, when the core assumptions became doubtful. A comparison between traditional methods and interdisciplinary methods with careful analysis of human and system behavior in traffic engineering and transport planning is underlined.

Chapter 2 presents an evaluation of walkability and pedestrian level of service. The authors of this chapter underline different points of views from planners' and engineers' sides to pedestrians and they point out diversity, and consequent inconsistency as a problem. The gap between the planning and the engineering approaches in evaluating PLOS (Pedestrian Level of Service) and walkability is also discussed.

Chapter 3 discusses mental barriers of decision makers in planning for cyclists with evidence from Austria. The authors explore the nature of these barriers, including the 'vicious cycle of mental barrier construction', and identify possible approaches to reduce, or even eliminate mental barriers. The authors use a list of factors that dissuade people from cycling from the literature to derive a classification of mental barriers in legislation, planning and infrastructure. Exemplary evidence from Austria is given.

Chapter 4 presents a comparative analysis of public transport users' perception targeting sustainable transportation. The authors argue that parameters on demand for public transport systems are varied by users' perception. Based on a survey, the authors investigate public transport users' perceptions in a case study in Denizli, Turkey. They further summarize and discuss the survey's results.

Chapter 5 addresses an assessment relationship between social exclusion and transportation. The chapter reviews the concept of social exclusion which has given impetus to the discussions of poverty and disadvantageousness since the 1980's. The authors investigate the relationship between social exclusion and transport has been evaluated as an attempt to identify socio-spatial segregation pattern of Ankara, Turkey.

Chapter 6 introduces equidistance concept with evidence of the influence of parking organization on mode choice. The authors provide empirical evidence to how parking organization influences the mode choice of transport system users. An overview of 4-stage models helps discuss flaws and errors in the transport demand and mode choice estimation. The chapter presents some transport policy formulations towards sustainability.

Chapter 7 analyses effects on car ownership rates resulting from increased parking lots in residential areas in the case of gated communities in Ankara. The author contends to analyze the relationship between development patterns of residential areas and supply of parking spaces by focusing on gated communities. The results of the study reveal some relationships between ratio of having garages and housing prices.

Chapter 8 discusses regional parking ordinances in Austria. The authors of this chapter contend that provision of parking space is a constant challenge in urban transport planning because streetscape is a limited resource and therefore highly contested between different uses. The chapter reviews the differences in parking regulations in nine different provinces of Austria. The authors then identify long-run goals that parking organization in cities should reach.

Chapter 9 presents a discussion for sustainable parking planning for smart growth of metropolitan cities with a comparison of parking plans of Ankara and Vienna. The authors draw a critical perspective for the parking planning policies for these city cases. The results of the comparison highlight similarities and differences in parking planning processes of the two cities.

Chapter 10 reviews the alternatives of transport network design of a metropolitan city using hierarchical cluster analysis in a case study of Istanbul's 2023 plan networks.

In this chapter the author develops a methodology to model urban structure (identifying and classifying centers and subcenters) based on urban travel data (interzonal urban passenger flows via urban rail modes during morning peak-hour) was developed using hierarchical cluster analysis. The travel patterns at the metropolitan level are analyzed.

Chapter 11 presents a comparison of transport modes in terms of energy consumption. The authors make a comparison of transport modes in terms of energy consumption is given for two chosen routes in the case study of Ankara. The chapter addresses the differences with respect to chosen travel mode in terms of energy use and travel cost reflected to traveler for certain OD couples.

Chapter 12 presents the effect of a motorway on development of accidents in a big city. The author contends that transport systems with higher speed have a higher accident risk than the others. Some empirical evidence is presented from Vienna. The author lists the necessary conditions to have a more responsible transport planning.

Chapter 13 presents a multi-modal method for the assessment of highway performance. The author defines an appropriate performance target and explains why a multi-modal concept has to consider transport demand in persons and goods rather than traffic volumes in vehicle units.

Chapter 14 presents a specific issue on sustainability of transportation planning in an urban region in ambulance location problem, which is a significant research field in transportation in healthcare. The authors assume that travelling time of vehicles between two demand points is uncertain and this uncertainty is represented as a crisp and as well as convex set and in order to deal with this uncertainty they developed a robust counterpart approach.

Chapter 15 discusses logistics for city and regional planning. The author criticizes urban and regional planning discipline for not taking into account effects of transport logistics. The chapter concludes that the introduction of principles of logistics into the early stages of land use and city planning would change the "given conditions" and open the path for a more sustainable development, with more pressure for innovation and fairness in the market.

REFERENCES

Allen, P. (1998). Cities as Self-Organising Complex Systems. In C. S. Bertuglia, G. Bianchi, & A. Mela (Eds.), *The City and Its Sciences*. Physica Verlag. doi:10.1007/978-3-642-95929-5_2

Butera, F. M. (1998). Urban Development as a Guided Self-Organisation Process. In C. S. Bertuglia, G. Bianchi, & A. Mela (Eds.), *The City and Its Sciences*. Physica Verlag. doi:10.1007/978-3-642-95929-5_6

Mossello, M. T. (1998). The Possibilities and Limits of Self-Organisation. In C. S. Bertuglia, G. Bianchi, & A. Mela (Eds.), *The City and Its Sciences*. Physica Verlag. doi:10.1007/978-3-642-95929-5_4

Chapter 1
The Role of Engineers and Their Tools in the Transport Sector after Paradigm Change:
From Assumptions and Extrapolations to Science

Hermann Knoflacher
Vienna University of Technology, Austria

ABSTRACT

On the contrary to other fields of human progress, transport is creating more problems with increasing wealth. Congestion, traffic victims, noise and air pollution are direct unwanted effects, not been solved so far. The cause is the lack of system understanding in the professional world. the existing paradigm in Transport is based on assumptions and extrapolations and not on solid science. The new scientific based paradigm, need also new tools for the sustainable future of transport and cities.

INTRODUCTION

Engineering tools in transport came from craftsmanship experiences to handle a great variety of materials such as wood, water, stone, earthwork, iron and steel till the 19th century. When railway came in use, new tools had to be developed and implemented to provide the society with a reliable and safe "modern transport system". Responsibility for the operation of the new system became an engineering issue not only for the infrastructure and the vehicles, but also for the customers and the environment. Railways had strong effects on urban, industrial and social structures. But the interrelationship was not in the awareness of transport engineers. Their tools were focused on vehicles, all kinds of infrastructure, provision of energy and a smooth operation of the system. With the mass motorization these tools, dealing with physics, traffic flow and speed of cars were applied to this new and fascinating individual mode. Core hypotheses of this Paradigm were "Growth of Mobility", "Travel Time Saving by Speed" and "Freedom of Modal Choice", leading thinking and acting of transport engineers. Forecast models, Cost Benefit

DOI: 10.4018/978-1-5225-2116-7.ch001

Analysis and a big set of Standards were developed to build and operate the Transport System for the motorized society. But the interventions, based on this Paradigm, produced a lot of adverse effects:

- Traffic accidents,
- Air pollution,
- Damage of local business,
- Urban sprawl,
- Climate change,
- Corporate domination, and
- Unemployment.

A critical analysis of the foundation of this paradigm was necessary – and possible, by using methods from other disciplines, like system dynamics or evolution theory to understand not only the behavior of the transport system, but also the behavior of people, especially the gang of people in the fields of transport. The outcome of this research was a scientific sound background for the Paradigm Change, with core hypotheses in accordance with the real behavior of the transport system of today. Using this tools, adverse effects of the old paradigm and contradiction between the promises of transport engineers and the real effects can be prevented. Paradigm Change is not a superficial new buzzword, as someone might hope, it change the understanding of terms, it change the unit of scale from car to man, it change the borders of responsibility of transport engineers by widening them into other disciplines and force for other solutions.

This chapter gives an introduction into the issue, the history of the transport system, the problems to cope with the needs of the new system, the dynamic and overlapping of inventions in technology and operation till today. The adverse effects, unwanted and unplanned by transport engineers and the struggle to find new supporting hypotheses, when the core assumptions became doubtful are then described.

Traditional assumptions, which are the core hypotheses of the traditional paradigm, are confronted with the empirical facts of the system and the reality of human behavior in this artificial environment. The new principles and tools are briefly described with references to existing textbooks. Since the new paradigm is already in use in some cities, some practical examples are added, as well as the critical questions from representatives from the "old Paradigm collected over the last decades. (The author himself was an excellent representative of this Paradigm 45 years ago. Otherwise he would never become Professor at the Technical University) Paradigm Change has an "incubation time" of decades or even generations and is not an issue of understanding the new hypotheses only. Between understanding and acting, the mental environment must change also. And this is the case today. There are so many seekers in the professional world and even more in the society, to find a way out of the problems, we as transport engineers have created with the old Paradigm.

The chapter is subdivided into

- History and genesis of the traditional paradigm and the tools of transport engineers.
- Contradictions between promises, forecast and reality of the system.
- Research results about system behavior and real people behavior – the foundation of the new paradigm.
- The new principles and tools.
- Can the new paradigm meet the challenge?

- Conclusions and Future Research directions.

10,000 YEARS EXPERIENCE WITH A SUSTAINABLE TRANSPORT SYSTEM

"Without knowledge of the past there is no future": this popular wisdom is not only valid for the financial sector, as cited in the Civis Mundi magazine for political philosophy and culture 2012. It is important for all fields of practice and science, particularly for the fields dealing with new innovations and new dimension of depth of engagement into the natural and social environment. The transport sector was a challenge for technicians and engineers since millennia. More than 10,000 years of engineering experience for sustainable transport systems in urban and rural societies existed until the 18th century. Boats, ships, harbors, canals, bridges, roads, the wheels, vehicles, elevators, water management systems, multimodal hubs and many others were built and operated in coherence with the local and global environment, within the limits of available solar energy. The burden of transport and work in rural and urban environments was eased by countless technical innovations developed, tested in practice and improved over the time. Historical cities are one of these products. Cities are complex living structures of relationships of societies, built environment, variety of functions and form, safe and origins of art and urban culture.

A NEW CHALLENGE FOR TRANSPORT ENGINEERS

19th Century: Abundant and Cheap Energy for the Transport System

Invention of transport system on rails – controlled by the society – responsibility for the safety and reliability of a system of big mass and high speed but without an inbuilt brain to cope with these new dimensions. Engineering tools from design and safety standards to operation manuals, knowledge of new materials for a new technology, thinking in networks etc. had to be developed, tested and optimized. Railway engineering became a special discipline in universities for technology. Engineering innovations in the transport sector had effects on many other disciplines from law to economics and political science. Railway acts specified the frame of privileges and obligations of railway engineering and railway operation. Indirect effects on land use, economy, social system and environment occurred, but were not perceived in the context of "engineer´s responsibility". Planning tools were developed under these conditions. Safety became an important issue in railway operation, since the new traffic mode had the power to move big masses with high speed, but not the brain to assure an accident free operation. Human brain and senses were not developed for such high speeds. So the railway engineers hat to develop the "artificial brain" for the railway system to ensure safe and reliable transport. This "brain" comprises physical structures, like signals, railway crossing, switches, stations etc. rules, contracts between railway operators and railway users, special regulations, safety equipments and many others. Many thousands of well trained and responsible employees were needed to compensate the missing brain of this new, artificial transport mode. Sustainability, or better durability and reliability, was important for the vehicles, the railway infrastructure and economic viability but not in the systemic context of today.

20th Century: How to Cope with High Speed of Cars and Mass-Motorization?

The United Kingdom Stage Carriage Act 1832 first introduced the offense of endangering the safety of a passenger or person by 'furious driving'. The first numeric speed limits were created in the UK by a series of Locomotive Acts (1861, 1865 and 1878); the 1861 Act introduced a UK speed limit of 10 mph (16 km/h) on open roads in town, reduced to 2 mph (3 km/h) in towns and 4 mph (6 km/h) in rural areas by the 1865 'red flag act'. The Locomotive on Highway Act 1896 which raised the speed limit to 14 mph (23 km/h) (being the estimated speed of a horse being driven 'furiously') is celebrated to this day by the annual London to Brighton Veteran Car Run..

The first person to be convicted of speeding is believed to be Walter Arnold of East Peckham, Kent, who on 28 January 1896 was fined for speeding at 8 mph (13 km/h). He was fined 1 shilling plus costs. (Wikipedia, https://en.wikipedia.org/wiki/Speed_limit)

2 mph in urban and 4 mph on "open country" for motorized vehicles were not a big challenge for road traffic engineers of the 19th century. The situation changed when higher speed limits were permitted under the pressure of rich and influential people from the automobile clubs; 30 mph in urban areas after 1934 and in principle no speed limit on rural roads. New design standards had to be elaborated, partly on analogy to experiences with existing roads, elementary physics and geometry. The person-car-unit became the scale unit for transport and urban planning.

At this time scientific knowledge about this new, fast developing high speed transport system for individual use, was not available. Engineers and planners used plausible assumptions based on personal experience and extrapolated their experience to the system. The "system" for engineers was restricted to the carriageway, the infrastructure, and the vehicles they had to built operate and maintain for the state or private companies. But when the number of cars increased, new problems occurred: How to manage the transport flow and the fast increasing number of vehicles in the streets? The field of Traffic Engineering became a discipline focused on car traffic flow. (Highway Capacity Manual, 1965) Highways and Expressways were planned in USA and copies as Motorways (in analogy to railways) in Europe. New terms were invented or "imported" from other disciplines. All other modes from the past disappeared from the screen of transport engineers in the 1960's of the last century. The way into the future was the motorway. Transport engineers were educated and trained to design, build, maintain this new transport system, which has so convincing advantages – for everybody. Pedestrians, cyclists and public transport modes were out of fashion and no option for the future of the 1960's.

The Age of mythology in transport science began: The term "mobility" was adopted from social science (Sorokin and modified into "car trips". Increasing car ownership – "growth of mobility" – became a pillar of traditional transport planning. "Speed saves time", is an obvious and objective fact on the individual level.) Cost-benefit analyses as a tool for decision support are based mainly on the benefits from time-savings by faster transport systems and it became the second pillar. Accessibility in combination with free modal choice is the third one. A new working method was established in the engineering field, implemented and casted in regulations and standards to cope with the growing demand for carriageways and space for cars, to satisfy the needs of the society, economy or political decisions. Behind these activities in transport engineering, new structures of thinking and common sense represented subconsciously a paradigm for the disciplines in transport. A common sense spread in engineering, transport economy, transport policy, land use and urban planning reinforcing each other, that "mobility" is growing, high speed is better than low speed an must given priority in traffic and transport and the more cars the people have, the better their lives are (pfister, 1994). Based on the paradigm, which determine terminology, indicators, values, professional world view, engineers, economists, land-use and urban planners built the new environment in cities, villages, settlements in the last two centuries. Their intention was the creation of undisturbed free flow of vehicles, especially cars, between every origin and every destination for a better economy, better life and accessibility for the future. Every place of human activities was provided with enough parking places due to the regulations of the building code. Guidelines for road design became the tools for transport, land use and urban planning. Traffic safety, quality of traffic flow (level of service), and parking provision was the aim of planning. Sustainability in the context of ecology, economy and sociology was not a common goal. It was an external issue and not a transport engineering responsibility.

TOOLS FOR THE TRANSPORT SECTOR OF THE 19TH CENTURY

In the 19th century railways were planned and built on engineering experience and knowledge about physics, thermodynamics, electricity, geology within a framework of legal contracts and responsibilities. Transport safety standards were implemented and monitored. An environment with clearly defined system boarders and clearly defined tasks were well known by engineers since centuries. The goals and tasks for transport engineers were specified: fast and cheap access to all kinds of resources and strategic military needs; transporting armed forces fast and in big quantities where needed. For these purposes railways were given special privileges e.g. to use private land for their infrastructures. Transport engineers had to cope with high speeds for a system of big mass but without an adequate brain and information system. This had to be developed and implemented in and around the railways.. Railways are "closed systems" with controlled access are operated by well-trained personal under defined operational standards and defined decision chains of responsibility. Tools to design, build and operate railways were developed, implemented and optimized and became standards for the engineers and other professionals, working in this new transport field of high speed vehicles on rails. On urban and rural roads the speed was mainly the speed of pedestrians and horse-drawn vehicles. Traditional civil engineering skills and tools to build carriageways, bridges adjusted to the local terrestrial and material environment existed since centuries. Speed on the road system was not an issue at this time (Knoflacher, 2015), transport volume was limited, congestion became an issue from time to time on roads of greater cities. Traffic regulations were implemented, the first "traffic light" called semaphore was established at the intersection of George and Bridge Streets in the 1860's (Lance & McNeil, 1996) to separate conflicting traffic streams.

20th CENTURY: TOOLS ADAPTED TO THE NEEDS OF CAR TRAFFIC

Standards and Legal Tools

The term "traffic" became a synonym for car traffic. The car became the unit of scale for transport engineers, land use and urban planners. Architects, land use, urban and transport planners internalized the private car into their work subconsciously. Traffic engineers developed tools and methods to provide the society with sufficient convenient usable parking facilities. They made standards in accordance with the dimension of existing car size and adapted these standards according to the new car design and dimensions. Car design was made for profit of car companies; traffic engineers followed the input requirements of the car industry and agreed on standards about the number and the size of parking places.

A powerful legal tool is the road traffic act, in Austria was introduced in 1961(Straßenverkehrsordnung, 1961). This law was written to organize road traffic but was executed by traffic engineers and other professionals in transport. They all followed the thought of the traditional paradigm for the needs of car traffic. The far reaching consequences for all other road users were not recognized. Public transport had some few privileges, but pedestrians and cyclists are subordinated to the needs of free flow of cars and their need for space in public areas. It is not the law which formed the behavior of people and the society in the transport system, but it is much more the manner, how the law was read and translated into action by engineers, administrations and lawyers.

Tools to Cope with High Speed

Speed on roads, faster than walking, were only available for horse riders or horse-drawn carriages. Experience with speed was: it saves time. Like in railway planning, speed of the road transport system became the leading parameter for transport engineers – till today. "The faster the speed is, the better the transport systems", became the credo. How to shape the environment for the needs of car traffic? When the first carriageways were designed for cars, transport engineers copied principles from railways and adapted and simplified it for the needs of car traffic. The intention was to build roads for fast speed and high traffic volume. Curve linearity has to be designed in accordance to law of physics, assumptions about human behavior, characteristics of material of road surface and tires and casted into road design standards during the first half of the 20th century. Lane widths were standardized in accordance to traditional principles of civil engineering: "the wider, the better". European motorway lane width was "imported" from USA where 12 ft wide lanes were used and adapted to 3.75m standard lane width (more could not be wrong). The high design speed made the road alignment rather stiff and forced traffic engineers to build much more bridges and tunnels than ever before in the transport engineering history. Construction industry recognized their chance for a new and profitable business area – and access to tax-payers money. This again had an influence on the dimensions of roads through the machinery. This new physical and legal environment encouraged more and more people to buy and use the cars. Car owners had benefits on the roads, comfort on their trips, opportunities for new recreation and holiday destinations. These benefits were amplified when the service industry became aware the chances for its business in this new transport sector. The increasing number of accidents and traffic victims were not a problem for transport engineers, since the legal system put the blame in "human errors", the people were the "95%-cause" of accidents. There was – and is – no reason to doubt about the own role in the system and the effects of the structures planned, built and operated by transport engineers, lawyers, economists and decision makers. Each of them do the best, within each narrow system view. Transport engineers checked the quality of their work against the planning and operation standards released and made compulsory by ministries or professional associations. "Somebody had the wisdom to make them so perfect". Logical and doubtful thought has been done somewhere else. Engineers have not been trained to think in a philosophical way, but they are trained to act. And action was really necessary since fast and unexpected car traffic "growth" had to be tackled. An unexpected problem for transport engineers appeared, when car ownership increased: congestion. Congestion creates "loss of travel time" and has negative effects on the economy and also negative effect on the joy of driving cars. That is quiet opposite to what transport engineers wanted. Therefore it must be a really serious problem. The number of publications, research project and technical and technological suggestions has a nearly endless range from more roads over more management to more ITS (Intelligent Transport Systems). The self-confidence of traditional educated transport engineers is shocked, when regularly congestion "emerges from something" on their well designed infrastructures with abundant capacity. They concentrate their efforts to make car traffic flowing smoothly again. They develop tools handle the traffic volume under the conditions of unlimited growth.

Tools to Cope with Car Traffic Volume

Since the 1950s of the last century, the term "mobility", invented in social science[1], was modified into "driving cars". Traffic engineers had the obligation to provide the society with "mobility", which means "free flow of car traffic". Something has to be done to prevent the system from congestion. The Americans had the solution: they released the first edition of the "Highway Capacity Manual" in 1950. It was the first document to quantify the concept of capacity for transportation facilities. HCM became a kind of a Bible for transport engineers since then. HCM 1965 introduced by definition the concept of "Level of Service", which has "become the foundation for determining the adequacy of transportation facilities from the perspective of planning, design, and operation". In the HCM-Version 2000 (Manual, 2000) capacity is defined as:

The capacity of a facility is the maximum hourly rate at which persons or vehicles reasonably can be expected to traverse a point or a uniform section of a lane or roadway during a given time period under prevailing roadway, traffic, and control conditions. Vehicle capacity is the maximum number of vehicles that can pass a given point during a specified period under prevailing roadway, traffic, and control conditions. This assumes that there is no influence from downstream traffic operation, such as the backing up of traffic into the analysis point. Person capacity is the maximum number of persons that can pass a given point during a specified period under prevailing conditions. Person capacity is commonly used to evaluate public transit services, high-occupancy vehicle lanes, and pedestrian facilities.

If somebody expects that the last version of HCM will convey a system view, he will be disappointed, since the methods and principles are still related to the private car as the unit of the transport engineer´s universe. Traffic volume has to be operated under best possible conditions. Solutions have to be found between levels of service A (free flow under ideal conditions) and level D (dense traffic flow). Level E and F are qualifies as "unacceptable". E is traffic density at maximum capacity. Level F is density above maximum capacity flow, the condition where congestions occur. These principles determine the scale of values and priorities – the thinking of traditional transport engineers: speed and car traffic volume are the predetermined parameters for their work. A level of service between A and D for car traffic is their goal.

In summary transport tools of the (still) existing paradigm are:

- Society has the need for more mobility.
- Traffic demand is given from outside. (By the society, economy, etc.)
- Mobility means mobility of cars.
- Trips are generated in "cells" (small units of area with people, working places, recreation, school etc. inside) and "attracted" to the destinations, again "cells".
- Tools have been developed for car traffic flow.
- Use "acceptable" levels of service A – D.

- Tools are available to optimize speed of car traffic.
- Travel time saving in the system is the main goal in models for traffic planning
- Congestion is seen as a problem and has to be prevented.
- Sufficient road space has to be provided in time for future mobility needs.
- Standard unit is the person car unit, PCU. Other road users are converted into PCU´s.
- Parking is a task for urban planning and building code and traffic law.
- Cause of traffic accidents is the people (95% human errors).
- Environmental problems have to be solved by technical means within the Eigendynamic of the transport system.
- Pedestrian, cycle and public transport modes are part of the transport system but have to be treated in a separate way (with the same principles like car traffic).

Traffic engineers have used these tools to make for the best transport system for people and the economy since the 1950's. If problems appear, the therapy is: "more of the same". But the reality of the world is not following these expectations. There is more congestion, less liveable cities, less local employment, less local economy, more air pollution after more than 50 years of traditional traffic and transport engineering.

"UNINTENDED EFFECTS"

- Traffic accident,
- Noise,
- Air pollution,
- Awareness of dependency on imported energy,
- "Growth of congestion",
- Urban sprawl,
- Damage of local economy,
- Long distance commuters,
- Financial problems in municipalities and cities, and
- Decline of public transport.

The transport system consumes more space, energy and resources than ever before in human history. These are clear indicators of an unsustainable system. Unintended effects are indicators showing us that the system does not behave as expected or promised. The modern transport system is an artificial one: it is made by the tools and how they are used by engineers and planners. If these unintended effects pop up, there are two explanations:

1. People planning, building and operating the system do not understand the system.
2. The core hypotheses of the paradigm, which determine their work, are not in accordance with the real system behavior.

The first explanation is the common criticism from self-proclaimed "experts" from transport sciences and other disciplines and has mainly an idealistic or ideological – and sometimes also a competitive background. The profession of transport engineering, which is a rather new and not clearly defined discipline, is a favored playground for many kind of miracle healer. Such miracle healer are very popular by politicians who want to solve problems with promises instead of necessary actions.

If we have a scientific look on the HCM, we find a collection of assumptions, based on personal experience, a composition of different mathematical formula with variables and parameters, a (road) one dimensional cross-section view of traffic and tools how the transport system should be adapted to the future needs of car traffic. The new version of this handbook is extended also to other modes of transport. The methods are the same: collection and analysis of data plus plausible explanations. The tools are dealing with the symptoms of the system, but not with causes, which are outside of the engineering domain. Many open questions remain. The transport system seems to be a system which can be treated without external causes and effects? The goals for planning and design are internal: Level of service, speed etc.? Transport engineers are closer to car and building industry than to urban and land use or landscape planning, psychology or environmental science. Can transport be separated from all these fields? Is there no interdependency with all of these disciplines? Not everything which is published under "science" must be scientific.

Contradicting Practical Experience

Practical solutions in transport management and urban planning, such as traffic calming, pedestrian areas on former urban carriageways, red waves instead of green waves for car traffic, priority for public transport etc., based on ideas of some planners or local situations, which are contradicting traditional paradigm, have been implemented in many cities with success. Well-disposed professionals are collecting these examples and experiences to write articles and even books with promising titles with nice photos from nice solutions. Networks of professionals are working with the promise to naïve politicians "Make me an omlet but don't break any eggs". The EU-commission spent millions of Euro for "research" of "best practice". International teams are specialized for "Best-Practice-Picking" and make a living with training courses. The effect on the system is limited and the understanding of the system behavior even much more. But it has become a business-model for planners and miracle-healer-groups, replacing the position of transport engineers, by blaming them for what they have planned. This symptom treatment is mainly based on believe, decorations and ideology and not on a new paradigm. The tools for this "practice of anything goes" are nearly arbitrarily. Story-telling is only one of many tools of planning. But many story-telling books about transport and urban planning are successful on the market. The common message is the promise: If we copy the good examples, we will have the same wonderful results, and it is much easier, if we forget the sleeping dragon, which makes all the unwanted effects. "We can go heaven with our cars". This are really good news and worth to be supported.

This kind of transport experts are travelling with a great collection of selected examples from elsewhere, to show what has been done and what is possible. Many cities and governments like these kind of good-looking "expertise" which can be very helpful in the short term – at least till the next election. Specialized tools have been developed, since the traditional transport engineers are fighting with increasing strong headwind.

Beside this wide empirical data base, new methods have been developed for a better understanding the human and system behavior. Based on sound science, theoretical background with empirical validation a paradigm change is on the way, since decades.

The paradigm of prevailing transport planning (not to confuse planning with design of vehicles or infrastructure or transport logistics) is based on plausible assumptions and extrapolation of individual experience as participants in an artificial environment.

An engineer is per definition a practitioner, concerned with applying scientific knowledge, mathematics, and ingenuity to develop solutions for technical, societal and commercial problems. Today we have to add environmental problems to this list. If the paradigm of his daily work is not based on scientific knowledge, as it was the case in the fast developing technical transport sector, the risk of mistakes is inevitable. And if it is a dynamic system, the effects can be fatal. The work of engineers forms the link between scientific discoveries and their subsequent applications to human needs and quality of life. Engineers can become a danger for the society, if they don`t know the limitations and preconditions of their formulas, which are part of their tool-box. Social problems are the general factors that affect and damage society. A social problem is normally a term used to describe problems with a particular area or group of people in the world. Social problems often involve problems that affect real life. Transport planning is an involvement into real life!

THE PARADIGM CHANGE IN TRANSPORT

Lewin wrote, "There is nothing more practical than a good theory" (1951, p. 169).

Careful analysis and examination of data from household surveys, a multitude of studies and scientific papers, field observations, before and after studies of implemented measures have proved, that the core hypotheses of the paradigm of prevailing transport planning are not only wrong but don´t exist at all on the system level. If core hypotheses have to be replaced by new hypotheses in line with the reality of the system behavior, a paradigm change is happening. The system view is changing. The previous principles are still valid for the individual level, but cannot be extrapolated to a system outside of human evolutionary experience.

The new core hypotheses for planning of transport and urban systems are the result of careful scientific examination of data, the implementation of evolutionary epistemology (Riedl, 1984 a,b) and empirical evidence from practical solutions. Engineering tools have been developed on this background and tested in practice. The results have been published and discussed in seminars with the scientific community and experienced practitioners in the last three years.

- Mobility, if it is defined in a purposeful manner, is not growing with car ownership. The average number of trips per person and time (day, year) is a constant. If car trips are increased, pedestrian, cycle and public transport rips are decreased simultaneously. The average number of trips remains constant, if the number of people does not change (Figure 1).

People are at least as smart as transport engineers. They are intelligent and selfish. They behave in accordance to the built, legal, organizational and financial structures, provided. They choose the most convenient mode, they find under their personal and environmental circumstances. Since roads have been converted into carriageways for cars and the environment became dangerous, polluted and noise, public transport nor easily accessible, the most attractive mode became the car. Instead of walking, cycling or public transport, car-mobility was enhanced, all other modes suppressed. Transport engineers feel only responsible for car traffic and planned and built the environment for the most convenient use of this

Figure 1. "Mobility-growth" in a car-oriented view – left; mobility in the real system – right
Knoflacher & Kloss, 1980.

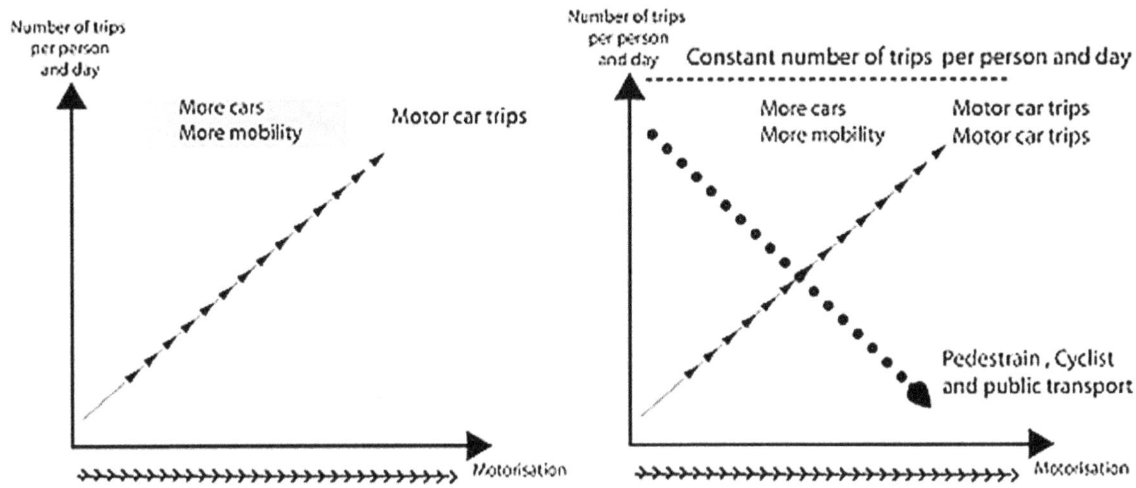

mode – and produces the demand, they recognized as "growth of mobility". The first pillar of the old paradigm was the consequence of the narrow view of transport engineers, not taking into account the reality of mobility in the system. This pillar doesn´t exist! (Figure 2).

- Faster traffic systems do not save travel time. If the system speed enhances, travel time stabilizes rather soon on the same level as before. Only travel distance increase with increasing speed of the system (Figure 3).

Figure 2. Structures determine behavior – and the effects of it. Transport system is man-made. If it produces accidents, man-made structures, as well as human "structures" must be the causes.
Knoflacher, 1985.

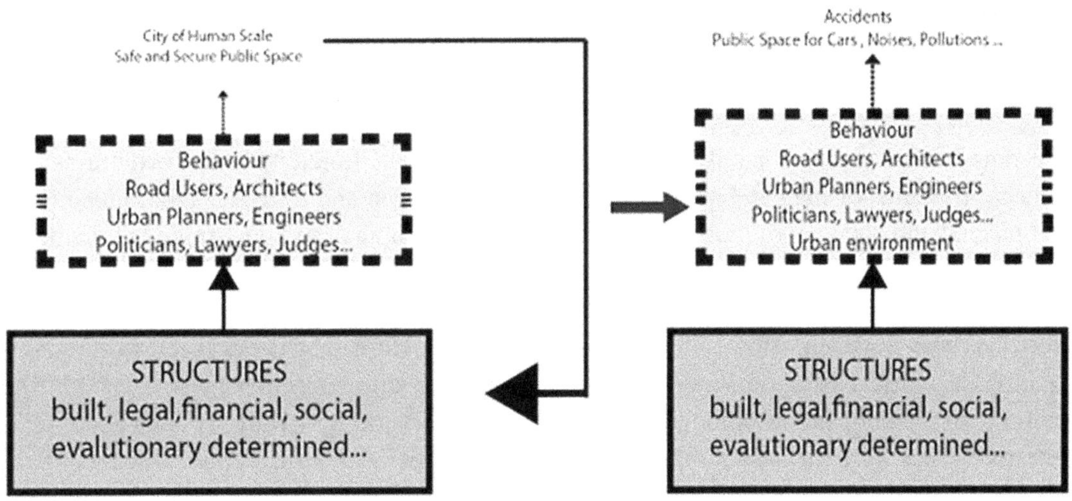

Figure 3. Increasing speed in the transport system, enhance distance. Travel time is constant.
Source: Travel surveys Austria, Germany, Italy, GB.

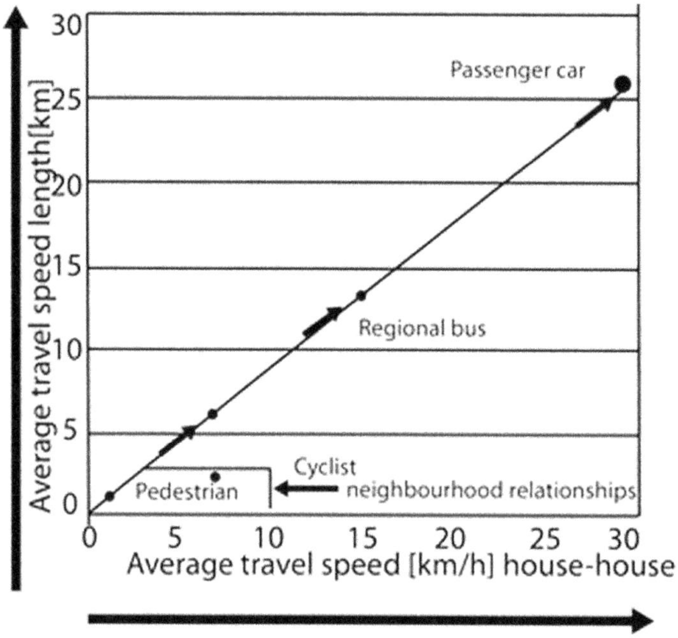

This system effect is a little difficult to understand and there were some obstructions to accept these facts (Hupkes, 1982). In each case we move faster between an origin and a destination, the travel time is shorter. If this would be true also for the whole system, societies with faster transport system must have plenty of time, they must become convenient. This is obviously not the case. Empirical data from surveys (Marchetti, 1994; Schaefer, 2000) show a constant travel time globally, independent of degree of motorization or income. A multitude of surveys on municipality level show that event the travel time distribution of pedestrian, cyclists and car users are practically identical (Knoflacher, 1985). Therefore it is nonsensical to calculate benefits from time savings by speed, which don´t exist et al and has the figure zero (Knoflacher, 1986).

The second pillar of traditional transport engineering – and science – was also a myth, caused by extrapolation of personal experience to a system, mankind has no evolutionary experience.

- Freedom of modal choice is only a slogan, a wishful illusion. In reality inner (evolutionary determined) structures of man and outside structures of the urban and transport environment, as well as financial and legal conditions, are constrains for "free choice" and determine human behavior.

But we have to explain, why people are addicted to cars, even if they have other opportunities to move. This secret has been unveiled in the 1970's (Knoflacher, 1981). Human behavior is following sensations and irritations. Empirical data from human behavior in our artificial environment by Walther (1975) about value of time for walking and in riding in busses, and the observations of Karl v. Frisch about information systems of bees were the base for the discovery what happens in human brain, when using the car (Knoflacher, 1986). It is the human body energy which determines our behavior. The relationship between sensations and irritations is not linear. It follows a logarithmic function as E. H. Weber (1795–1878) and Fechner (1801-1887) have discovered. The input variable occurs as an exponent in the inverse function is our sensation (Knoflacher, 1982). This sensation is related directly to our "body energy-calculator" deep in our brain, the hypothalamus. This is the place in which the car sits in our brains and control our world view, our value system and how we think and how we behave (Figure 4, Figure 5).

Consequences for Transport Planning

Traditional educated transport engineers try to solve transport problems on the level of symptoms, far away from the causes. Transport engineers have not been educated to understand human behavior and the nonlinear feedbacks in the artificial world planned and built on assumptions, hopes and promises, but without a sound scientific background. Therefore the used instruments make the problems even worse. Besides this ignorance about system and human behavior they made severe methodological mistakes, without knowing it. To solve problems, it is necessary not only to understand the cause of it, but also to have goals from outside and not from inside the system to fix a negative feedback, which is necessary to mitigate and solve any problem. Goals are expressed by indicators.

Typical indicators of traditional transport engineering were:

- Person car as the unit of scale;
- Growth of mobility is inevitable;
- Level of Service A – D;
- Speed of 100 km/h (HCM) on rural roads;

Figure 4. The effect of the car on human behavior
Riedl & Knoflacher, 1985.

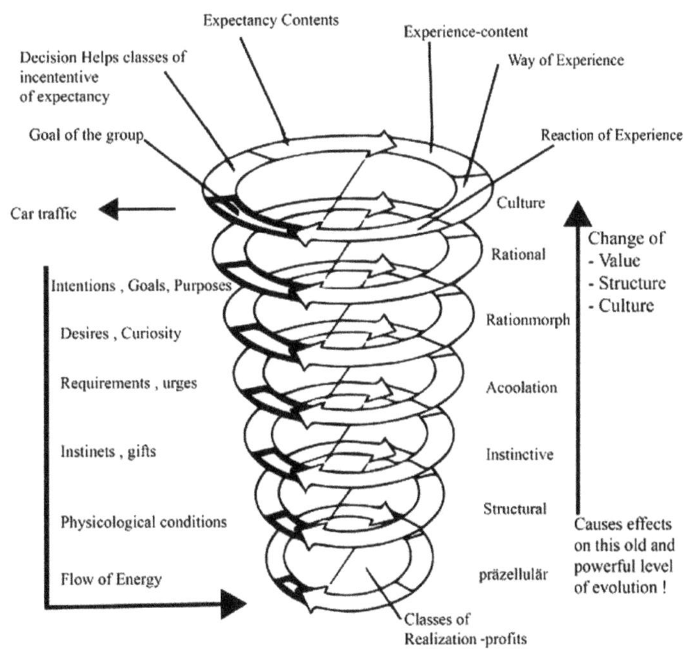

Figure 5. a) Relationship between walking-distance and acceptance; b) acceptance of walking distance: 100% acceptance of the nearby parked car and 10% chance of choice for a PT-stop 420m away. Quality of the environment has a strong influence on the acceptance of walking-distance.
Knoflacher1981, Peperna 1982.

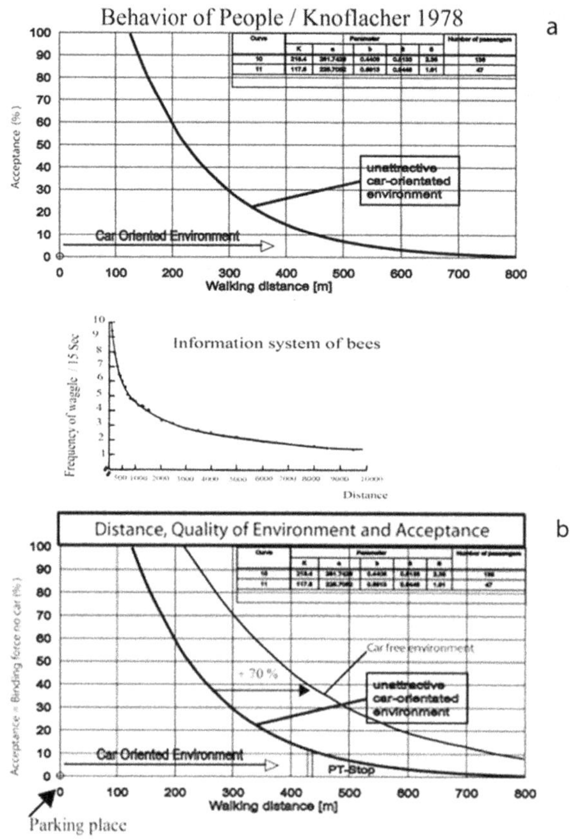

- 50 – 70 km in urban areas;
- Parking provisions on every place of human activity;
- Traffic signals for green waves; and
- Minimizing travel time within given borders of the system.

Behind these indicators was the faith for infinite need of physical mobility, unlimited space and resources and unlimited growth and technology will solve all the other problems (which were created by technology?). Transport infrastructure of the 20th century was built on such ideas. Transport engineers were educated and trained to act as servants of the mystic machinery for technical mobility, necessary for the society and the continuously expanding and growing economy. They have never controlled the system, although they promised to do. They did not recognize the etymologic content of “level of service” which made them to servants of the road traffic system. Their position was and is that of servants in the system. Therefore they were replaced from most of the leading positions in administrations and decision bodies by members of other “academic tribes”.

THE ROLE OF TRANSPORT ENGINEERS IN THE NEW PARADIGM: FROM IGNORANT SERVANT OF AN UNKNOWN SYSTEM TO QUALIFIED AND RESPONSIBLE EXPERT

Engineers and planners cannot act anymore like helpless servants of a mystic "Eigendynamic" (momentum) of the society and the transport system, which push them into unsustainable future as it is the case in the prevailing transport paradigm.

The scientific background of the new paradigm gives the transport engineers the opportunity to understand the reality of the system and help to distinguish between personal experience and the behavior of the system, they are responsible for. The perception of the reality changes therefore. To be able to have control on the transport system, transport engineers must have access to the origins and destinations of trips and not only on movements between them. Transport engineers of the new paradigm must therefore be able to understand the urban fabric and humans on the level of evolution, which is crucial for their behavior in traffic and transport system.

- Instead of believing an infinite "growth of mobility" traffic engineers have to decide about the modal split. They determine the modal choice substantially by the structures and infrastructures they plan, build und organize. They "produce" the kind of mobility that people will choose, but not the number of trips. Transport engineers have now the full responsibility for the modal split.
- To control the modal choice, transport engineers must control the origins and destinations of all trips. They must get into urban planning, where the cause of the transport problems are located.
- They are not anymore the people for high speed physical mobility, but much more the expert for intelligent transport solutions.
- They have to understand and work for people, for children, adult, aged people, handicapped and not for the best "life" of vehicles.
- If they want to control the transport and traffic system, they must know how to implement, operate and maintain negative feedbacks.
- They have to check, whether their brain is infected by the car-virus (Knoflacher, 2009). If this is the case they cannot work as a responsible person for the society. A good exercise is to use a wheelchair for one day to get some impression and knowledge, what kind of environment has been planned in the transport sector.
- Transport is a mean for something and not a goal.

Goals and Tools for Transport Engineers in the New Paradigm

External goals are necessary to solve a problem. For the transport sector following indicators (goals) and tools are important and have to be used: To each goal basic facts and simple tools are added to justify the necessary measures. First of all the scale must be applicable to every mode in the transport system.

- **Unit of Scale in Personal Transport:** The man from childhood to age.
- **Unit of Scale in Good Transport:** Weight (ton) or volume (m^3) and not the container or truck.
- **Mobility:** "Barrier-free-Mobility".

Barrier free mobility means not only "mobility for all", the term used for barrier-free access to public transport, its definition goes much wider and include mobility for poor, mobility at eye level, children, elder people on all of their trips also. The least in the society has to be provided with safe and convention opportunities for mobility to participate in the society.

Different Tools on Different Levels

In general transport engineers have to work and act on three levels:

1. Planning level,
2. Project level, and
3. Level of urban and traffic management.

Planning Level

On the level of transport planning, transport engineers have to provide the base for decisions for strategies for the transport system in the widest context.

The tools today are:

- Knowledge about behavior of dynamic systems and how to handle it in a responsible manner to define appropriate system borders and key indicators.
- Implement negative feedbacks to control the transport system and prevent positive feedbacks which make the system uncontrollable.
- Since the causes of the transport problems are located in urban planning and building codes, as far as parking is concerned, transport engineers have to change these codes and rules. Parking cars have to be separated from human activities physically, financially and organizationally in order to control the modal split.
- Travel time saving on the planning level don´t exist. The value is zero.
- Speed is not a precondition for the project level any more. Speed is determined by the natural or built or social environment of the transport system.
- Congestion in car traffic is not the problem; congestion is an indicator for mistakes in land-use and urban planning and/or the indicator for excessive attractive car traffic.
- Computer models have to be used to reproduce not only the intermodality of the transport system, but also the interactions with urban and regional structures and external effects, as far as necessary;
- Models to determine effects of different time scales, short to long term behavior and interdependencies have to be used instead of assignment-models.
- **Laws and Regulations:** They are important tools for practical work. Existing legal and administrative structures have been written with the old paradigm in mind. Sometimes the problem lies in the interpretation, sometimes the content itself can be cause of the problem. If the tool is not appropriate it has to be replaced. If existing laws and regulations force the engineer to do wrong, he must inform the government about the effects to change the law according to the goals.

Indicators to characterize the *real behavior* of the complex system are:

- Modal split including all kinds of modes, from pedestrian to public transport.
- Constancy of travel time or loss of travel time as a positive indicator for good solutions.
- Acceptance functions for car parks and public transport stops.
- Risk analysis techniques, not only within the transport system, but also for all effects from the transport system on society, environment and economy; and
- Decision support methods.

Project Level

Methods to define the system borders in accordance to the project effects:

- Laws, regulations and ordinances, e.g. building code, design guidelines, coherent with the reality of human behavior;
- Intermodal methods and guidelines based on person-unit for mobility of people and volume or weight for good transport.
- Data about local circumstances, physical, social and legal environment, material, landscape, history, sensibility of the ecosystem, societies, etc.
- Zero victims in the transport sector: Tool: Speed below 30 km/h, car free environment. (See Figure 6.)

Traffic Management Level

There are many reasons to keep the speed in urban areas below 30 km/h: fatality risk, energy consumption, noise, and shared space. If a traffic engineer as a consultant or an expert permits a speed on urban roads higher than 30 km/h he has to carry the consequences of such a decision.

Figure 6. The fatality rate of pedestrian in crashes with passenger cars as a function of the collision speed *Rosen et al., 2011.*

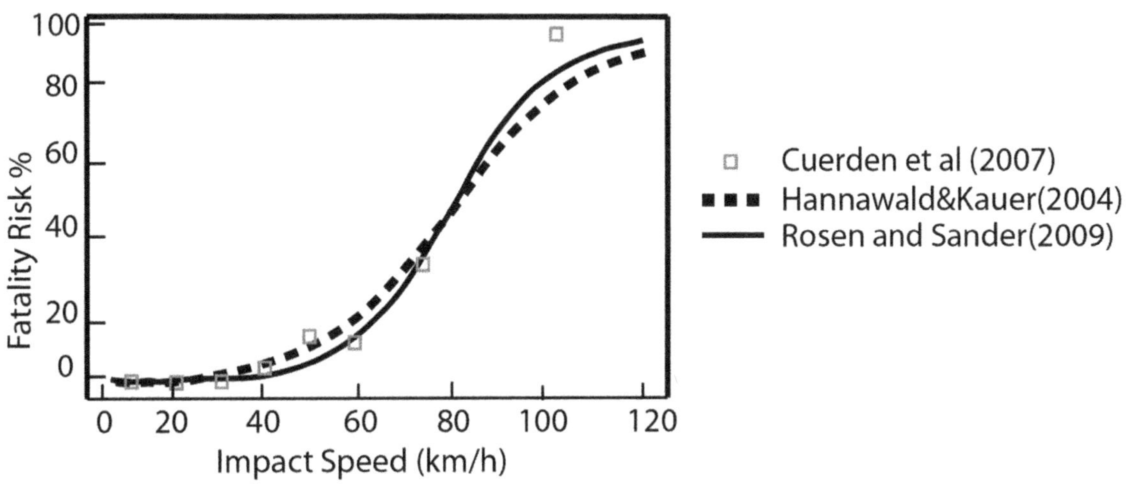

- **Minimize Space for Transport:** Use the diagrams shown in Figure 7.

The diagrams show apparently the big difference between the car and the nonmotorized and public transport (Figure 7). It is not difficult to find the right decision. To agree with this goal, cars in urban areas should not have separate lanes or subsidized parking space on roads.

- **CO^2-Reduction:** Reduction of car use and speed.

CO^2 emissions from pedestrian and cycle mobility is negligible. Emissions per passenger kilometer from cars are 5 – 10 times higher than from public modes. Rails and light rails (trams) are driven by electricity. If electricity comes from solar or water, CO^2 emissions are very low. CO^2 emissions come also from building transport infrastructure and maintenance. Reducing car use is the most efficient reduction of CO^2 in transport.

- **Active Noise Reduction:** Reduction of speed, traffic volume and some technological means.
- **Save and Support Local Economy and Employment:** Promotion of walking, public transport and cycling, car free cities, capturing travel time in the cities by reduction of speed.

Figure 7. a) Space, occupied by pedestrian, cyclists, PT-user and car-user in the urban transport system; b) dynamic spatial efficiency of urban transport modes

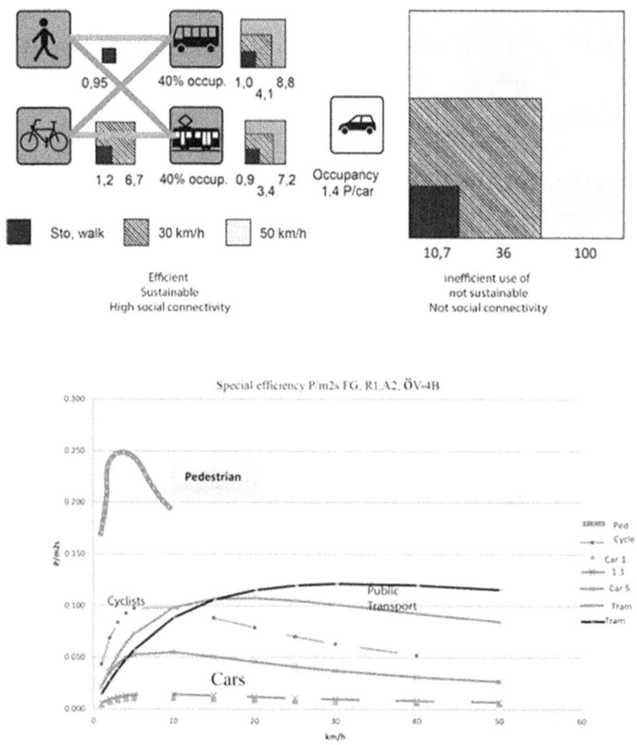

a

b

- **Provide Opportunities for Social Contact and Social Cohesion:** Pedestrian areas, public places, local shops and workshops, use of streets for walking and meeting place for residents especially children.
- **Prevent Unwanted Positive Feedback Loops in the Transport System:** Keep car traffic controlled by limited space for driving, parking fees, reduction of lanes, use of car lanes for public modes, busses or trams or cycling, convert parking places into green areas etc.
- **Respect the Dignity of People and the Life-Supporting Nature:** Environment for "mobility of all", shelter for pedestrian, seating accommodation for pedestrian. Dignity of man is related to walking. Minimizing paved surface whenever possible. Use trees and plants as air conditioners in public space. prevent boring building facades.
- **Take Care on Landscape and History:** Adapt travel speed and alignment in accordance to the topography and the landscape.
- **Check if the Defined System Borders Are Not Exceeded by the Effects of Your Projects and Implemented Measures:** Diagrams showing spreading of noise and air pollution from traffic, water pollution, extension of distance along motorways. Accessibility has positive effects for the big centers and corporations, but negative effects on local structures, which become the looser in fast systems.
- Doubt about the content of guidelines and standards and understand their limitations. Recommendation: standards and guidelines are in general compromises between stakeholders in the transport systems and represent therefore not a kind of scientific "truth". In general they are years, even decades behind the existing "state of the art". Look at the background, history and evolution of standards and guidelines.
- Urban planning and management is part of transport engineering work. Comment: Parking places are not an individual issue. Parking in the system has the same or even stronger effect on transport than planning, building and organizing public transport stops. Organizing parking in accordance with the goals of sustainability, social coherence, quality of life is only possible if parking is totally separated from urban structures, physical, financial and organizational. Tools. Legal instruments to repair mistakes in existing structures. Financial instruments to support solutions and fees on misbehavior of car users. Enforcement and monitoring bodies.
- Every solution and action should become a Synergetic Satisfier. A Synergetic Satisfier is not only satisfying the purpose for which it is implemented, but has other positive – synergetic – effect on other purposes. The strongest positive synergetic satisfier in the transport system is the parking reorganization above. If parked cars are removed from human activities, the whole system capture travel time, public space became safe for everybody, noise and air pollution disappear, social networks appear, local shops get their customers back, local employment enhance, opportunities for entrepreneurs pop up, public transport operate on time and become profitable, public space can be used as an extension of apartments in house blocks, security is improving, since people open the windows to the streets, elder people can participate and meet, space for trees and nature is available, financial situation of cities and municipalities can heal etc. This measure has also a negative flipside: international corporations are forces to compete in eye level with local providers, urban and land-use planning is not so easy any more, much more mental mobility is necessary for planners and decision makers, if they cannot put the burden of their professional ignorance on people's shoulders and easy, cheap and comfortable car traffic.
- Quality, humanity and beauty should be the outcome of transport and traffic engineering work.

CHALLENGE FOR THE NEW PARADIGM

What about the Cost?

If these goals are satisfied, the solution must be the most cost-effective and cost efficient one. A very simple diagram can demonstrate this: the so called "fundamental diagram" of traffic engineering, the relationship between density and traffic volume in person-units:

Four to five times more space is needed to transport the same amount of people on a road in cars instead with busses or trams (Figure 8). It is obvious, that cities put a lot of effort into the transport system to reestablish a fair balance between sustainable modes, pt, cycling and pedestrian and the unsustainable car traffic to save money and to prevent a further damage of the local economy. If speed is reduced, traffic safety increases, noise and air pollution goes down and quality of life is enhanced. Direct and indirect cost saving are one of the effect of the new paradigm.

What about Sustainability?

- Concerning sustainability, they have to provide the society with solutions in favor of sustainable modes.
- Sustainable modes are walking, cycling, use of rickshaws and public transport.
- Car traffic is not sustainable due to its huge amount of resources, energy, space and the adverse effects on environment and society.
- The former "unit of scale", the Person Car Unit, has to be replaced by the person-unit, the human being.

Figure 8. The diagram show the inefficient use of precious urban space if planners use the PCU as unit of scale
Knoflacher, 2014.

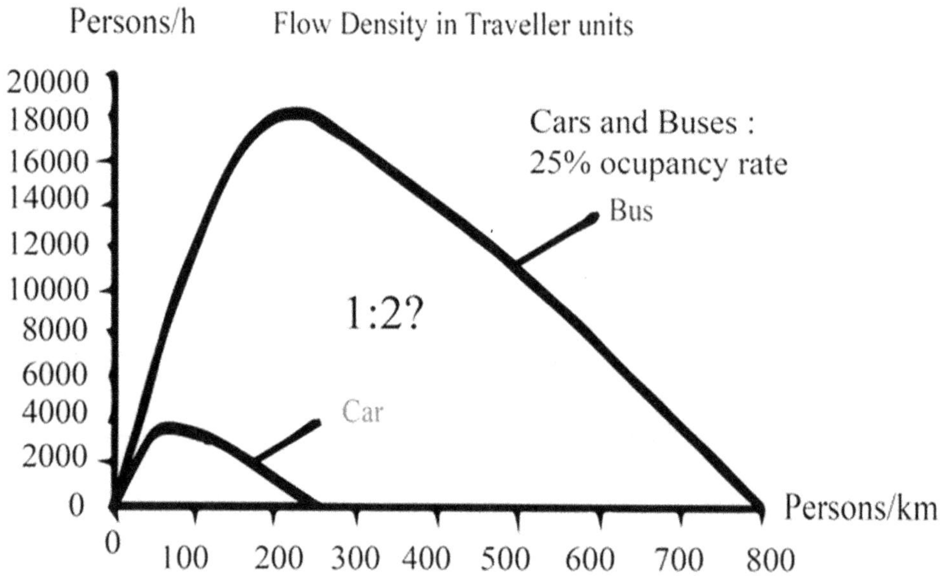

- The unit for transport of goods is not anymore the truck; it is the weight or the volume of goods: ton, cubic meter.
- Multimodal thinking is the consequence and a must.

What about Planning Models?

Assignment models, to calculate traffic demand, flows and indicators fit not into the new paradigm, since they are either static or dependent on assumptions about the urban or regional structures. In the meantime they are easy to handle and produce more or less plausible pictures and even simulations whatever needed. Models for all kinds of modes are on the market today. But models are only devices for transport engineers, to play with them to get a feeling how the system behave. They can reproduce the transport system behavior rather well if they are used in a professional manner. But they are not sufficient for the new system view.

Since structures at the origins and destinations of trips are affected by changes in the transport, they have to be included into the planning model structure. Several kinds of such models have been built in the last decades. Some of them are successful in practical use. One chapter in the book is dealing with the MARS model, an offspring of the new paradigm.

What about Time Saving by Speed?

Travel speed is not anymore the core-determinant for transport planning. “Speed” in this context mean “speed of the transport system”.

- Cost-benefit calculations based on travel time saving cannot be used any more, since there is no “travel time saving” possible in the system. It doesn´t exist at all. Its mathematical value is zero!
- Since the system cannot save time it cannot lose time by reducing speed, e.g. congestion. Congestion is not a problem. Congestion is a (not very elegant but effective) tool for traffic management (Don’t forget: it is important for the negative feedback effect).
- Concerning sustainability and laws of physics: sustainable transport modes determine the level of speed. In urban areas pedestrian speed is the most important one.
- Speed of technical traffic modes is related to risk and accidents, energy consumption, air pollution, noise and land use. Engineers are responsible for these adverse (unintended) effects.
- Speed changes land use and urban structures: Not only the location, but also the form and content of buildings, the patterns of economy and the quality of the environment changes with the speed of the transport system. Engineers in transport an urban planning are responsible for these effects.
- The guidelines for planning have to be rewritten as well as tools for transport, land use and urban planning.

What about Freedom?

“Freedom of modal choice” has been challenged in the 70’s of the last century when the discrepancy between estimated and measured travel time became an issue of science. The problem was solved by using the methods of evolutionary theory. This approach made it possible to close the gap between change of structures and human behavior. It is the body energy for mobility, which determines the behavior in the (artificial) environment. The relations and feedbacks are nonlinear. The fascination of cars is based on this fact, deep rooted in the human brain and binding people and cars on the level of physics – plus all psychological and social effects above this level. Position of car park is therefore the decisive element if we want to transform our unsustainable transport system into one toward sustainability.

- Building code must be rewritten. Car parks and human activities have to be separated in such a way, that sustainable transport modes become more attractive to use than cars.
- Parking management is the key for sustainable transport. If parking is organized in the proper way, outside of settlements or at least further away from locations of human activities than stops of public transport, car-traffic-flow problems disappear.
- Land use and urban planners become an important and crucial role, since they have to develop car-free urban and settlement structures with parking cars outside or at locations further away than pt-stops.
- When parked cars are removed from roads and houses and put into central garages, the environment become socially attractive, economically viable, safe, and healthy.
- Parking organization under the premises of sustainability stop urban sprawl and force engineers and planners to implement their tool-box in accordance to the human scale.
- To handle the system for transition toward sustainability financial tools and tools of organization are necessary. Today parking is heavily subsidized direct and indirectly by tax-payer´s money. Parking charges in accordance with the real cost have to be implemented – and work effective in practice, where they are introduced.
- Parking charges have to be introduced on all parking places outside allocated garages or parking lots.
- Financial support with money from parking fees can now be used for public transport, cycle or walk.

What about Guidelines?

Guidelines to copy, without thought, to build a uniform cloned environment should not be used by transport engineers. Tools to develop innovative solutions within the borders from laws of physics, evolution, human behavior, chemistry, and social sciences are now necessary. Transport engineers must not be educated in all these fields of science, but they must understand human behavior since they change the environment of the human society, its economy, its opportunities with the technical means much more than ever before in human history. Transport engineers have converted the world people have built for them over 10,000 years into a world for cars. They have now the obligation and the duty to repair the well minded, sometimes fatal mistakes of the transport engineers from the last century. Guidelines, not based on scientific background have paved the carriageways and the way into the transport and adjacent problems of today. The way in the future, toward the cities for people are not so easy as promised by some architects, just to change the surface. There solutions looks nice, but the cause of the problems lies much deeper in human and urban structures of the last century. It is the engineer´s responsibility also, to treat the problem at the source, if he wants to have a successful and satisfying professional future. Traditional standard of a cross-section of an urban road provide space for cars, parking and driving as the most important elements. Public transport is not considered, pedestrian and cyclists are separated.

These kinds of standards and guidelines have destroyed the public space in cities in the last century. They are a professional imprinting for transport engineers thinking and acting

If the new principles of transport engineering are used we get a cross section in which pedestrians, cyclists and public transport have priority in the public space. Cars have to share the street space with public transport and have no special lanes for themselves. Even in historical cities this principles encourage people to walk, cycle, restaurants and shops are reopened, the city start to live again. Cities livability is dependent on livability of streets. Livability of streets depends on the presence of pedestrian in public space. This organization of urban space is in agreement with the constitution, the scientific background and the goals of a sustainable city and sustainable transport system.

SOLUTIONS AND RECOMMENDATIONS

Is the New Paradigm an Illusion or of Practical Use?

There are numerous examples, where the tools of the new paradigm have been implemented successfully. Car free areas in historical cities are easy to revitalize. They have an urban structure, which is tailor made for pedestrians and public transport plus cycling. If these parts of the city become car free, public space soon become alive, shops are reopened and houses remake. Today these centers of cities, depraved in the past, have become the most attractive – and expensive – living, working and shopping areas and in many cases tourist attractors. Vienna, the global "quality of life city number one" has pedestrianized not only parts of its center, but also many streets in the districts surrounding the center.

Architects show fascinating colored photos from pedestrianized streets inspiring promises for solutions of urban problems, which will not be achieved if the not visible causes of problems are not tackled. This approach looks plausible, because it deals with the visible symptoms, but it will not heal an ill city. Of course it is of importance to build an attractive and nice environment for people, the urban landscape (Gehl, 2010). But it is not enough to repair the damage to the cities, which was done in the last century. A brain tumor cannot be treated by a coiffeur, although he can make a person looking nice. A lot of politicians and system-ignorant planners are impressed by the nice looking pictures, but the real solution has to treat the cause of the transport problems of today: the behavior of people in the car oriented environment: the new organization of parking. Parking places have to be removed from streets as well as from private ground into central public garages. In these storybooks this bloody business is not visible and not touched. But this is the future field for transport engineers in the urban environment: development of strategies to become more parts of the cities car-free. (Car free is not a city without cars, it is a city, in which people are free from the necessity to use cars).

In Vienna, thousands of parking places have been removed from the streets, provide space for people, to plant trees and build cycle lanes instead. Public transport on the surface was given priority at traffic signals and separate lanes, taken away from car traffic. Anyway the car traffic is much more relaxed than 40 years ago, when the urban streets were congested much more. Public space is now reshaped following the principles of the new paradigm. Cars have to share the space with – and behind – public transport. And the Viennese people have changed their behavior, following the change of structures. Physical structures tell them, that it is better to walk, cycle or use public transport. Beside the greater pedestrian areas, thousands of mini-pedestrian areas at stops of trams, busses and metro stations have been implemented. Parking on streets is not free of charge anymore, neither for residents, nor for commuters or visitors, in most of the districts of the city. 1.400 km of cycle lanes have been introduced into the street network. Information systems for sustainable modes were implemented, on streets as well in digital form. Shortcuts through building blocks were opened to reduce detour for pedestrians. Low floor trams provide easy access to all kinds of public transport for elder people, handicapped persons, children and people with child carts etc. These all are not only interested but also satisfied professional activities for the new generations of transport engineers.

Time is needed not only for a sound traffic-concept, but also for implementation. Car ownership in Austria is still increasing. In the City of Vienna, car ownership is in decline since 2002. This change of system behavior has its roots decades before:

In the city of Vienna the new goals have been implemented in the traffic-concept, elaborated between 1975 and 1981. Tools were developed and tested since then. Car use started to decline since the 1980s. Since 2002 car ownership is declining in the city continuously till today while the number of residents in the city is increasing. Vienna has become the number 1 City in quality of life globally, since 8 years. Without the contribution of transport engineers, using the new goals and tools, this development could not happen.

THE KEY FOR SUCCESS: REORGANIZATION OF PARKING

If we have direct access to cars and affordable energy modal choice is determined by these inner and outer structures. The key are the outside structures, the artificial world, transport engineers plan, build and organize. If the car is parked directly adjacent to every human activity, we are forced to use it. We are in the trap (Figure 9).

The key for success is the reorganization of the parking regime. The physical separation of parking and human activities is inevitable to give the people the freedom od choice between private car and public transport (Figure 10). The financial inequity between the transport modes has to be removed step

Figure 9. The urban structure, which cause risk, air pollution and damage in urban areas
Knoflacher, 1985.

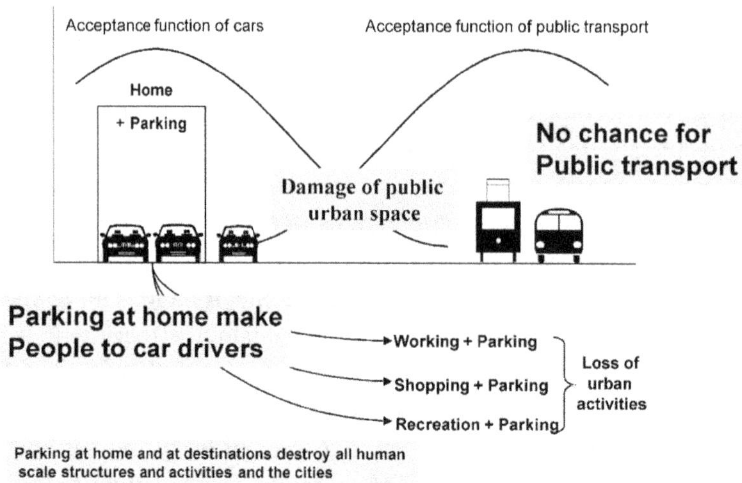

Figure 10. Reorganization of parking is the key to solve most of the traffic and urban problems of today. *Knoflacher, 1985.*

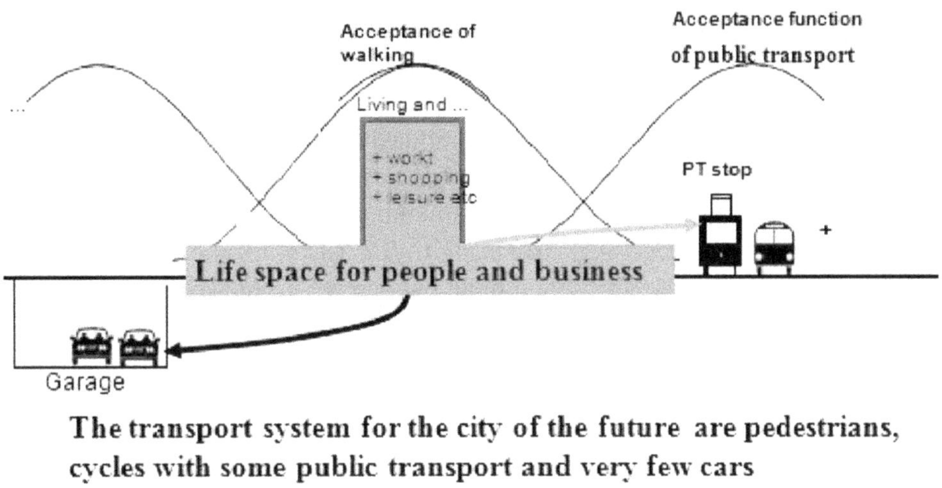

by step by introducing public charges for parking on public space and private parking facilities, like in Vienna (2016).

After this brief introduction into the basics of the new paradigm, we can now draw the consequences for transport engineering practice.

1. The base of traditional transport engineers were myths and not science.
2. Captured by these myths they have not recognized the adverse effects of their tools and work.
3. The tools were developed to realize the imaginations of these myths.
4. Instead of solving the problems they promised, they are producing and enhancing not only transport problems but also environmental degradation, urban damage and social isolation.
5. They have never understood the dynamic of the interrelationship between transport and the urban system.
6. Transport is a mean and need therefore goals from outside and not from the transport system itself!
7. Since they have not understood the system behavior, they don´t feel guilty for traffic accidents, damage of local economy, urban sprawl, congestion maximization, environmental degradation.
8. Environment and urban settlements were shaped in order to save time by high design speed. Like societies have built temples to worship their gods, the cult of mobility built the temples of elevated motorways, cloverleaf intersections, tunnels to serve the goddess of speed.

FUTURE RESEARCH DIRECTIONS

The tools presented in the chapter have an interdisciplinary scientific background and stimulate the discussion across the borders of traditional scientific and professional disciplines. In practical use they have been tested since more than three decades successfully. The tools are supporting the emerging trends toward a sustainable urban and transport future. They provide planners and decision makers with the core indicators needed to prevent the bias of the past. The viability of the paradigm has been tested in cities, regions and countries. Models and methods have been developed on these theoretical and practical foundations, applied and implemented in many cities by closing the gap between transport planning and urban planning. The new paradigm approach the problems of today not at the symptom level, it is focused on the level of causes in and outside of the human being. Using the tools of the new paradigm in the process of implementation of proposed programs, strategies or project helps not only to reach the goals, but also to control the process of change continuously. This saves not only time and money but prevent undesirable trends on an early stage. But nevertheless future research is needed.

The new tools are much more sensitive to local situations than the traditional global planning tools which created the unintended effects on environment, traffic safety, local economy and loss of beauty of urban settlements. Research is needed to calibrate parameters for the specific cultural, climate and topographical as well as social conditions. New standards for the "design of public space" have to be developed to replace the old car-oriented road design standards. Research has to be carried out for the transition from the existing parking regulations and practice to the new parking regulations for parking facilities which fit into a sustainable future. Existing trends in the legal structures toward more equality in the transport system need a new and sound scientific background, which has to be elaborated. Since the transport and traffic system is changing economic structures research is needed in order to prepare a smooth transformation from global to much more local economy and more employment. How it works, can be seen in cities where the tools of the new paradigm are used since years. Further research is needed to provide a wider knowledge- and database by collecting data and experiences in a much more systematic way, which is now possible. Since the core-assumptions of the traditional traffic engineering have changed with the paradigm change, terms used so far, change their meaning, new guidelines and methods have to be applied.

CONCLUSION

The solution of the problems in the traffic- and urban-system, cannot be found by applying the principles and methods, which have caused and contributed to the problems. Traditional traffic planning was not based on scientific knowledge of the behavior of people in the totally new man made technical environment of railways and cars. By using interdisciplinary methods and careful analysis of human and system behavior tools for traffic engineers are available to help them to organize the traffic and urban system in a sustainable way. The tools are based on the real behavior of the people in the traffic system of today:

There is no "growth of mobility", the average number of trips per person and day is constant. We can only change the modal split.

There is no time saving possible in the system if speed is changed. Travel time of the system is constant. Speed change, change settlement and economic structures.

There is no free choice of mode. Structures – inner human and outside manmade structures – determine human behavior.

Bases on these three facts, new tools were developed and described briefly in the chapter.

REFERENCES

Fancher, R.E. (1996). Pioneers of Psychology. New York: W.W. Norton & Company.

Gehl, J. (2013). *Cities for people*. London: Island Press.

Highway Capacity Manual. (1965). *National Research Council (U.S.) Highway Research Board*. Committee on Highway Capacity.

Highway Capacity Manual (HCM). (1950). *Practical Application of Research*. Washington, DC: By the Committee on Highway Capacity Department of Traffic and Operations Research Board.

Hupkes, G. (1982). The Law of Constant Travel Time and Trip-Rates. *Futures*, *14*(1), 38–46. doi:10.1016/0016-3287(82)90070-2

Knoflacher, H. (1981). *Human Energy Expenditure in different modes, Implications for town planning*. Paper presented at the International Conference on Surface Transportation System Performance, Washington, DC.

Knoflacher, H. (1982). Zum Problem der subjektiven Widerstände in Transportsystemen. *Int Verkehrswes, 34*(6).

Knoflacher, H. (1985). *Katalysatoren für Nichtmotorisierte*. Eigenverlag.

Knoflacher, H. (1986). Kann man Straßenbauten mit Zeiteinsparungen begründen. *Internationales Verkehrswesen*, *38*(6), 454–457.

Knoflacher, H. (1986). Verkehrsplanung für den Menschen: Vol. 1. *Grundstrukturen*. Orac Verlag.

Knoflacher, H. (2007). *Grundlagen der Verkehrs- und Siedlungsplanung*. Wien, Köln, Weimar: Böhlau Verlag.

Knoflacher, H. (2013). *Virus Auto: die Geschichte einer Zerstörung*. Wien: Böhlau.

Knoflacher H. (2015). Road safety in urban areas, transport planning and traffic safety: Making cities, road and vehicles safer. *CRC 2016*.

Knoflacher H. Kloss H. (1980). *Verkehrskonzeption Wien. Konsulentengutachten – Radverkehr*. Magistratsabteilung 18 Wien.

Lance, D., & McNeil, I. (1996). *Biographical Dictionary of the history of Technology*. Taylor & Francis.

Lewin, K. (1951). *Field theory in social science: Selected theoretical papers* (D. Cartwright, Ed.). Harper and Row.

Manual, H. C. (2000). *Transportation research board*. Washington, DC: National Research Council.

Marchetti, C. (1994). *Anthropological Invariants in Travel Behavior*. Technological Forecasting and Social Change. IIASA Laxenburg.

Peperna, O. (1982). *Master Thesis:Die Einzugsbereiche von Haltestellen öffentlicher Nahverkehrsmittel im Straßenbahn und Busverkehr*. Wien.

Pfister, C. (1994). *Das 1950-er Syndrom, GAIA*. Trnsdiziplinäre Zeitschrift, Oekom Verlalg München.

Riedl, R. (1984a). *Biology of Knowledge: The Evolutionary Basis of Reason*. Chichester, UK: John Wiley & Sons.

Riedl, R. (1984b). Evolution and Evolutionary Knowlege. In F. M. Wuketits (Ed.), *Concepts and Approaches in Evolutionary Epistemology* (pp. 35–50). Dordrecht: Reidel. doi:10.1007/978-94-009-7127-1_2

Rosen, E., Stigson, H., & Sander, U. (2011). Literature review of pedestrian fatality risk as a function of car impact speed. *Accident; Analysis and Prevention*, *43*(1), 25–33. doi:10.1016/j.aap.2010.04.003 PMID:21094293

Schafer, A. (2000). Regularities in travel demand: An international perspective. *Journal of Transportation and Statistics*, *3*(3), 1–31.

Walther, K. (1975). Die fahrzeugäquivalente Reisezeit im öffentlichen Personennahverkehr. *Verkehrs und Technik*, 7.

Wien. (2016). *Parking regulation*. Retrieved from https://www.wien.gv.at/verkehr/parken/

KEY TERMS AND DEFINITIONS

Congestion: Maximum density of system users in a part of the traffic system. Speed in congestion is zero. Congestion is not a problem and can be successfully used as a tool to change modal split. In general congestion is the symptom of exaggerated good conditions for the part of traffic in which it occure.

Mobility in Traffic: The number of trips between origins and destinations in public space and related to a purpose. In the system, the average number of trips per person per day is constant. Therefore, no "growth of mobility" exist.

Modal Split: The share of trips of different modes: percentage of trips on foot, cycle, car, public transport.

Parking Space: The space to leave cars for a certain time. The location of parking space determines the modal choice of people. To give people the freedom of modal choice, Parking space has to be outside of human settlements, villages or cities. Parking has to be totally separated, spatially, financially and organizational, from places for residents, working, shopping and recreation.

Speed: An indicator to characterize the traffic system. Since travel time in the system does not change with the travel speed of the system, speed has no importance anymore in the new paradigm. Well organized systems have less need of high speed traffic and they are therefore more sustainable. "High speed" in the new paradigm: everything faster than walking.

System: In the context of traffic engineering is the society, the natural and artificial environment (buildings, infrastructure, vehicles etc.), the legal system including administration and management with all interactions and relationships.

Travel Time: The time between origin and destination. On the operational level of public transport, travel time is important. On the system level for planning travel time is a constant and cannot be used to support faster modes, since the overall travel change is always zero.

Unit of Scale for Traffic Engineers: The person (Person-Unit, PU) and for goods the weight (kg, ton) or volume (cubic centimeter, cubic meter).

ENDNOTE

1 Sorokin Pitirim, Sociologist 1889 – 1968.

Chapter 2
Evaluation of Walkability and Pedestrian Level of Service

Hediye Tuydes-Yaman
Middle East Technical University, Turkey

Pinar Karatas
Middle East Technical University, Turkey

ABSTRACT

Due to decreasing resources, living in urban regions focus on sustainability in many aspects, including transportation. Sustainable transportation encourages non-motorized modes of walking and cycling as well as public transit (which also relies on walking while accessing a station), as well. However, walking as a mode is still a big mystery itself that needs further attention and research effort especially in the evaluation part. So far, the planners have discussed the concepts of walking and walkability, while engineers have mostly focused on Pedestrian Level of Service (PLOS). The scope of the problem is reflected in the diversity, and consequent inconsistency, in the available PLOS methods, which is one of the problems addressed in this chapter. The second and the bigger problem is the gap between the planning and engineering approaches in evaluating PLOS and walkability producing no consensus or clear relationship between the two, even though they overlap greatly.

INTRODUCTION

Walking can be defined as one of the individual transportation modes, which is the consequence of personal preferences (Rodrigue et al., 2006). It was the dominant mode during the 1800-1890s due to small sizes, compactness and mostly circular shape of the cities offering walking distances less than 5 km (Kaiser et al., 1995). First, invention of the railways changed the urban morphology, which was further reformed by the rapid increase of automobile usage after the1930s that marked the start of "the automobile era". Since then, the threat by the ever increasing popularity of automobile in the urban transportation not only required major investments in road infrastructure, but also in the traffic management systems to overcome its negative effects such as, congestion, air pollution and accidents. But, neither

DOI: 10.4018/978-1-5225-2116-7.ch002

has provided a real solution, and the "vicious circle" of automobile-based transportation problems still threatens city traffic networks, today.

A more realistic solution has been sought after within the concept of "sustainability" that gained popularity as early as the 1970s. However, the easiest way to define "sustainable" transportation starts with defining what is "non-sustainable" in transportation. According to a definition proposed by Black (2010) a sustainable transportation system is,

... One that provides transport and mobility with renewable fuels while minimizing emissions detrimental to the local and global environment and preventing needless fatalities, injuries, and congestion.

This can be further supported by the recommendation of selecting focus of urban transport policy as "reducing car trips and encouraging alternative modes, such as public transport, walking and cycling" (Babalik-Sutcliffe, 2013). Furthermore, walkability and biking are encouraged due to their positive impacts on human health (Brown et al, 2009; Rosenberg et al, 2009; Van Dyck et al, 2011).

Though easy to recommend, it is much harder to get people to choose "walking" as a mode in most cities, today. Because, despite its great capacity in sustainability, walking has the major disadvantage stemming from physical limitations of human body, in terms of walking distance and speed. While planning public transport services, guidelines often use one-quarter mile or 400 meters (or at most upto one-half mile or 800 meters) as "walking distance to access public transport" as key measure (Daniels & Mulley, 2013). There can be many factors pushing the walking distance to much lower or higher limits. But, even a range of a few kilometers is rarely enough to get all our travel needs (commute, shopping, etc.) satisfied in a city. Thus, plans to encourage walking as a mode have to be integrated by encouragement of public transit use in an intermodal way, as well. However, public transit has its own challenges when it comes to mode choice, with its stop-to-stop structure, waiting/dwelling times, problems in accessing to stops and making transfers. As a result, it is very difficult to evaluate which of these two modes, walking or public transit, creates (or to what extent) challenges, such as:

- Limits of effective capture zone of public transit stations,
- Impact of infrastructure aspects, road network design or land use measures on the usage of public transit stations, or
- Whether people do not chose public transit because they do not want to/cannot walk, or vice versa.

It should be noted here that, in developing countries, walking limits and use of public transit may go up to levels much higher than developed countries, especially among certain traveler groups that cannot afford automobile; but, such mode choices made based out of necessity are easily reversed towards the favor of automobile, when a traveler's socioeconomic demographics change over time, creating a setback in our efforts towards a more sustainable urban transportation management.

To succeed in reaching sustainable urban transportation goals, we have to start with the basic concepts of "walking as a mode", and understand the mode choice and modal attributes of walking, as they are evaluated in a pedestrian mind. Pedestrians can be described as "subjective utility maximizers", because they arrange their activities, activity areas and the paths between the activities synchronically to maximize the predicted utility of their efforts and walking (Hoogendoorn & Bovy, 2004). This is very similar to the utility concept behind the traditional four-step model, which included further model developments in trip generation and distribution, mode choice and network assignment steps for motorized modes, but not for non-motorized ones. On the other hand, examination of the interaction between travel mode choice and characteristics of the physical environment, especially local topography and sidewalk availability, revealed a great influence especially in non-motorized modes. Incorporating non-motorized modes into local transportation planning and built environment influenced travel behavior further (Rodriguez & Joo, 2004). While developing a conceptual model for walking in urban spaces, Yazıcıoğlu et al. (2011) drew attention to the individual, group, regional, and spatial environmental variables, which had a direct influence on walking choice. Also, an important task is to address the cost issues making a location or a region "walkable" or not. So far, the planning literature have mainly focused on the concepts and evaluation of "walking" and "walkability", while engineering literature have mostly dealt with "Pedestrian Level of Service (PLOS)", which is a more quantified concept evaluating pedestrian flows or design aspects. However, there is a big gap between the methods of evaluation of the two concepts, which is the main goal of this chapter. To analyze the main problem of disconnection between walkability and PLOS, we have to start with the definitions and discussions about these concepts presented below.

BACKGROUND

Walkability

While discussing the mode of walking and its benefits for society, Litman (2003) defined two key concepts, "Walking (activity)" and "Walkability (quality of walking)". Abley (2005) also discussed these concepts in detailed, which finally concluded with a very basic definition of "walkability" as "walking friendly". However, expanding the concept of "walkability" to include each new suggestion, makes it harder to define what walkability is in a concrete way (Park, 2008).

In an attempt to define aspects of walkability, some planners listed three major aspects of walkability as connectivity, quality of road network and proximity (Gori et al., 2014), while others listed density, capacity and directness as major aspects from land use perspective (Abley, 2005). While assessing walkability of routes to schools in different suburban regions, Gallimore et al. (2011) considered aspects such as traffic safety, accessibility, pleasurability, density of housing, but, finally concluded that land use patterns that created "shorter more direct" pathways were perceived to be "more walkable". However, walk purpose can affect the perspective of a pedestrian and perception of "walkability" significantly. Despite the real distance values, people can walk more/less depending on their perception of neighborhood characteristics.

Though there are no generalized results agreed on, some studies reported various aspects of land-use, trip and traveler characteristics affecting walkability as follows:

- Guo (2009) showed that walking environment significantly affected the one's walking experience and the utility of walking by the investigation of travel behavior, path choice decision and walking travel time of commuters after exiting the subway. The study concluded that improvements in the streets should be planned according to the utility of walking and justified through a cost-benefit analysis, without providing any concrete improvement recommendations.
- Studying a regional coastal Australian city, Humpel et al. (2004) showed that weather, aesthetics, accessibility and location were the aspects associated with "general neighborhood walking",

while weather and aesthetics were associated with "walking for exercise"; safety and accessibility were found important in "walking for pleasure".

- Analyzing the impact of walkability on transit users, Park (2008) listed 42 different aspects regarding street, road and land-use, which were called "path walkability indicators" and used to develop a composite index based on data from Mount View, California, USA.
- In Chicago, pedestrian friendliness was measured by a Pedestrian Environment Index (PEI) which used the data available for planners combining four components: Land-use diversity, Population density, Commercial density and Intersection density (Peiravian et al., 2014). The methodology investigated the areas for each component at each sub-traffic analysis zone level and produced a comparative evaluation between zones.
- Studying street characteristics (length, location and width) and pedestrian behavior characteristics (trip purpose, travel time information, trip frequency, etc.), Liu et al. (2014) suggested to increase walking in a neighborhood by a combined design (with business activity, landscape, circulation, transit services and beyond).
- A series of studies on walking behavior and pedestrian perception in Putrajaya, Malaysia, showed that three walking behavior features (speed, direction and experience) had very small differences among urban and rural region pedestrians and pedestrian's perception on walking time between their homes to the nearest community facilities, and were slightly affected by household income and gender. However, lighting of streets, signals for street crossing, safety during night walks, traffic volume and speeds along nearby street, existence of trees and awning, attractiveness of environment/building articulation, cleanness, and street furniture for resting, were listed as factors positively effecting walking from "safety and security" and "convenience and attractiveness" perspectives. (Azmi et al., 2012, 2013; Karim & Azmi, 2013).
- In the analysis of the walk trips beyond metro stations in Beijing, China, Sun et al. (2015) showed that mean walking time was 8 minutes and both perceived and measured travel times decreased with greater connectivity in the road network.

Walkability Evaluation Methods and Tools

Due to diverse nature of interaction between the pedestrians, built environment and walk trip characteristics, it has been very challenging to develop a commonly accepted tool to evaluate the walkability. Comparing different assessment techniques (such as reviewing, auditing and rating) Abley (2005) set their scopes such that "reviewing" was mostly qualitative and employed to analyze an existing condition, while "rating" was performed in a quantitative way performed for an existing or a proposed situation. "Auditing" can be performed either in qualitative or quantitative way for both existing and proposed conditions. Shorter and simpler evaluation tools have also been proposed, such as a simple checklist of the selected number of local destination points (Bias et al., 2010).

An auditing-based assessment tool, called Scottish Walkability Assessment Tool, (Millington et al., 2009) was designed to assess walkability according to: Functional elements (walking surface and permeability), Safety (personal and traffic), Aesthetic (streetscape, architecture, views) and Destinations (parking, land use mix, services, public transport, parks). In San Francisco Bay Area, three different surveys were conducted to develop a "perception-based composite walkability index" (Park et al., 2014). Among 42 "Path Walkability Indicators" defined (Related with curb-to-curb roadways, Pedestrian crossings, Buffer zones, Sidewalks and their facilities, Street scale and enclosure, Nearby buildings and properties), 22 of them were used in a final formula.

Fairnie et al. (2016) focused on walking with its direct effect on physical activity, and investigated London Travel Demand Survey (LTDS), Active People Survey (APS) and Health Survey for England (HSfE) to get the variables which could be the predictors of walking. Results showed small differences between the ethnic groups, age and car ownership.

As walkability may continuously change over space based on aspects in land use, road network and walkway characteristics, it inevitably calls for a spatial evaluation in Geographic Information System (GIS) environment. As early as the 1990s, GIS analysis was performed to gain insights about the effect of street configuration and walking distance on pedestrian access to transit mode and their mode choice (Hsiao et al., 1997). More attempts have been made to develop GIS-based "walkability index" concepts:

- Using TIGER data, Schlossberg (2006) proposed a GIS-based Pedestrian Assessment Tool to evaluate the pedestrian-environment areas by three visualization and quantification techniques; "street network classification (quality)", "pedestrian catchment areas (proximity)" and "intersection intensity (connectivity)".
- Owen et al. (2007) studied the relationship between the pedestrian access and the transit preferences at district level based attributes (such as dwelling density, street connectivity, land-use mix and net retail area) and participant's socio-demographic characteristics.
- Defining a GIS-based "Walkability Index", Leslie et al. (2007) studied the walkability according to the proximity and connectivity characteristics with a spatial data set in Australia.
- Hajna et al. (2013) performed GIS-based analyses of walkability, in addition to the street-level audits and participant-reported ones, where GIS was used to derive land use mix, street connectivity and residential density.
- More recently, Ellis et al. (2016) employed the six connectivity measures (intersection density, link-node ratio, pedestrian route directness, pedshed, metric reach and directional reach) using a footpath network in Belfast, North Ireland. Results showed that intersection density and metric reach were seemed as the best measures for connectivity.

A noteworthy application in evaluating walkability in GIS is an open source website, "Walk Score®", which has a patented system that analyzes hundreds of walking routes to nearby amenities for the given addresses and awarded points on a scale from 0-100 (Front Seat Management, 2011). While amenities within 5 minute walk (0.25 miles, which is approximately 400m) were given maximum points, a decay function was used in point awarding for the distances up to 30 minute walk. A location with less than 50 points was labeled as "car-dependent". To show the variability of walkability in space, results of the walkability evaluation for two locations in the Northwestern University Campus in Evanston, IL, USA by Walkscore ® are presented in Figure 1. When the analysis point was taken as the Graduate School building in south end of the Campus located very close to downtown Evanston, a very high walkability level as estimated with Walk Score 95 with many possible destinations within a very short distance. Though still very pedestrian friendly, a lower walkability level of Walk Score 70 was estimated for the Technological Institute located in the northern part of the Campus, which was only 0.6 miles away from the former. The website also produces scores for transit and bicycle modes, wherever possible. Pedestrian friendliness metrics included population density, average block length, intersection density, link/node ratio, and route directness. However, such precision in evaluation of walkability requires rich

Figure 1. Walkscore ® evaluation of a) the Graduate School and b) the Technical Institute in Northwestern University Campus, Evanston, IL

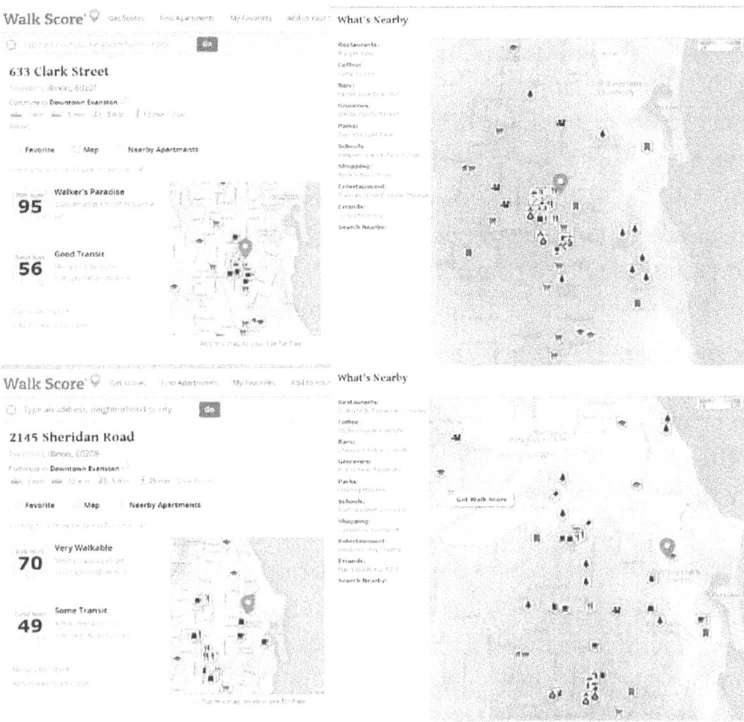

data sources which were listed as Google, Education.com, Open Street Map, the U.S. Census, Localeze, and places added by the Walk Score user community for Walk Score ® analysis (Walkscore, 2011). For developing countries, availability of such diverse data is rather challenging, thus, Cubukcu et al. (2015) presented a GIS-based method to measure the walkability on street level as an alternative way for developing countries. This method was used to evaluate Izmir, Turkey with 6500 street segment, and calculated the walkability score based on betweenness, centrality - and accessibility scores derived from the street network and land use.

Pedestrian Level of Service

The concept of "Level-of-service (LOS)" is merely an abstraction of quality of the considered phenomenon. If it is traffic we are considering, LOS shows the state of traffic in a qualitative way for the selected time and location. LOS is generally represented by levels of A through F, where A shows high levels of comfort and/or available capacity, while F stand for congestion and/or severe delays in the system. For different traffic flows (vehicular, pedestrian, bicycle, transit, etc.) different definitions of LOS are made based on the characteristics and factors affecting the flows, such as capacity usage, delay, fluidity of the flow, etc.

Pedestrian Level of Service (PLOS) got less attention than vehicular LOS in traffic engineering. As an early attempt, Polus et al. (1983) investigated features and properties of pedestrian flow on sidewalks in Haifa, Israel which reflected an inverse proportion between speed and density. Mori and Tsukaguchi

(1986) evaluated "ordinary sidewalks" based on pedestrian behavior and opinion. The need and lack of PLOS in design, maintenance and improvement of sidewalks, walkways and alleys for the purposes of safety, security, continuity and comfort was brought into attention by Yukubousky (1994), but, it fell short in providing any details about PLOS. A comprehensive evaluation of design and traffic flow conditions along sidewalks for the city of Gainesville by Dixon (1996) formed the basis of a design based PLOS rating method. Another PLOS methodology focusing more on the design aspects on/along sidewalks was proposed by Jaskiewicz (1999), and called as Trip Quality Method. Highway Capacity Manual (HCM) offered different formulations and thresholds for the PLOS on sidewalks, crosswalks and stairways (2000, 2010); but, all were based on the concept of "space per pedestrian", which was considered a measure of pedestrian comfort and mobility. Focusing on pedestrians' perception of safety and comfort in the roadside environment to provide a measure of PLOS, Landis et al. (2001) grouped the factors which had a significant effect on pedestrian environment under three perspectives: performance measures (sidewalk capacity), quality (enjoyment) aspects, perceived safety or comfort. As an output, the model provided guidance on how to better design pedestrian environments on presence of sidewalk, lateral separation between vehicle traffic, motor vehicle volume and speed. Muraleetharan et al. (2003) used a conjoint technique to analyze the PLOS by determining the importance of the sidewalk characteristics such as, flow rate of pedestrians, sidewalk width, existence of obstacles, crossing facilities, etc., and suggested that the most significant factors were the pedestrian flow rate and the sidewalk width, but not obstacles.

Tan et al. (2007) assessed the PLOS by analyzing the relationship between the pedestrian perceptions and the quality of road facilities, as well as, traffic flow operation. Pedestrian survey was conducted to investigate the primary factors which had an effect on PLOS, followed by a step-wise regression model. Factors significantly affecting the pedestrian LOS were the bicycle volume, the vehicle volume, the pedestrian volume, driveway access frequency and the distance between sidewalk and vehicle lane. Sponsored by the Danish Road Directorate, Jensen (2007) developed a method to measure pedestrian and bicyclist stated satisfaction for road sections based on the existing traffic operations, geometric conditions and some other variables. The results demonstrated that vehicular volume and speed, urban land uses, rural landscapes, type and width of pedestrian and bicycle facilities, number and width of drive lanes, pedestrian volumes, bicyclists and parked cars, presence of median, trees and bus stops significantly affected the level of satisfaction. Petritsch et al. (2008) developed a user-based method for pedestrians instead of a provider-based method on evaluating the quality of service. As part of the study; sidewalk width, separation of walkway from traffic, buffer width, traffic volume and speed, pedestrian volume and on-street parking effect was analyzed for pedestrian level of service on walkway segments. As a more special case, Asadi- Shekari et al. (2014) focused on the pedestrian level of service for campus streets to evaluation and improvement. 27 "pedestrian design indicators" based on 20 different guidelines were listed. The researchers composed a formulation depends on all 27 indicators and determine the PLOS for Universiti Teknologi Malaysia Campus.

With a slight change in focus, Florez et al. (2014) defined pedestrian's quality of service, named "Quality of Pedestrian Level of Service (Q-PLOS)" by interviews conducted at three matches of FIFA Confederations Cup at Maracana Stadium in Rio de Janeiro. The results showed that the most important attributes for the travel on foot were accessibility, comfort, reliability, convenience, safety, security. Similarly, Talavera-Garcia and Soria-Lara (2015) also focused on Q-PLOS evaluated based on the "quality of urban design" for pedestrians and walking need relations of it. Based on previously defined walking needs, the results grouped under four groups; accessibility, safety, comfort and attractiveness. Daniel et al. (2016) developed a method for pedestrian facilities, called pedestrian footpath level of service

(FOOT-LOS), provided both LOS determination and determines the factors dominantly affect pedestrian LOS. Method used factors such as pedestrian flow, traffic volume, footpath width, speed, obstructions, etc. under three categories: physical, location and user attributes. Kadali & Vegadiri (2016) defined a measure of effectiveness concept (depending on pedestrian safety, delay, pedestrian behavior, etc.) for evaluation of pedestrian environment which differed by the type of pedestrian facility in metro stations in China by comparison of available studies.

EVALUATION OF WALKABILITY AND PLOS

As shown for both walkability and PLOS concepts, different definitions and evaluation methods have been developed and tried, which had strength and weakness in different aspects. However, there has not been any consensus on either of them or any evaluation methods for them. At this point, there seems to be two problems blocking the progress in evaluating walking as a mode discussed below.

Dimensions of PLOS

Variety in the definition and scope of PLOS studied resulted in comparative analyses made to better illustrate the situation about evaluation of PLOS (Yukubosky, 1994; Singh & Jain, 2011; Sisiopiku, 2006; Hasan et al., 2015; Kadali & Vegadiri, 2016). These studies discussing the strengths and weaknesses of the existing PLOS methods based on the factors employed showed that these methods mostly had rather narrow scopes either focusing on one or two dimensions of the PLOS concept. Comparing six PLOS methods, Singh and Jain (2011) grouped the methods into two categories:

i) Capacity based methods and ii) Roadway characteristics based methods. Capacity based methods were found mostly useful in planning progress to give base information about pedestrian facilities, such as HCM method. Roadway characteristic based methods mostly had a focus on evaluation of pedestrian perception, characteristics of walkways such as Trafitec Model (Jensen, 2007) and SCI Model (Landis et al., 2001). Another classification of PLOS methods by Tan et al. (2007) was made in three categories: i) PLOS considering the pedestrian flow operation (as in HCM), ii) PLOS based on road environment quality (as in Trip Quality method) and iii) PLOS considering both Australian method (Gallin, 2001) and Landis method (2001).

Sisiopiku et al. (2007) reviewed, compared and contrasted five PLOS methods, more commonly used and accepted five methods in the literature (see Table 2). In a more comprehensive study, Christopoulou (2012) compared 11 methods which were created based on the USA conditions (see Table 3). Hasan et al. (2016), compared five different PLOS methods (Australian Method, HCM, Trip Quality Method, Landis Model and Conjoint Analysis), and emphasized their opportunities and limitations.

A more comprehensive classification of measures employed in PLOS evaluations can be made under five categories as shown in Figure 2 including:

1. Traffic,
2. Pedestrian,
3. Walkway,
4. Land use, and
5. Safety-Comfort dimensions.

Figure 2. Pedestrian Level of Service (PLOS) measures employed in current methods

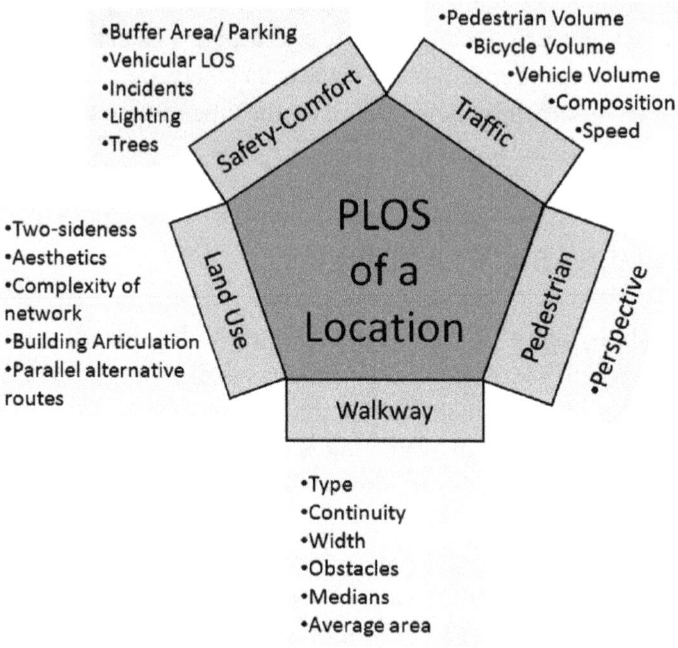

Traffic dimension represents evaluation of pedestrian traffic on the walkway and other modes found important, as well as the composition of vehicular traffic and speed variations. Besides the design features related to safety/comfort dimension (such as existence of buffer areas, lighting and shading opportunities, etc.), vehicular LOS, existence and prevention of incidents are also categorized in this group, as they are important in pedestrian's perception of safety or comfort while walking. Land use characteristics provide convenience to users by sidewalk on both side (two-sideness), complexity of network and parallel alternative routes; while having attraction by aesthetic, building articulation (for the sake of attractiveness) and such kind of design aspects. Another dimension which gets the highest attention in the PLOS literature is the physical infrastructure characteristics that include measures showing function (cross-section, sidewalks, intersections or walkways) and geometric design features (width, continuity, obstacles, etc.). Though included in the figure, "pedestrian" dimension is limited to include their perspective only in few studies (Tan et al., 2007). However, most of the available PLOS methods do not include any measures to include human factors affecting walkability in a region. To avoid duplication of measures in this compact form, each measure is listed under one dimension based on its major use. But, some of the measures can be used to evaluate more than one dimension. For example, in Christopoulou (2012), vehicle volume was interpreted as a factor which affects the pedestrians' perception of safety and comfort.

Problem of Inconsistency among PLOS Ratings

The variety in the scope and techniques used in PLOS methods bring the potential risk of inconsistency of outcomes. As a consistency test case was designed by Sisiopiku et al. (2007) using five methods (presented in Table 2) to evaluate 13 sidewalks, in the downtown Birmingham, AL. The results confirmed that same segment could receive varied PLOS results according to different methods. Christopoulou

(2012) focused on the five of the methods (Methods 6, 7, 9, 10 from Table 3 and HCM Method) to create a new model for Greek conditions. The results revealed that level of service results could greatly differ according to method, and the inclusion of both qualitative and quantitative parameters could reflect the actual condition. Furthermore, it showed that HCM (2000) overestimated the condition in general. To better illustrate the differences and potential inconsistency among PLOS rating, three methods, HCM, Gainesville Method (GM) and Trip Quality Method (TQM), were selected to evaluate PLOS at 84 walkway locations (shown in Figure 3) in the Middle East Technical University (METU) Campus, Ankara, Turkey (Karatas, 2015; Karatas & Tuydes-Yaman, 2016). The details of the selected methods are presented in detail in Appendix . But, PLOS calculations for two selected location, Point 1 and Point 2, are discussed here with the necessary data summarized in Table 1.

Pedestrian flow counts needed for HCM method were taken on October 11, 2011 with 10 observers at 6 different time periods, starting early in the morning (T1 period: 08:15-09:00) until the end of work hours in the evening (T6 period: 17:00-17:45), to capture the variation in pedestrian activity throughout the day. Pedestrian counts were collected for 3-minute periods as used by Parma et al. (2013) and Maddock et al. (2012), and, which were later converted to 15-minute volumes used in HCM, by simple proportionality. The results showed the highest 15-minute pedestrian volume of 260 pedestrians (during T5 period) and 455 pedestrians (during T6 period) at locations Point 1 and Point 2, respectively. Assuming no platooning on the campus walkways and due to lack of pedestrian speed data, PLOS rating was performed based on flow rate (p/min/ft.) criterion shown in Table 4 which produced the scores of

$$PLOS_{HCM\_1} = \frac{V_{15}/15}{w} = \frac{65/15}{1.90/3.28} = 0.70 \text{ p/min/ft} \rightarrow \text{Score} < 5;\; PLOS_{HCM\_1} = A$$

Figure 3. (a) METU campus layout and (b) PLOS analysis locations

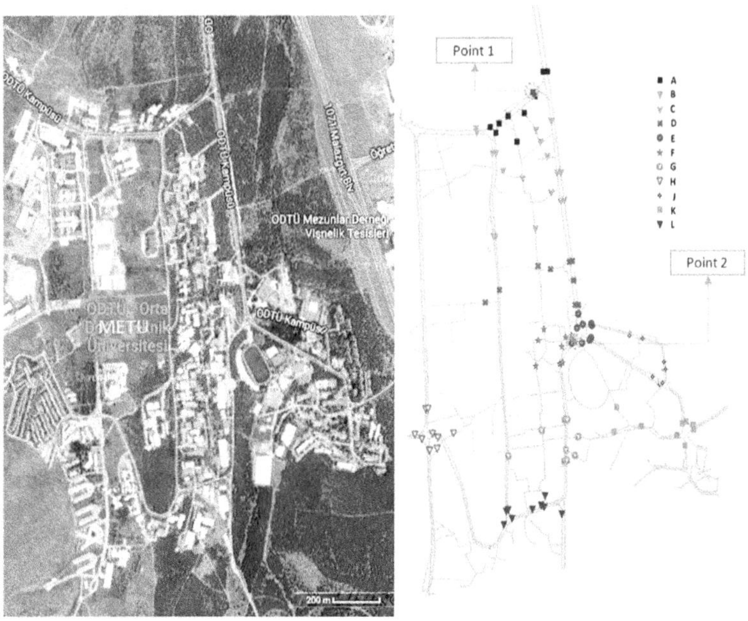

Table 1. PLOS data for two walkway locations at METU campus

<table>
<tr><td colspan="7">Sidewalk-Point 1 (Width=1.9 m)
</td><td colspan="7">Sidewalk-Point 2 (Width=1.57 m)
</td></tr>
<tr><td colspan="7">✓ Continuous on both sides
✓ Medians
✓ Buffer not less than 1 m
✓ Shade trees
✓ Vehicular LOS=A, B(off-peak)
✓ Vehicular LOS=F (morning-peak)
✓ No maintenance problems</td><td colspan="7">✓ Continuous on both sides
✓ Shade trees
✓ All campus vehicular LOS=A, B
✓ Minor or infrequent problems</td></tr>
<tr><td colspan="7">× No-off street parallel alternative route
× No lighting
× Public Transportation support</td><td colspan="7">× No-off street parallel alternative route
× No lighting
× No buffer
× Public Transportation support</td></tr>
<tr><td colspan="3"></td><td colspan="2">Rater 1</td><td colspan="2">Rater 2</td><td colspan="3"></td><td colspan="2">Rater 1</td><td colspan="2">Rater 2</td></tr>
<tr><td colspan="3">• Enclosure
• Network Complexity
• Building Articulation
• Complexity of spaces
• Transparency
• Buffer
• Shade trees
• Awnings
• Phy. Comp./ cond.</td><td colspan="2">=Very Poor
=Very Poor
=Very Poor
=Poor
=Very Poor
=Good
= Excellent
= Very Poor
= Good</td><td colspan="2">Poor
Average
Very Poor
Average
Very Poor
Excellent
Excellent
Very Poor
Excellent</td><td colspan="3">• Enclosure
• Network Complexity
• Building Articulation
• Complexity of spaces
• Transparency
• Buffer
• Shade trees
• Awnings
• Phy. Comp./ cond.</td><td colspan="2">=Very Poor
=Very Poor
=Very Poor
=Average
=Very Poor
=Very Poor
= Good
= Very Poor
= Good</td><td colspan="2">Very Poor
Average
Very Poor
Poor
Poor
Very Poor
Average
Very Poor
Average</td></tr>
<tr><td colspan="3">Rater Average =</td><td colspan="2">D</td><td colspan="2">C</td><td colspan="3">Rater Average =</td><td colspan="2">D</td><td colspan="2">D</td></tr>
<tr><td>Time Period*</td><td>T1</td><td>T2</td><td>T3</td><td>T4</td><td>T5</td><td>T6</td><td>Time Period*</td><td>T1</td><td>T2</td><td>T3</td><td>T4</td><td>T5</td><td>T6</td></tr>
<tr><td>$V_{15}(p)^{**}$</td><td>30</td><td>15</td><td>40</td><td>25</td><td>65</td><td>30</td><td>$V_{15}(p)^{**}$</td><td>265</td><td>175</td><td>360</td><td>215</td><td>195</td><td>455</td></tr>
<tr><td>$PLOS_{HCM}$</td><td>A</td><td>A</td><td>A</td><td>A</td><td>A</td><td>A</td><td>$PLOS_{HCM}$</td><td>A</td><td>A</td><td>A</td><td>A</td><td>A</td><td>B</td></tr>
<tr><td>$PLOS_{GM}$</td><td colspan="6">B</td><td>$PLOS_{GM}$</td><td colspan="6">D</td></tr>
<tr><td>$PLOS_{TOM}$</td><td colspan="6">D</td><td>$PLOS_{TOM}$</td><td colspan="6">D</td></tr>
</table>

* T1=08:15-09:00; T2=09:15-10:00; T3=12:15-13:00; T4=13:15-14:00; T5=16:00-16:45; T6=17:00-17:45

** V_{15} represent the 15-min pedestrian volumes

$$PLOS_{HCM\_2} = \frac{V_{15}/15}{w} = \frac{455/15}{1.57/3.28} = 5.89 \text{ p/min/ft} \rightarrow 5 < \text{Score} < 7;\ PLOS_{HCM\_2} = B$$

During the busiest time periods at Point 1 and Point 2, respectively.

To obtain $PLOS_{GM}$, infrastructure data was collected by one rater and evaluated according to the point scale presented in Table 5. At Point 1, the GM score was calculated as 15, which was the sum of points from six categories as

$$PLOS_{GM\_1} = C1 + ... + C6 = 9 + 0.5 + 1.5 + 2 + 2 + 0 = 15 \rightarrow 14 < \text{Score} < 17;$$

$$PLOS_{GM\_1} = B$$

$$PLOS_{GM\_2} = C1 + ... + C6 = 7 + 0 + 0.5 + 2 + 0 + 0 = 9.5 \rightarrow 7 < \text{Score} < 11;$$

$$PLOS_{GM\_2} = D$$

As seen in the calculations, majority of the loss of points at Point 2 was due to conflict, maintenance problems and Public transportation support.

$PLOS_{TQM}$ measures included collection of relative data, thus, two trained raters were employed to evaluate the campus as a city street, as the method required. For each location, raters decided the sufficiency of 9 measures (M1 to M9) produced rater scores (see Table 1) according to the score points in Table 6 and produced point based scores for Point 1 as

Rater 1:

$$PLOS_{TQM\_1\_1} = \frac{M1_1 + ... + M9_1}{9} = \frac{1+1+1+2+1+4+5+1+4}{9} = 2.22 \rightarrow 2.2 < \text{Score} < 2.7;$$

$$PLOS_{TQM\_1\_1} = D$$

Rater 2:

$$PLOS_{TQM\_1\_2} = \frac{M1_2 + ... + M9_2}{9} = \frac{2+3+1+3+1+5+5+1+5}{9} = 2.90 \rightarrow 2.8 < \text{Score} < 3.3;$$

$$PLOS_{TQM\_1\_2} = C$$

Before, making any decision about producing an overall $PLOS_{TQM}$ rating for Point 1, one has to make a decision about the consistency of the rater evaluations, which is called “inter rater reliability”. If the reliability among the evaluations of two raters were more than 70%, it was accepted as “reliable” and we could produce a way of averaging the two ratings somehow. But, for the evaluations of 84 locations at METU Campus, the similarity between PLOS values of two raters was 69.8%, which was slightly lower than the acceptable limit. An inter rater reliability analysis using the Kappa statistic resulted in a value of Kappa = 0.57 ($p < .001$). As a rule of thumb, values of Kappa from 0.41 to 0.60 are considered moderate agreement (Viera & Garret, 2005; Altman, 1990) and extinguished the re-evaluation process according to Fleiss (1981). Without any further evaluation, average of the ratings were calculated as

$$PLOS_{TQM\_1} = \frac{PLOS_{TQM\_1\_1} + PLOS_{TQM\_1\_2}}{2} = \frac{2.22 + 2.90}{2} = 2.56 \rightarrow 2.2 < \text{Score} < 2.7;$$

$$PLOS_{TQM\_1} = D$$

The analyses by the selected three methods reveals enough to detect potential inconsistency both within and among themselves. HCM method was able to capture time-dependent change of pedestrian activity, but, could not reflect other pedestrian friendly aspects of the neighborhood or sidewalk design. As pedestrian volumes were very small compared to the sidewalk width at Point 1, PLOS ratings were always at "A" level in the HCM rating method; however, GM based ratings were able to reflect the differences in safety-comfort or sidewalk design aspects, but none of the pedestrian flow conditions, such as average space, walking speed, etc. TQM ratings showed that PLOS can differ even among raters for the same location.

Problem of Disconnection of Walkability and PLOS Research

Even if the infrastructures are designed adequately and similarly, utility of walking may vary greatly between cultures and locations. PLOS, either from planning or capacity perspective, is a step that evaluates only current available conditions (flows, capacity or design). Because, contrary to HCM that looks mostly at average space per pedestrian, Gainesville Method evaluates the locations for existing of sidewalk alongside the vehicular roads as a couple and assign higher PLOS score. Trip Quality Method focuses on building articulation and architecture for campuses. Partially, they all reflect conditions or outcomes about "walkability", thus, could be a sub-task in evaluation of walkability.

To establish a link between the more general concept of walkability and more specific concept of PLOS, we investigated the dimensions and measures used in the former, as they relate to those proposed for the latter above and showed the overlapping parts in Figure 4. It can be easily seen that there is a great match between the measures used in PLOS and walkability evaluations, but walkability studies mostly aimed at evaluation of a region (such as a neighborhood, vicinity of transit stations, etc.), whereas PLOS evaluations mostly focused on a location. However, planners have not made much use of traffic related data, such as pedestrian volumes, vehicular traffic volumes, mostly likely because it is very difficult to collect and monitor for urban locations. While land use measures, such as complexity of the network, existence of parallel alternative routes were referred in PLOS methods, similar concepts are considered within connectivity, accessibility, land use density under the "neighborhood surrounding" in walkability studies. Also, most of the issues regarded under safety/comfort in PLOS studies (such as existence of buffer areas, lighting, etc.) are also considered within the scope of this heading in walkability literature. Beyond the walkway characteristics used in PLOS, some walkability studies included walking surface characteristics, directness, etc. which took the aspects as the "road network characteristics" level. What is majorly not covered in the PLOS evaluations is the pedestrian characteristics, such as gender, age, group size, as well as walk purpose or proximity issues such as intended time or destination.

Figure 4. Overlaps of measures employed in current PLOS methods and walkability studies

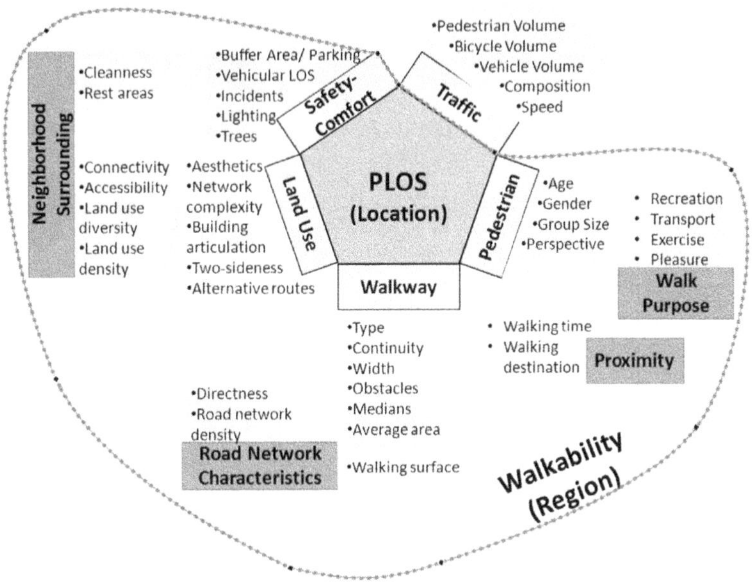

Finally, the existing disconnection between the two concepts, and researchers focusing on them, not only causes loss of energy, but also prevents formation of consensus on basic concepts and procedures necessary to study walking as a mode. At this point in literature, there is a wide enough perspective describing the dimensions and measures. The next step should be the selection of key factors affecting the walkability strongly and some standard procedures to evaluate their impact. To weed the local and common elements, similar studies must be conducted at different locations (such as CBDs, transit station areas, campuses, etc.) repeatedly and comparatively.

SOLUTIONS AND RECOMMENDATIONS

A Consistency Check for PLOS Evaluations

As seen in the PLOS evaluations of two points using different methods, there can be big differences among the ratings and there is no possibility of converting a PLOS rating by one method to the rating of another. However, ratings by different methods for the same locations can be compared in a pairwise fashion to gain insights about their consistency while evaluating the walkability in a region. For this purpose, choosing two PLOS methods (such as TQM versus HCM), we have to form PLOS rating pairs such as (C, A) and (D, B) for study locations Point 1 and Point 2, respectively. When the ratings at every study location are paired, frequencies of 6x6 possible rating pairs $\left(f_{(A,A)}, f_{(A,B)}, \ldots, f_{(F,E)}, f_{(F,F)}\right)$ should be determined. Depicting of these frequencies in a 2D graph as shown in Figure 5, where size of the circles represents the frequency of the possible rating combination, can visually reveal the consistency/inconsistency among the selected methods: If the larger circles (higher frequencies) are located on or around the diagonals, this means selected PLOS methods produced consistent results. For example, for the case

Figure 5. Examples of frequency graph for pairwise evaluation of PLOS ratings

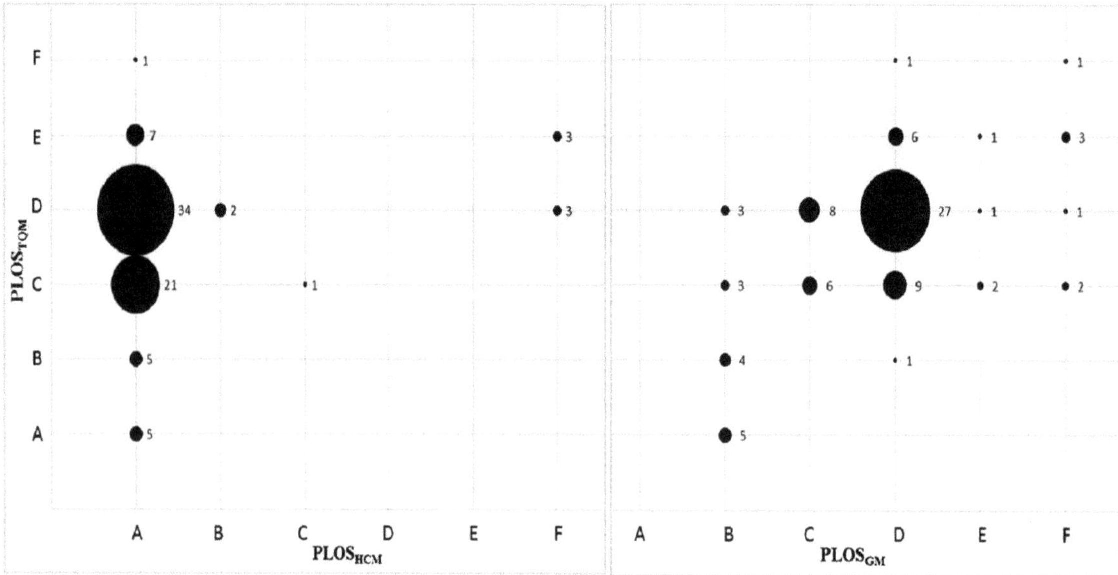

of METU Campus PLOS ratings, $PLOS_{TQM}$ ratings were very inconsistent with the ratings of $PLOS_{HCM}$, considerably consistent with those of $PLOS_{GM}$. This was not unexpected as both TQM and GM focused on evaluation of infrastructure from design measures mostly, while HCM focused on flow characteristics, only.

Integrating PLOS in Walkability Evaluation

Though there is no simple form or method for the evaluation of walkability, certain associations are almost universal, such as connectivity and the directness within the walkway network can promote walking within a neighborhood, making it "pedestrian friendly" in terms of network design. Diversity in the land use is a good, but an indirect, indicator of capacity of walk trip demand in a neighborhood, creating a "pedestrian friendly" region in terms of origins/destinations. While existence of trees, wide walkways, lighting, etc. contributes walkability by creating a "pedestrian friendly" environment in terms of safety and comfort. Furthermore, existence of good public transit and bike sharing services encourages the walkability within proximity of their stations.

On the other hand, walkability can change rapidly even within a short distance, as walkable distance concept itself is short (mostly no more than 1 km) and any of aforementioned encouraging measure can deteriorate within urban design and built environment. Thus, evaluating walkability needs a spatial-temporal evaluation in GIS environment, ideally. Analysis of pedestrian counts at 102 locations (84 of which were walkways) in METU Campus at 6 different time periods in a day, showed such dramatic changes over consecutive segments as well as change over time in Figure 6.

GIS-based walkability evaluations must include different sub-task evaluating different dimension of walkability that can only be achieved, if relevant data is available. While it is possible to create a GIS-based evaluation system with many dimensions considered easily in major cities for most of the

Figure 6. Pedestrian flow volumes and directions in the a) morning (08:15- 09:00) and b) evening (16:45-17:00) hours at METU campus

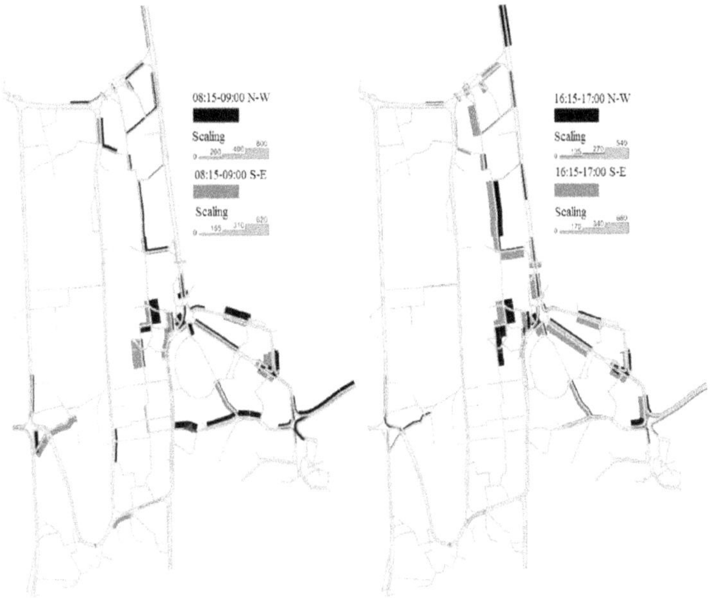

developed countries, lack of data would be the bottleneck of the process for most of the developing countries allowing only simple analyses. A simple depiction of such a modular walkability framework is presented in Figure 7. Within this new nature of walkability evaluation, PLOS should be used not only to evaluate design features, but also to verify the walkability assessments. If everything is done according to the "pedestrian friendly" perspective, then people should be walking on the streets, thus, should be verified by high pedestrian volume.

FUTURE RESEARCH DIRECTIONS

Whether designed as "pedestrian friendly" or not, analysis of walking as a mode should include vigorous measurement of pedestrian activity within a region, as was done for motorized traffic starting in the 1950s. But, this is rather challenging, as walk routes can be very flexible as any available area, paved or not; can be used by pedestrians for walking and in any direction needed, unless restricted by landscaping or land use. While people can move in groups partially or throughout a route, they can also break from one whenever they prefer, which creates much complex flow characteristics that can change rapidly, in terms of speed or density, over short time of periods.

The possibility of route change within very small distances during walking brings the challenge of monitoring many locations within a region, which is very costly, especially if performed manually. Automated pedestrian count technologies via video recording, do exist. However, they are too expensive to monitor a mode, such as walking, used by millions, but for very short distances. Though some simple pedestrian count technologies exist (i.e. Eco-Counter), they mostly do not include details about speed, direction of the movement or any user characteristics that would be more helpful in walkability evalu-

Figure 7. A GIS-based modular walkability framework integrating PLOS

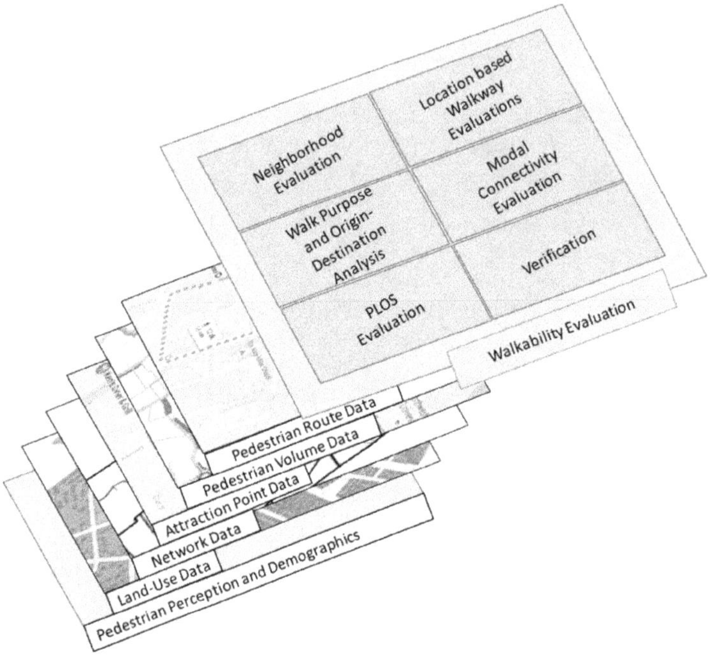

ation methods. Any advancement in sensor technologies that would provide fast, cheap but relatively precise data about pedestrian movements without violating any privacy issues, could create a big leap in understanding and modeling walking.

Another way to look at the problem is to provide services and opportunities that would not only promote walking, but also collect and archive walking data from pedestrians. Having route-based walking data is more revealing in terms of network and infrastructure usage, especially, when digitized in or directly transferred to GIS environment, which can be supported with land use and network information (traffic volumes, roadway type, pavement type, etc.). Increasing availability of smart phones can provide an effective medium to collect such data, but, researchers or agencies must find further promotions than standard reports of burned calories, number of steps, etc. Moreover, mobile application should be developed to get traveler opinion on walkway or neighborhood characteristics.

PLOS concept has to be addressed from a wider perspective to include both qualitative and quantitative measures, which somehow reflects the user's perception and route choice measures. This may even yield to determination of network-level PLOS concept, which would be beyond the current point- or corridor-based perspectives.

CONCLUSION

Though very important and critical in the road map towards more sustainable urban transportation, walking as a mode has not been fully understood, yet. To create a modal shift towards walking, it is important to change travel behavior of an individual that is a result of their perception of walking. While it is possible to start with the "expected utility" concept for walking, which is the first step of most of the mode choice models, it is still not clear what the cost items are that lead to quantification of "walkable/not walkable" conditions.

A popular term in the planning literature is "walkability" whereas there is no consensus on the scope or evaluation methods of it. Similarly, engineers mostly focused on developing tools to evaluate conditions of pedestrian via Pedestrian Level of Service (PLOS) concept, and yet came short of reaching a consensus on the limits or methods for it, either. Though coming from different backgrounds, both groups have been trying to conquer the same domain of parameters affecting walking. But, there are two major problems blocking the progress in modeling walking:

1. Diversity in the PLOS definitions also brought the problem of inconsistency among ratings.
2. Concepts of PLOS and walkability have been developed in a very disconnected way.

As a quick solution to the first, it is possible to use multiple PLOS rating techniques for the evaluation and perform a consistency check via a pairwise comparison. But, for the solution of the latter, a comprehensive definition of walkability has to be made that include and take advantage of PLOS ratings with pedestrian flow measurements as a verification tool. Evaluations of walking must also take advantage of advancement in data collection and GIS-based analyses.

REFERENCES

Abley, S. (2005). *Walkability scoping paper*. Academic Press.

Altman, D. G. (1990). *Practical statistics for medical research*. CRC Press.

Asadi-Shekari, Z., Moeinaddini, M., & Shah, M. Z. (2014). A Pedestrian Level of Service Method for Evaluating and Promoting Walking Facilities on Campus Streets. *Land Use Policy*, *38*, 175–193. doi:10.1016/j.landusepol.2013.11.007

Azmi, D. I., Karim, H. A., & Ahmad, P. (2013). Comparative study of neighbourhood walkability to community facilities between two precincts in Putrajaya. *Procedia: Social and Behavioral Sciences*, *105*, 513–524. doi:10.1016/j.sbspro.2013.11.055

Azmi, D. I., Karim, H. A., & Amin, M. Z. M. (2012). Comparing the walking behaviour between urban and rural residents. *Procedia: Social and Behavioral Sciences*, *68*, 406–416. doi:10.1016/j.sbspro.2012.12.237

Babalik-Sutcliffe, E. (2013). Urban form and sustainable transport: Lessons from the Ankara case. *International Journal of Sustainable Transportation*, *7*(5), 416–430. doi:10.1080/15568318.2012.676152

Bias, T. K., Leyden, K. M., Abildso, C. G., Reger-Nash, B., & Bauman, A. (2010). The importance of being parsimonious: Reliability of a brief community walkability assessment instrument. *Health & Place*, *16*(4), 755–758. doi:10.1016/j.healthplace.2010.01.008 PMID:20303317

Black, W. R. (2010). *Sustainable transportation: Issues and solutions*. Academic Press.

Brown, B. B., Yamada, I., Smith, K. R., Zick, C. D., Kowaleski-Jones, L., & Fan, J. X. (2009). Mixed land use and walkability: Variations in land use measures and relationships with BMI, overweight, and obesity. *Health & Place*, *15*(4), 1130–1141. doi:10.1016/j.healthplace.2009.06.008 PMID:19632875

Christopoulou, P., & Pitsiava-Latinopoulou, M. (2012). Development of a model for the estimation of pedestrian level of service in Greek urban areas. *Procedia: Social and Behavioral Sciences*, *48*, 1691–1701. doi:10.1016/j.sbspro.2012.06.1144

Cubukcu, E., Hepguzel, B., Onder, Z., & Tumer, B. (2015). Active Living for Sustainable Future: A Model to Measure Walk Scores via Geographic Information Systems. *Procedia: Social and Behavioral Sciences*, *168*, 229–237. doi:10.1016/j.sbspro.2014.10.228

Daniel, B. D., Nor, S. N. M., Rohani, M. M., Prasetijo, J., Aman, M. Y., & Ambak, K. (2016, January). Pedestrian Footpath Level of Service (FOOT-LOS) Model for Johor Bahru. In *MATEC Web of Conferences* (Vol. 47). EDP Sciences.

Daniels, R., & Mulley, C. (2013). Explaining walking distance to public transport: The dominance of public transport supply. *Journal of Transport and Land Use*, *6*(2), 5–20. doi:10.5198/jtlu.v6i2.308

Dixon, L. (1996). Bicycle and pedestrian level-of-service performance measures and standards for congestion management systems. *Transportation Research Record: Journal of the Transportation Research Board*, (1538), 1-9.

Ellis, G., Hunter, R., Tully, M. A., Donnelly, M., Kelleher, L., & Kee, F. (2015). Connectivity and physical activity: Using footpath networks to measure the walkability of built environments. *Environment and Planning. B, Planning & Design.*

Fairnie, G. A., Wilby, D. J., & Saunders, L. E. (2016). Active travel in London: The role of travel survey data in describing population physical activity. *Journal of Transport & Health*, *3*(2), 161–172. doi:10.1016/j.jth.2016.02.003

Fleiss, J. L. (1981). *Statistical Methods for Rates and Proportions* (2nd ed.). New York: Wiley.

Florez, J., Muniz, J., & Portugal, L. (2014). Pedestrian quality of service: Lessons from Maracanã Stadium. *Procedia: Social and Behavioral Sciences*, *160*, 130–139. doi:10.1016/j.sbspro.2014.12.124

Front Seat Management, L. L. C. Walk Score® Methodology. (2011). Retrieved from https://www.walkscore.com/methodology, at July 30, 2016.

Gallimore, J. M., Brown, B. B., & Werner, C. M. (2011). Walking routes to school in new urban and suburban neighborhoods: An environmental walkability analysis of blocks and routes. *Journal of Environmental Psychology*, *31*(2), 184–191. doi:10.1016/j.jenvp.2011.01.001

Gallin, N. (2001). Quantifying pedestrian friendliness--guidelines for assessing pedestrian level of service. *Road & Transport Research*, *10*(1), 47.

Gori, S., Nigro, M., & Petrelli, M. (2014). Walkability Indicators for Pedestrian-Friendly Design. *Transportation Research Record: Journal of the Transportation Research Board*, (2464), 38-45.

Guo, Z. (2009). Does the pedestrian environment affect the utility of walking? A case of path choice in downtown Boston. *Transportation Research Part D, Transport and Environment*, *14*(5), 343–352. doi:10.1016/j.trd.2009.03.007

Hajna, S., Dasgupta, K., Halparin, M., & Ross, N. A. (2013). Neighborhood walkability: Field validation of geographic information system measures. *American Journal of Preventive Medicine*, *44*(6), e55–e59. doi:10.1016/j.amepre.2013.01.033 PMID:23683990

Hasan, T., Siddique, A., Hadiuzzaman, M. D., & Musabbir, S. R. (2016). Determining the Most Suitable PLOS Methodology for Dhaka City through Synthesis of Subjective and Objective Measurements. *TRB 2015 Annual Meeting*. Transportation Research Board of the National Academies.

Hoogendoorn, S. P., & Bovy, P. H. (2004). Pedestrian route-choice and activity scheduling theory and models. *Transportation Research Part B: Methodological*, *38*(2), 169–190. doi:10.1016/S0191-2615(03)00007-9

Hsiao, S., Lu, J., Sterling, J., & Weatherford, M. (1997). Use of geographic information system for analysis of transit pedestrian access. *Transportation Research Record: Journal of the Transportation Research Board*, (1604), 50-59.

Humpel, N., Owen, N., Iverson, D., Leslie, E., & Bauman, A. (2004). Perceived environment attributes, residential location, and walking for particular purposes. *American Journal of Preventive Medicine*, *26*(2), 119–125. doi:10.1016/j.amepre.2003.10.005 PMID:14751322

Jaskiewicz, F. (1999). *Pedestrian Level of Service Based on Trip Quality, TRB Circular*. TRB.

Jensen, S. (2007). Pedestrian and bicyclist level of service on roadway segments. *Transportation Research Record: Journal of the Transportation Research Board*, (2031), 43-51.

Kadali, B. R., & Vedagiri, P. (2016). Review of Pedestrian Level of Service: Perspective in Developing Countries. *Transportation Research Record: Journal of the Transportation Research Board*, (2581), 37-47.

Kaiser, E. J., Lawrence, D. F., Peter, O. E., & Thomas, L. S. (1995). *Health and Community Design*. Urbana, IL: University of Illinois Press.

Karatas, P. (2015). *Determination of Pedestrian Level of Service for Walkways: METU Campus Example* (MS Thesis). Middle East Technical University, Ankara, Turkey.

Karatas, P., & Tuydes-Yaman, H. (2016). A pairwise Comparison of Different Pedestrian Level of Service (PLOS) Ratings. *Transportation Research Board (TRB)95th Annual Meeting*.

Karim, H. A., & Azmi, D. I. (2013). Convenience and Safety of Walking Experience in Putrajaya Neighbourhood Area. *Procedia: Social and Behavioral Sciences*, *101*, 318–327. doi:10.1016/j.sbspro.2013.07.206

Landis, B., Vattikuti, V., Ottenberg, R., McLeod, D., & Guttenplan, M. (2001). Modeling the roadside walking environment: pedestrian level of service. *Transportation Research Record: Journal of the Transportation Research Board*, (1773), 82-88.

Leslie, E., Coffee, N., Frank, L., Owen, N., Bauman, A., & Hugo, G. (2007). Walkability of local communities: Using geographic information systems to objectively assess relevant environmental attributes. *Health & Place*, *13*(1), 111–122. doi:10.1016/j.healthplace.2005.11.001 PMID:16387522

Litman, T. (2003). Economic value of walkability. *Transportation Research Record: Journal of the Transportation Research Board*, (1828), 3-11.

Liu, B., Wang, Z. W., & Zhou, Y. B. (2014). Investigating Pedestrian Behavior Differences of Walking Streets and their Effect Factors with Three Wuhu Cases. *Mechanics of Materials*, *587*, 1879–1883.

Maddock, J. E., Ramirez, V., Heinrich, K. M., Zhang, M., & Brunner, M. A. (2012) Statewide Observational Assessment of the Pedestrian and Bicycling Environment in Hawaii, 2010. *CDC- Preventing Chronic Disease Control and Prevention, 9*(1), 1–9.

Manual, H. C. (2000). *HCM* (Vol. 1). Washington, DC: National Research Council, Transportation Research Board.

Manual, H. C. (2010). *HCM* (Vol. 1). Washington, DC: National Research Council, Transportation Research Board. (Original work published 2009)

Millington, C., Thompson, C. W., Rowe, D., Aspinall, P., Fitzsimons, C., & Nelson, N. et al.SPARColl—the Scottish Physical Activity Research Collaboration. (2009). Development of the Scottish walkability assessment tool (SWAT). *Health & Place*, *15*(2), 474–481. doi:10.1016/j.healthplace.2008.09.007 PMID:19041275

Mōri, M., & Tsukaguchi, H. (1987). A new method for evaluation of level of service in pedestrian facilities. *Transportation Research Part A, General*, *21*(3), 223–234. doi:10.1016/0191-2607(87)90016-1

Muraleetharan, T., Adachi, T., Uchida, K. E., Hagiwara, T., & Kagaya, S. (2003). *A study on evaluation of pedestrian level of service along sidewalks and at intersections using conjoint analysis*. In Annual Meeting of Japanese Society of Civil Engineers (JSCE) Infrastructure Planning, Toyohashi, Japan.

Owen, N., Cerin, E., Leslie, E., duToit, L., Coffee, N., Frank, L. D., & Sallis, J. F. et al. (2007). Neighborhood Walkability and the Walking Behavior of Australian Adults. *American Journal of Preventive Medicine*, *33*(5), 387–395. doi:10.1016/j.amepre.2007.07.025 PMID:17950404

Park, S. (2008). *Defining, Measuring, and Evaluating Path Walkability, and Testing Its Impacts on Transit Users' Mode Choice and Walking Distance to the Station* (PhD. Dissertation). University of California, Berkeley, CA.

Park, S., Deakin, E., & Lee, J. (2014). Perception-Based Walkability Index to Test Impact of Microlevel Walkability on Sustainable Mode Choice Decisions. *Transportation Research Record: Journal of the Transportation Research Board*, (2464), 126-134.

Parma, K. D., Koprowski, Y., & Zhang, W. (2013). Origin-Destination and Travel Time Measurement Using Bluetooth Technology: Results and Lessons Learned. ITE Technical Conference and Exhibit.

Peiravian, F., Derrible, S., & Ijaz, F. (2014). Development and application of the Pedestrian Environment Index (PEI). *Journal of Transport Geography*, *39*, 73–84. doi:10.1016/j.jtrangeo.2014.06.020

Petritsch, T., Landis, B., McLeod, P., Huang, H., Challa, S., & Guttenplan, M. (2005). Part 2: Pedestrians: Level-of-service model for pedestrians at signalized intersections. *Transportation Research Record: Journal of the Transportation Research Board*, (1939), 53-62.

Polus, A., Schofer, J. L., & Ushpiz, A. (1983). Pedestrian flow and level of service. *Journal of Transportation Engineering*, *109*(1), 46–56. doi:10.1061/(ASCE)0733-947X(1983)109:1(46)

Rodrigue, J. P., Comtois, C., & Slack, B. (2006). *The Geography of Transport Systems*. Hofstra University, Department of Economics & Geography.

Rodríguez, D. A., & Joo, J. (2004). The relationship between non-motorized mode choice and the local physical environment. *Transportation Research Part D, Transport and Environment*, *9*(2), 151–173. doi:10.1016/j.trd.2003.11.001

Rosenberg, D., Ding, D., Sallis, J. F., Kerr, J., Norman, G. J., Durant, N., & Saelens, B. E. et al. (2009). Neighborhood Environment Walkability Scale for Youth (NEWS-Y): Reliability and relationship with physical activity. *Preventive Medicine*, *49*(2), 213–218. doi:10.1016/j.ypmed.2009.07.011 PMID:19632263

Schlossberg, M. (2006). From TIGER to audit instruments: Measuring neighborhood walkability with street data based on geographic information systems. Transportation Research Record: Journal of the Transportation Research Board, (1982), 48-56.

Singh, K., & Jain, P. K. (2011). Methods of Assessing Pedestrian Level of Service, Journal of Engineering. *Research Studies (Pullman, Wash.)*, *2*(1), 116–124.

Sisiopiku, V. P., Byrd, J., & Chittoor, A. (2007) Application of Level of Service Methods for the Evaluation of Operations at Pedestrian Facilities, In *TRB 2007 Annual Meeting*. Transportation Research Board of the National Academies. doi:10.3141/2002-15

Sun, G., Zacharias, J., Ma, B., & Oreskovic, N. M. (2016). How do metro stations integrate with walking environments? Results from walking access within three types of built environment in Beijing. *Cities (London, England)*, *56*, 91–98. doi:10.1016/j.cities.2016.03.001

Talavera-Garcia, R., & Soria-Lara, J. A. (2015). Q-PLOS, developing an alternative walking index. A method based on urban design quality. *Cities (London, England)*, *45*, 7–17. doi:10.1016/j.cities.2015.03.003

Tan, D., Wang, W., Lu, J., & Bian, Y. (2007). Research on Methods of Assessing Pedestrian Level of Service for Sidewalk. *Journal of Transportation Systems Engineering and Information Technology*, *7*(5), 74–79. doi:10.1016/S1570-6672(07)60041-5

Van Dyck, D., Cardon, G., Deforche, B., Owen, N., & De Bourdeaudhuij, I. (2011). Relationships between neighborhood walkability and adults physical activity: How important is residential self-selection? *Health & Place*, *17*(4), 1011–1014. doi:10.1016/j.healthplace.2011.05.005 PMID:21596613

Viera, A. J., & Garrett, J. M. (2005). Understanding Interobserver Agreement: The Kappa Statistic. *Family Medicine*, *37*(5), 360–363. PMID:15883903

Walkscore ®. (2011). Retrieved from https://www.walkscore.com/

Yazicioglu Halu, Z., & Yürekli, F. (2011). Yürünebilirlik Kavramı ve Kentsel Mekanlarda Yürüme. *İTÜ Journal Series A: Architecture, Planning, Design*, *10*(2), 29–38.

Yukubousky, R. (1994). *Level of Service Standards* (Vol. 7926). Academic Press.

KEY TERMS AND DEFINITIONS

Connectivity: Level of directness of links and density of connections in a network.

Level of Service (LOS): A qualitative ranking of the traffic condition under flow and capacity characteristics experienced by users (HCM, 2000, 2010).

Pedestrian Level of Service (PLOS): A LOS concept customized to evaluate pedestrian flow or environment.

Proximity: Level of distance or time desired in walking.

Quality of Road Network: Capacity of the road network showing in terms of "pedestrian friendly" features such as width and continuity of walkways, existence of parallel routes, medians, etc.

Sidewalk: Walkways allocated alongside the vehicular roads.

Walkability: Quality of walking. Also defined as "walking friendly".

Walkway: Any road capacity, paved or unpaved, used by pedestrians (such as sidewalks, alleys, crosswalks, etc.).

APPENDIX

Comparisons of PLOS Studies

See Tables 2 and 3.

PLOS Evaluations by Three Different Methods

HCM-Based PLOS Method

The main steps of PLOS determination are given as:

Step 1: Determination of effective walkway width, w_E,
Step 2: Calculation of pedestrian flow rate, v_p,
Step 3: Calculation of average pedestrian space, A_p,
Step 4: Determination of LOS (for walkways with and w/o platooning, and stairway)

The effective walkway width determination is rather cumbersome, which require measurement of total walkway width, fixed-object (such as trees, signs, etc.) effective width and shy distance. This effective width has to be calculated at every point of change. Pedestrian flow rate per unit width, $v_p = \frac{v_{15}}{15xW_E}$ (p/ft/min) is taken as 15-minute pedestrian counts. The average pedestrian space, A_p, is a surrogate measure for density calculated by dividing pedestrian speed, S_p, by pedestrian flow, v_p, if speed is

Table 2. Comparison of factors included in pedestrian sidewalk assessment methods

Issue	HCM 2000	Australian	Trip Quality	Landis	Conjoint
Geometry	Ped. Space; v/c ratio	Path width	Ped. Path components	Motor path width; On-street parking	Width and separation
Flow	Pedestrian flow; Speed	Pedestrian volume; Mix of users	Not considered	Vehicle flow; Speed	Flow rate
Path	Not considered	Obstructions; Connectivity; Environment	Route; Buffer; Trees/ Overhangs	Sidewalk an buffer widths	Obstructions
Vehicle Conflicts	Not considered	Potential for conflicts; Crossing opportunities	Not considered	Not considered	Bicycle events
Security	Not considered	State of security	Buffer; Transition to other spaces	Not considered	Not considered
Support facilities	Not considered	Exist or not	Not considered	Not considered	Not considered
Quality of path	Not considered	Surface quality	Path condition	Not considered	Not considered

Source: Sisiopiku et al., 2007.

Table 3. Comparison of the pedestrian level of service methodologies with regard to their criteria

	Methodologies	Volume (veh, ped, bic)	Traffic Incidents	Safety/Comfort	User's Aspect
1	Mozer D.	Veh volume and speed, ped volume	-	Buffer zone	Taken into account
2	Dowling R et. al.	Veh volume and speed, ped volume	Ped crossing	Buffer zone, on street parking	Taken into account
3	Landis B. et al	Veh volume and speed	-	Buffer zone, on street parking	Taken into account
4	FDOT	Veh volume and speed	-	Buffer zone, on street parking	Taken into account
5	TRB	Ped volume	-	-	-
6	Jaskiewicz F.	-	-	Buffer zone, accessibility	-
7	Gallin N.	Ped volume, users' categories		Personal Safety	-
8	Jensen S.	Veh, ped and bic volume	-	Buffer zone, land uses, trees	Taken into account
9	Muraleetharan T. et al	Ped volume	-	Trees, parked cars etc	Taken into account
10	Tan D. et al	Ped, veh and bic volume	Roadway crossing	Distance between sidewalk-outside traffic lane	Taken into account
11	Dixon L.	Veh volume	Taken into account	Facilities easing pedestrian movement	-

Source: Christopoulou & Pitsiava-Latinopoulou, 2012.

measured. The guidebook also suggested a capacity value of 23 p/min/ft. value for walkways with random flow. PLOS limits from A to F are decided using Table 4. The table allows different measures for PLOS determination according to: data availability; average space, flow rate, average speed and v/c ratio.

Gainesville Method

This method assumes a grading point system of 1 to 21, and evaluates the PLOS of a walkway segment by six categories (see Table 5). Category I, pedestrian facility provided, evaluates the characteristics of the pedestrian walking area according to existing of sidewalks in sides of vehicular lanes, sidewalk width and off-street parallel alternatives and gives up to 10 points. This category gets approximately half of 21 points and the most important indicators of the method. Following category, conflicts, gets up to 4 points and gives information about interaction between the motor vehicles and pedestrians. Category III, IV and V give up to 2 points. Amenities stated the condition of sidewalk in the right-of-way, comfort and convenience for pedestrians, and Motor vehicle LOS determine the effect of the motorized LOS on pedestrians. Category V evaluate impact of the obstacles (i.e. holes, tree root intrusion, patching) which have a big impact on pedestrian flows. Category VI represents the existence of a bus stop on the connection of intermodal accessibility. For each sidewalk/walkway segment, category points are determined, and then the summation of all category points gives the total grade for segments. According to Pedestrian

Table 4. Average flow level of service for walkways and sidewalks w/o platooning in HCM (2010)

LOS ($PLOS_{HCM}$)	Average Space (AS) (ft^2/p)	Related Measures		
		Flow Rate (p/min/ ft.)[a]	Average Speed (ft./s)	v/c Ratio[b]
A	AS >60	Score ≤5	4.25<Score	Score≤0.21
B	60≥ AS >40	5< Score ≤7	4.17<Score≤4.25	0.21<Score≤0.31
C	40≥ AS >24	7< Score ≤10	4.00<Score≤4.17	0.31<Score≤0.44
D	24≥ AS >15	10<Score≤15	3.75<Score≤4.00	0.44<Score≤0.65
E	15≥ AS >8	15<Score≤23	2.50<Score≤3.75	0.65<Score≤1.00
F	8≥ AS	Variable	Score≤2.50	Variable

Notes: [a] Pedestrians per minute per foot of walkway width; [b] v/c ratio= flow rate/23. LOS is based on average space per pedestrian.

Table 5. Pedestrian level of service (PLOS) categories in Gainesville method

($PLOS_{GM}$)	Score* (Points)
A	Score >17
B	14< Score ≤17
C	11< Score ≤14
D	7< Score ≤11
E	3< Score ≤7
F	Score ≤ 3

Point Categories and Criteria

Category	Criterion	Points	Category	Criterion	Points
C1: Pedestrian Facility Provided (Max Value=10)	No Continuous or Non-existing	0	C3: Amenities (Max Value=2)	Buffer not Less Than 1m(3.5')	1
	Continuous on One Side	4		Benches or Pedestrian Scale Lighting	0.5
	Continuous on Both Sides	6			
	Min. 1.53m(5') Wide& Barrier Free	2		Shade Trees	0.5
	Sidewalk Width>1.53m(5')	1	C4: Motor Vehicle LOS (Max Value=2)	iv. LOS E,F or 6 or More Travel Lanes	0
	Off-Street/ Parallel Alternative Facility	1		v. LOS D and <6 Travel Lanes	1
C2: Conflicts (Max Value=4)	Driveways & Side streets	1		vi. LOS A,B,C and <6 Travel Lanes	2
	Pedestrian Signal Delay 40Sec. or Less	0.5	C5: Maintenance (Max Value=2)	iv. Major or Frequent Problems	-1
	Reduced Turn Conflict Implementation	0.5		v. Minor or Frequent Problems	0
	Crossing Width 18.3m(60') or Less	0.5		vi. No Problems	2
	Posted Speeds	0.5	C6: TDM/ Multi- Modal (Max Value=2)	No Support	0
	Medians Present	1		Support Existing	1

Source: Dixon, 1996.
*Score: C1+ C2+ C3+C4+C5+C6.

LOS ratings given in Table 5 and the total segments grade; LOS for pedestrians are defined in the scale of A to F. Finally, segment score could be turned into corridor score by using weighted averages by the ratio of segment length over corridor length and the available segment scores.

Trip Quality Method

There are nine specific qualitatively evaluated measures as listed in Table 6. A 5-scale rating is applied as 5= excellent, 4=good,3=average, 2=poor and 1=very poor. The scores of each measure averaged and overall LOS was determined according to Table 6. The method mostly suitable for city street design aspects such as shopping areas and working places through a street.

Enclosure / definition symbolizes the degree to which the edges of the street are defined, when buildings are constructed side-by-side along the sidewalk in the focus of minimizing the empty places between in and in front of buildings. Good enclosure increases the positive effect on safety and also on aesthetics. Complexity of path network indicates the route choice alternative in a well-designed land use pattern; and the alternative paths between each origin-destination pair. Building design, materials, color and décor has a good impact on pedestrian walking choice, speed and interest under the title of building articulation. The varied, interesting and rapidly changing areas increases the level of interest on walking such as, natural elements, spaces, plazas and parks shows the complexity of spaces.

Overhangs/ awnings/ varied roof lines has importance on both appearance and functional perspectives. Like building articulation, it increases the level of interest, besides that it works as a protection against to sunlight and rainfall. Buffer is the barrier between pedestrians and vehicular traffic and it is very important on safety. Shade trees improve the sun protection and also the aesthetic of the sidewalks. Transparency determines the interaction between public and private by the use of windows, outdoor displays and sidewalk cafes with a smooth interface. The last measure defines the physical component/ condition, both structural and functional view, concurred by sidewalk configuration and condition, vehicular speed and lighting characteristics.

Table 6. Overall pedestrian LOS rating according to Trip Quality Method

LOS ($PLOS_{TQ}$)	Averaged Score* (Points)			Pedestrians' Pleasantness	
A	$4.0 \leq$ Score ≤ 5.0			Very Pleasant	
B	$3.4 \leq$ Score ≤ 3.9			Comfortable	
C	$2.8 \leq$ Score ≤ 3.3			Acceptable	
D	$2.2 \leq$ Score ≤ 2.7			Uncomfortable	
E	$1.6 \leq$ Score ≤ 2.1			Unpleasant	
F	$1.0 \leq$ Score ≤ 1.5			Very Unpleasant	
Evaluation Measures and Point Levels					
Category	**Very Poor**	**Poor**	**Average**	**Good**	**Excellent**
M1: Enclosure/ Definition	1	2	3	4	5
M2: Complexity of Path Network	1	2	3	4	5
M3: Building Articulation	1	2	3	4	5
M4: Complexity of Spaces	1	2	3	4	5
M5: Overhangs/ Awnings/ Varied Roof Lines	1	2	3	4	5
M6: Buffer	1	2	3	4	5
M7: Shade Trees	1	2	3	4	5
M8: Transparency	1	2	3	4	5
M9: Physical Component/ Condition,	1	2	3	4	5

Source: Jaskiewicz, 1999.
*Average Score Points= (M1+M2+ …M9)/9.

Chapter 3
Cycling Related Mental Barriers in Decision Makers:
The Austrian Context

Tadej Brezina
Vienna University of Technology, Austria

Alberto Castro Fernandez
Independent Researcher, Switzerland

ABSTRACT

Cycling is increasingly being implemented as a viable option for sustainable mobility in transport master plans from the local to the regional and even national level. However, when the implementation of actual measures is nearing the final stage, often mental barriers of decision-makers surface. These mental barriers refer to the resistance to innovation adoption, a phenomenon well documented in the social sciences. In our policy debate we use a list of factors that dissuade people from cycling from the literature to derive a classification of mental barriers in legislation, planning and infrastructure. Mental barriers have an influence on out-turn cycling levels and we provide exemplary evidence from Austria. The nature of these barriers, including the 'vicious cycle of mental barrier construction', is explored, before identifying two possible approaches to reduce, or even eliminate mental barriers. These approaches are: social pressure upon and self-experience by decision-makers.

INTRODUCTION

After some decades of car dominance captivating the thinking and actions of decision-makers in urban mobility, in most European countries cycling is again regarded as a vital ingredient of a sustainable urban transport system. This is shown in a plethora of policy papers and master plans that mostly agree on cycling as being crucial for the improvement of society and their transport regimes – ranging from local to national levels, for example (Koch, 2006; Land Salzburg, 2006; Rosinak & Partner & Besch+Partner, 2006; Strele, 2010; Winkler & Oblak, 2003). These documents also unanimously state that efforts for more cycling need to be pursued. However, problems arise when measures need to be put into practice to

DOI: 10.4018/978-1-5225-2116-7.ch003

meet the ambitious goals and actually increase bicycle use. Vigar (2000) has found, that while a definitive shift in transport planning rhetoric has been observed, shifts towards alternative modes and demand management regimes have been very slow partly due to political difficulties.

Two types of obstacles to increase the level of cycling can be identified: 1) dissuading factors for citizens to actual bicycle use and 2) mental barriers of decision-makers to implement bicycle oriented transport policies.

There have been numerous in-depth studies on the factors influencing bicycle use and cyclists' behaviour, which have helped to provide a clearer picture (Alrutz, Bohle, & Willhaus, 1998; Aultman-Hall, 1996; Christmas, Helman, Butress, Newman, & Hutchins, 2010; Hagemeister, Schmidt, Seidel, & Schlag, 2004; Handy & Xing, 2011; Jones, 2001; Kirner Providelo & de Penha Sanches, 2011; Menghini, Carrasco, Schüssler, & Axhausen, 2010; Sener, Eluru, & Bhat, 2009). Heinen, van Wee, and Maat (2010) for example have carried out an extensive literature review to show the most relevant factors influencing the level of cycling. The factors can be classified in the following categories: built environment,

psychological factors, socio-economical factors and natural environment[1]. The built environment appears to be named among the most prominent factors. Table 1 shows a summary of factors mentioned by Heinen et al. (2010) along each category and their tested influence.

Natural environment factors such as temperature, rain and wind cannot be influenced. However, others like hilliness, beauty of the route and darkness have a potential to be improved, e.g. by finding better routes, plants and installing street lights respectively. These are actions of decision-makers, i.e. politicians, planners and technicians at national, regional and municipal levels. Psychological factors and socio-economic factors are linked to collective and personal frameworks and attitudes, but they can also be modified through soft measures of decision-makers such as specific campaigns and traffic

Table 1. Factors influencing the level of cycling, modified after Heinen et al., 2010.

Built Environment		Psychological		Socio-Economic		Natural Environment	
Trips distances	-	Attitude	+	Car ownership	+	Hilliness	-
City size	-	Public image	+	Bicycle ownership	+	Beauty of the route	(+)
Population density	+	Ideological beliefs	+	Costs of transport	+	Temperature	+
Mixed land use	+	Perceived behavioral control	+			Rain	-
Secured bicycle parking	+	Habits	+			Wind	-
Safety	+					Darkness	-
Density of cycle network	(+)						
Segregation grade of cycle ways	(+)						
Quality grade of cycle ways	(+)						
Continuity of cycle ways	(+)						
Road width	(-)						
Traffic lights and stops	(-)						
A positive operator means that the stronger pronounced the factor, the stronger the level of cycling. A negative operator means that the higher the factor, the lower the level of cycling. Operators in brackets mean that evidence with some contradictory results was found.							

Source: Heinen et al., 2010.

education. Ultimately, factors influenced by the built environment are exclusively caused by decision-makers. Therefore, decision-makers are crucial for minimizing the negative factors impacting on bicycle use among citizens and can have a big influence on the level of cycling. Decision-makers' actions can be a driving force for cycling when implemented correctly, but they can also discourage cycling when policy results are not satisfactory. We focus in this paper on the barriers of decision-makers that appear to be the reason for inconsistencies, avoiding consequently a correct implementation of cycling policies. The objective is to explore what mental barriers in decision-makers are, why they exist and how to abolish them.

In the next section we classify and adumbrate types of mental barriers. Section number three shows exemplary evidence from Austria. Before the conclusions, we discuss ways to overcome these mental barriers.

LITERATURE ON MENTAL BARRIERS

When planning for cycling many decision-makers enter new territory, because education and planning in the past was concentrating mainly on car traffic. Decision-makers therefore often lack sufficient knowledge for essential bike planning, which mainly also stems from little hands-on experience by them. By mental barriers we understand the resistance to innovation adoption that cause reservations and limiting beliefs hindering decision-makers to introduce fully suitable solutions for everyday and every-use cycling.

Decision-maker's misinterpretations of user needs in transport have been documented extensively by Brög (1991, 1994, 1996). Davies, Brook-Carter, and Gardner (2000), Ausserer and Risser (2001) and Daley & Rissel (2011) with their surveys give ideas on organisational and political attitudes towards and images of cycling. Mental barriers have been previously analyzed in different publications. Banister (1996) suggested strategies (based on principles, objectives, key concerns and actions) for four sustainability fields: existing resources, land use, energy and transport. Regarding transport the author pointed out as required actions: the restriction of car use and parking, the promotion and improvement of environmentally friendly conditions, as well as awareness and information. However, he identified the following barriers that hinder the decision of implementing these measures:

- Inappropriate planning for the given context,
- Price of parking not showing the real value,
- Reluctance of users to use public transport, to cycle or walk,
- The strong wish of citizens to travel by car,
- Fragmentation of public transport connections, and
- Lack of coordination in municipal departments of planning and transport.

To evaluate barriers and drivers of the increase of active modes policies, Cole, Burke, Leslie, Donald, & Owen (2010) carried out structured interviews of managers of public authorities and private companies and associations in Queensland, Australia. The result was that respondents mainly agreed with issues such asconstruction of infrastructure, improvement of public transport service, support of walking and cycling groups, political leadership and coordination and cultural issues have to be addressed. Additionally, the responders pointed to a limit of technical knowledge, a cultural influence and a lack of identification with the role of active mobility in decision-makers and technicians as a likely cause for the limited success of policies promoting active modes. Low, Gleeson, and Rush (2003) analyzed infrastructure policies and institutional arguments in the two largest cities of Australia, Sydney and Melbourne, in the past four decades. In order to reach a more sustainable transport scheme, they concluded that understanding "not only of what should be done, but of what stops it being done" should be increased. The inertia of old beliefs and individual preferences are part of society and are consequently also embedded in persons working in public authorities. This causes resistance to change and implementation of new, more sustainable policies. A questionnaire survey among Swedish traffic planning authorities shows that 47% of respondents recognised the lack of a professional traffic planning culture within their own organisations as a significant problem hindering the proposal and implementation of improved cycling facilities (Envall, 2009).

Envall surveyed the decision support for cycle planning as well. He found that 47% of decision-makers in authorities pointed out that generally bad planning culture (as above), a lack of up-to-date planning documents and a lack of in-house cycle planning expertise prove to be more problematic in cycle facilities delivery than lack of interest among politicians. (Envall, 2012)

In the OPTIC project barriers to implementation of transport policy packages were studied. The strategies that were identified from empirical analysis comprised of: using good examples, trials to legitimacy and acceptance, preparation of windows of opportunity and the installation of funding on higher levels to instigate investments on a lower administrative levels (Kth et al., 2011).

Some economy and psychology concepts can also be applied to transport research in order to increase the understanding of individuals' behaviors facing a new approach. At the end sustainable policies are being introduced to existing older standards and can be seen as "new products" that need to be adopted by decision-makers and planners. Wiedmann, Hennigs, Pankalla, Kassubek, and Seegebarth (2011) studied the reluctance of customers to use alternative fuel vehicles. The authors interviewed 240 customers in Germany and performed a cluster analysis taking into account exogenous variables such as: the financial risk of buying an alternative fuel vehicle, the technological risk, the infrastructural risk, the physical risk, the time risk,the social risk and the psychological risk. Additionally the reluctance to new technologies was selected as endogenous variable. As a result of the cluster analysis four types of customers were identified in increasing order of acceptance to buy an alternative vehicle the risk adverse, the status-oriented, the performance oriented and the ecologically minded customers. While the first and the fourth group need intensive communication measures, the intermediate groups rather require pricing and distribution strategies. Kleijnen, Lee, and Wetzels (2009) studied the resistance of consumers to innovations by means of a literature review and a qualitative focus-group in the Netherlands. The authors identified three main stages of consumers' resistance: postponement, rejection, and opposition. In the first one consumers may accept to buy the product but later, in the second stage after consideration, decide not to buy and in the third one the consumers adopt a radical position against the product (in our case that would be sustainable transport policies). The study discussed two determinants of resistance: the degree of change required and the conflicts with the consumer's prior belief structure.

Very often too, cycling is not only considered as generally inferior but plainly as dangerous per se. Horton (2007) identifies that fears of cycling are wide spread and vary socially, geographically and historically. There are three basic underlying ways in which cycling is constructed as dangerous, and thus a contemporary fear of cycling is produced: through road safety education, helmet promotion campaigns and the separation of cycling from motorised traffic when not needed. The vicious, negatively self-enforcing cycle of cultural marginalization of cycling derived from Horton's notion is shown in Figure

1. The more cycling is seen as dangerous the more spatial marginalisation takes place and moreover cyclists are not only pushed towards society's margins but also geographically within.

There also appears to be a self-spreading effect. Too often national-level decision-makers, incorporating only a limited view on cycling, produce legal regulation and guidelines that leave only limited options for local decision-makers to progressively construe the latitude of judgement.

Our paper is important because it identifies such mental barriers and adds to their systematic addressing. Because such mental barriers are a limitation of possible policy success over all scales of appearance. In the next section we explore the concept of mental barriers by discussing six Austrian examples.

AUSTRIAN EVIDENCE

Decision-makers carry out cycling policy relevant actions that we classify in the style of Heinen et al. (2010) in three main categories: legal regulation, planning and infrastructure. The results of these actions may show evidence of mental barriers of the decision-makers involved. Table 2 gives examples that have been a point of discussion in Austria in the recent past.

The following subsections show an overview of Austrian exemplary evidence of mental barriers of decision-makers and provides qualitive "highlights" grouped by Table 2 categories.

Legal Regulation

In Austria, legal regulations are enacted at two levels of society: the national assembly and nine provincial parliaments. Legal matters concerning cycling are regulated as well on the federal level (cycling in road code) as on the provincial level (bicycle parking in building codes). Three examples from the legal field illustrate mental barriers and their consequences follow.

Figure 1. The vicious control cycle of marginalisation of cyclists
Derived from Horton, 2007.

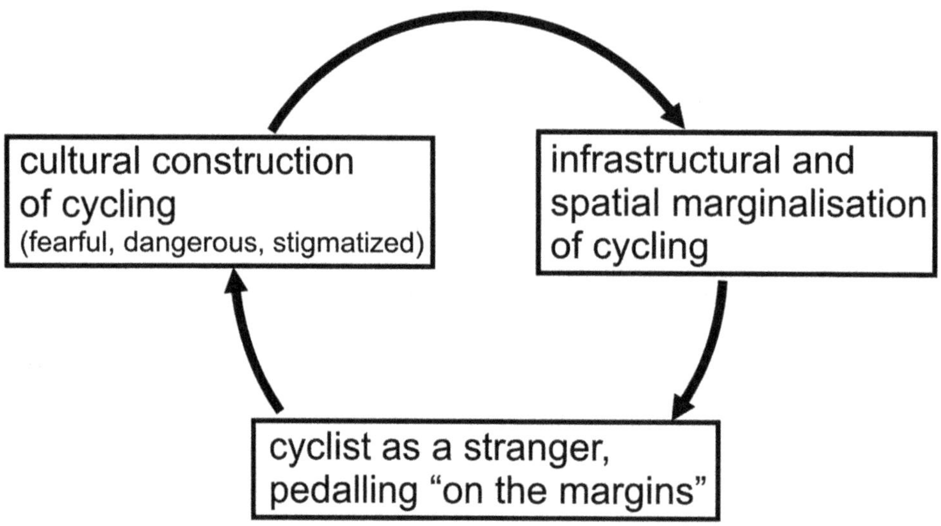

Table 2. Mental barriers in cycling policy classified in three categories – actual points of discussion in Austria

Infrastructure	Planning	Legal Regulation
• Hairpin curves: Bikes are vehicles, they have vehicular dynamics. Too often engineers still treat them like pedestrians with sharp turns and radii close to zero • Limited usability: infrastructure that's unable to meet user's needs and is therefore used little or not at all • Detours, horizontal and/or vertical • Lack of visibility and priority in junctions • Lack of maintenance of surfaces and installations • Under-estimation of safety distance between cycle paths and car lanes • Inaccessibility and lack of safety of bike parking • Inappropriate cycle path surface coatings	• Under-funding of cycling: The share of funding for cycling infrastructure and campaigns does not meet the aspired bicycles modal share objectives • Over-dimensioning of public space allocated to cars • Lack of integrated planning of cycling and pedestrian facilities	• Speed inconsistencies: limits for cars in cities are still much higher than achievable bike speeds • Genesis and boundary conditions of bicycle-related laws • Obligatory use of helmets • Obligatory use of cycle ways • Lack of regulation for sufficient bike parking provision

- **Genesis of Bicycle-Related Laws:** The federal road code is the competency of the national assembly. The federal ministry for transport sends law amendment drafts to civic associations for evaluation, before submitting them to the national assembly's discussion and decision procedures. These drafts are always accompanied by an evaluation sheet, where the ministry points out its view on what kinds of impact the proposed law amendment will have. Interestingly, the evaluation sheets of recent cycling related drafts (e.g. helmet obligation for children up to 14 years) always stated that there will be no financial or environmental impacts whatsoever triggered by the road code amendment. In Austria there are currently 1,220,600 children aged 14 or less. Prior to the amendment the helmet wearing rate in this age group was about 65% and rose foreseeable to 86% following the introduction[2]. So an obligatory helmet law leads directly to an increase in helmet wearing rates (which is the actual intention of course) and therefore is very much a cost relevant regulation. Then again, the health effects of obligatory helmet laws on cycling population health are a point of ongoing debate (eg. Jacobsen, 2003).
- **Obligatory Use of Bicycle Paths:** The Federal Road Code knows a general rule of obligatory use of cycling infrastructures. Only in very exceptional cases, e.g. for racing bikes in training, or bikes with trailers, cyclist are exempt from this obligation and may use the road lanes instead. Municipalities – e.g. the City of Vienna – frequently join in the calls to the National Parliament for the end of this obligatory rule. On the other hand, these very same municipalities concurrently proclaim to proceed with the construction of classic cycling ways as hitherto or even want to amplify it. The obligation produces three consequences:
    - A negative public opinion (car drivers expecting bicyclist to move away from the road lanes),
    - Increasing numbers of people cycling on sidewalks where there are no bike paths in the vicinity, and finally
    - It forces cyclists to use the existing, deficient infrastructure in places and cases where many would opt for the road lanes due to quality or safety concerns.

This leads to cycling structure not being built to the highest standards. Meanwhile the Federal Road Code's use obligation has been reformed. Once obligatory-use bicycle paths may now be marked as free-choice paths, if a local authority actively choses to do so. Otherwise a cycling infrastructure keeps its original status. The last two years since the amendment's introduction show, most bicycle infrastructure remains in obligatory use, no matter how good or bad its design may be.

- **Provision of Bicycle Parking:** Building codes in Austria are enacted by provincial parliaments. The provision of car and bicycle parking falls into the competence of the building code. The Lower Austrian provincial government's notion is that bicycle parking provision should not be regulated mandatory by building codes but real estate developers should voluntarily provide sufficient spaces. This idea is in total opposition to existing, obligatory and highly detailed car-parking regulations in the Lower Austrian building code. Upon the regional parliament's unanimous resolution the responsible regional minister responded in a letter from June 30th 2009 that obligatory bike parking cannot be adopted (Rosenkranz, 2009). The given reasons are: detached houses do not need such provisions on an objective basis; the location of businesses is important and could not be considered in such regulation; lack of legal practicability.

It has to be kept in mind, that all those alleged reasons do not matter at all for car parking obligation. The provincial planning department's public servants give detailed recommendations to developers and municipalities for voluntary bike parking provisions (Westhauser & Hofecker, 2011). Meanwhile in 2014 a building code amendment has been passed, which defines target values – in comparison to obligatory values for car parking – which can be even reduced by the municipal council. Figure 2 and Figure 3 show two bike parking situations in Vienna and Lower Austria, results of insufficiently regulated bicycle parking in building codes. The first figure illustrates the incorporation of physical discrimination for bicycles over favouring car parking. The previous bike parking was substituted by a car parking space and the new bike parking (covered cage) was located further from the main entrance and beyond a steep slope. The second figure shows a residential setup with obligatory parking lot and self-made, no-standards bike racks located dispersedly, because there is only a limited access bike locker room in the building's basement.

Planning

We bring two examples that show in detail the lack of practical understanding of actual cycling needs and the big potential in competence development with involved decision-makers

- **Over-Dimensioning of Car Traffic Space:** The assignment of sufficient envelope space for cyclists (Jones, 2001) is a benchmark whether policies are executed seriously or just half-heartedly at best. For example, the intersection of federal road 210 and Waltersdorferstraße in the town of Baden (Lower Austria) hosts a combined cycling and walking path with insufficient width right next to a road with 3 lanes. This is an important access route to the main train station for cyclists and pedestrians alike. The provided envelope width is breaking engineering rules but was nevertheless implemented by the public administration. Austrian road design rule RVS 03.02.13 (FSV, 2001a) asks for a minimum clearance (incl. Safety distances) of 2.5 m for cycle tracks, as does the design rule RVS 03.04.12 (FSV, 2001b) for mixed bidirectional cyclist and pedestrian surfaces

Figure 2. Bike parking design at the Campus Vienna Biocenter
Drawing: A. Castro-Fernandez; map source: City of Vienna.

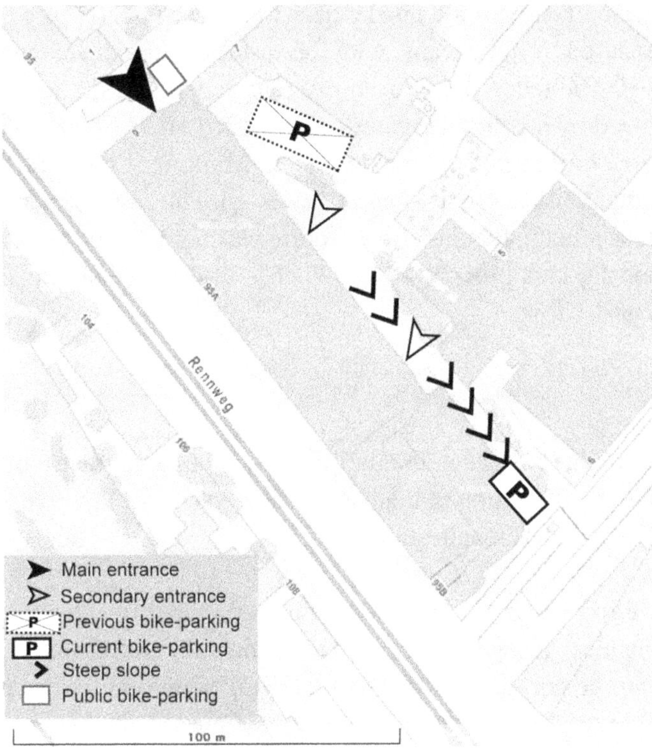

Figure 3. Lower Austrian bicycle parking vs. car parking requirements in effect
Photo: T. Brezina.

at 2.5 m (Figure 4). Upon making the faults aware to the regional road authority, an in-situ appraisal session of administrative transport experts was scheduled. This lead to the bike path sign being moved from the outside to the inside, next to the car lane (Figure 5). The lawful signposting needed to be ensured by any means, with the result that the envelope was cropped even more by the new signpost. The reduction in usability for cyclists and the ignorance of guidelines was unanimously approved by the present experts (Ferstl, 2013).

- **Lack of Integrated Planning:** A calm side road in Mödling, Lower Austria, receives a separate infrastructure treatment instead of an integrative design. Thus pedestrians and cyclists need to figure out their yields where the regular cross-section was not adapted to local constrictions (Figure 6): the cross-section for cars prevails at the expense of cyclists and pedestrians over a potential adaption to local constrictions.

Infrastructure

As laid out in the introduction, infrastructure quality is crucial as an attractor of or deterrent for cycling and it is the most tangible manifestation of decision-makers' actions. We give three Austrian examples illustrating such bad infrastructural manifestations.

- **Lack of Visibility and Priority in Junctions:** Infrastructure often discloses the inherent planning mindset. Where planners fail punctually in providing safe and usable designs, these shortcomings are attempted to be corrected with accompanying signposts commanding cyclists to behave in an un-cyclist style to prevent any harm resulting from this bad design – e.g. by dismounting the bicycle. In the south of Vienna, the Euro Velo 9, one of the main international cycling routes of the European network is crossed by many local roads. These intersections are designed either

Figure 4. Cross section of existing combined walking and cycling path at odds with minimum envelope requirements according to Austrian road design rules RVS 03.02.13 (FSV, 2011)

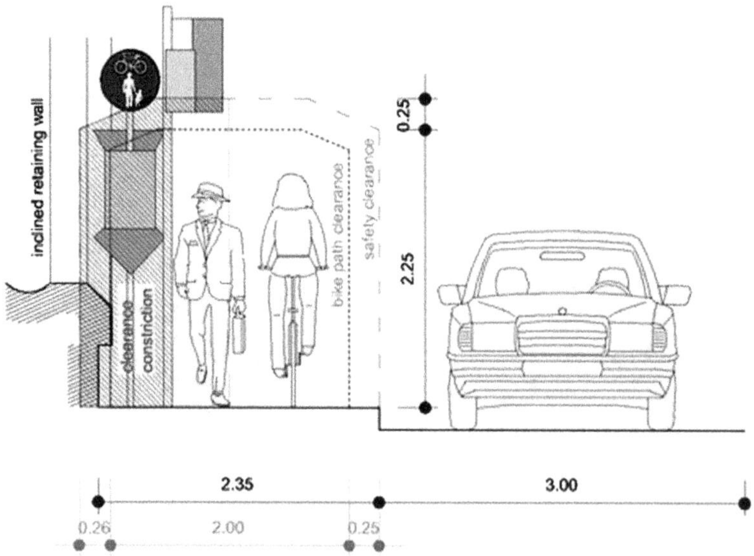

Figure 5. left (before): Insufficient space for pedestrians and bicyclists cramped together on the sidewalk with three car lanes adjoining; right (after): the importance of law-abiding signposting increases the administrative neglect of envelope requirements
Photos: T. Brezina.

Figure 6. Mödling: subliminal discrimination of active transport modes

as de-prioritized level crossings or as very narrow underpasses. Such sub-optimal solutions are frequently accompanied with signs like "intersection of death, cyclists dismount" or "pedestrian subway, cyclists dismount" (Figure 7 and Figure 8).

- **Winter Services:** Communities usually plow the snow from car lanes first and then, if at all, from bicycle lanes. The Austrian road code states that obligatory-use bike lanes that cannot be used, e.g. due to lack of cleaning, loose the obligation and the neighbouring car lanes may be used instead. And communities leave bike lanes quite often unplowed either for cost reduction reasons or for way-side storage of surplus snow from car lanes. Due to separation under snow-free conditions,

Figure 7. Euro Velo 9 in Lower Austria with a sign commanding "pedestrian underpass, bicyclists dismount"
Photo: T. Brezina.

Figure 8. Euro Velo 9 in Lower Austria with a sign commanding "intersection of death, bicyclists dismount"
Photo: T. Brezina.

cyclists and car drivers now meet in the road lane under conditions that ask for special attention and handling skills (snowy, wet, freezing, reduced widths). This does not appear to be the best precondition for mutual exposure, as there is hardly any culture of sharing the lane at perfect conditions.

- **Intersections with Railroads:** The regional railway Badener Bahn in the south of Vienna runs along the federal road B17 during long sections. On four locations the two-tracked alignment changes from the eastern to the western side of the road or vice versa. One of the railroad crossings in the town of Wiener Neudorf, was rebuilt in 2002 from a grooved rail type to a railway crossing with a regular rail groove for the wheel flanges. As the rail line crosses the road in a very acute angle, the rectangular rail groove gap increases significantly by the inverse sinus of the crossing angle. Thus, making the crossing along the road's side line a risky undertaking for cyclists. After a cyclist's crash and strong efforts by other cyclists on this route to improve the situation, the responsible regional traffic authority's sole concession was to install a street sign in both directions saying "cyclists, attention grooved rail". On the contrary, within the same authority's jurisdiction, just a few kilometers away in the town of Perchtoldsdorf another railroad crossing received a simple but effective treatment: a safer crossing geometry for cyclists including explanatory signs (see Figure 9) that reduces the dangers of wide grooves and direct lines of riding was installed.

Figure 9. Railroad crossing with safer bicycle crossing geometry in the town of Perchtoldsdorf, Lower Austria
Photo: T. Brezina.

These selected examples clearly indicate a consolidated state of insecurity of authorities with how to deal with cyclists properly and consistently. The next section addresses possible tools for the reduction of these mental-barriers-based insecurities with decision-makers.

OVERCOMING MENTAL BARRIERS

It is the mixture of various decision-maker's attitudes towards actual cyclists' needs that hinders cycling to actually become a major mode of transport. In most cases it is inobservance, often neglect, and sometimes even open, direct opposition conflicting with the goals set in transport master plans. Such decision-makers lack hands-on experience with cycling combined with a lack of understanding across effect levels from law making to construction details. At best they are random leisure cyclists only and have limiting beliefs about cycling. Still too often their decisions are based on the deeply-rooted idea that only car-emphasizing policies solve transport problems and lead to social and economic development (eg. Vigar, 2000). For example the perception that shopping ultimately needs a car is still common in society, although empirical research shows different results. A survey from the city of Graz (Schmitt & Landgraf, 2009) reveals that actually only a low fraction of shoppers (6%) objectively need a car for transporting their shopping goods. The remaining 94% of shoppers buy goods small and light enough to be carried away on foot or by bike, or they did not buy anything at all.

Because perceptions of reality shape concepts of transport systems design, a self-referential system can easily emerge where beliefs about transport solutions and the normative forces of perceived built reality feed each other (Rauh & Maierbrugger, 2003). Thus, many decision-makers with mental barriers cannot imagine and therefore do not produce consistent regulation and planning for sustainable mobility. In Figure 10 we improved the original control cycle from Figure 1 for decision-making in bicycle policies based on the feedback loop "infrastructure – cultural construction – decision-makers". It identifies the point of action for improving cycling conditions. The opinions, attitudes and actions of decision-makers sculpt the laws, the planning and infrastructure design for cycling. Laws and infrastructure in turn influence cycling's cultural construction, which feeds back to the decision-makers' attitudes. How to break this vicious circle of self-enhancement against cycling in favour of cycling? We suggest that the most promising spot within this cycle for triggering improvement are the decision-makers' attitudes. We further suggest two options that we amended to the control cycle:

- The first one is an increasing social pressure from the general public on decision-makers. For example when the numbers of cyclists have increased so much in recent history that decision-makers simply can't ignore them and need to take care of cycling matters more detailed and more consistently.
- The second one is self-experience: Either by scale effect – the more people cycle the more likely more decision-makers are amongst them – or with specially organized field trips (e.g. by cycling advocacy groups) where decision-makers experience the limitations of existing regulations, planning and infrastructure at first hand. Cycling decision-makers due to the scale effect will tend to improve cycling conditions as they themselves want to experience the benefits of improvement.

Figure 10. The control cycle of bicycle policy decision-making is based on the control cycle derived from Horton in Figure 1

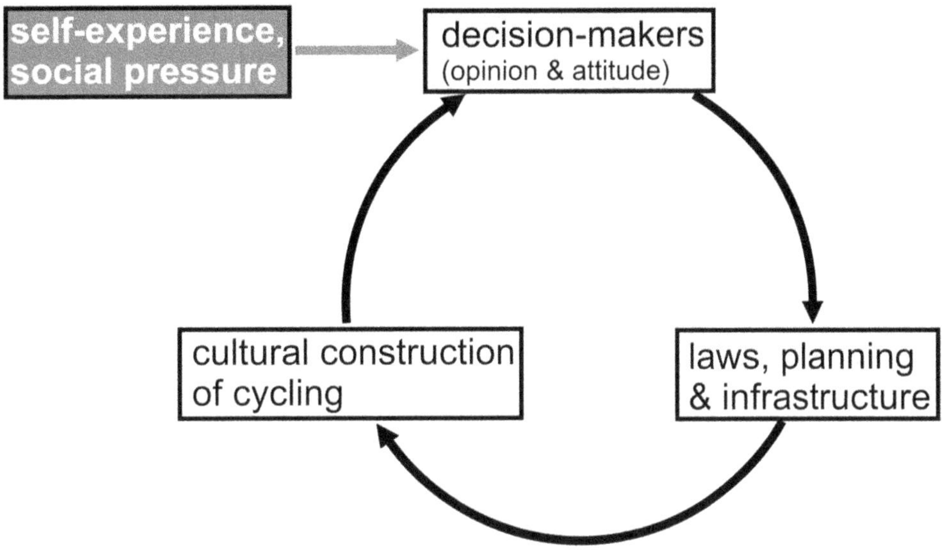

Steenberghen et al. (2008) discuss policy options to overcome barriers to implementation of a technological system and market penetration. These policies include and are strongly founded on issues of pricing and taxation as well as vehicle efficiency and regulatory standards. Similar strategies do not appear well applicable for the problem at hand. On the one hand, higher costs do not appear as an implementation barrier in the light of the very low costs of cycling infrastructure in comparison to road project budgets. On the other hand, a lack of standards does not appear to be of primary concern, as regulatory standards exist, e.g. for infrastructure design. But they denote the standard of engineering and do not pose an obligatory rule for wit in implementation of the very same standards on any level of decision making – at the planning or political level. Financial instruments as well do not appear suitable at first hand for the decrease of mental barriers. A controlling-based funding system may only move mental barriers from local decision-makers to those centralized institutions that control funding quality. We therefore think that the best point of access of the control cycle are decision-makers, as indicated above, because the leverage of policy interventions appears to be bigger with them than with the other two cycle entities.

Austria traditionally has had rather low levels of cycling with a recent average yearly distance of around 180 km per capita (Rauh et al., 2006) and just a handful of cities show significant cycling shares, e.g. Graz (14%) and Salzburg (16%)[3]. Therefore it is very likely that increasing numbers of cyclists will also increase the amount of cycling decision-makers in the future and by thus decrease the extent of social marginalisation. This may lead to cycling being increasingly recognized as a viable transport alternative. The examples of Salzburg and Graz evidently support this notion. Mayor Heinz Schaden and City councillor Erich Edegger, who are/were both everyday cyclists themselves, pushed the cycling agenda in Salzburg and Graz and helped to make their cities stand out in terms of cycling modal-split in Austria.

CONCLUSION

A factually strong fraction of mental barriers remains amongst ambitious ideas about the increased future importance of cycling. Although published objectives aim high, the legislation and planning complex has not yet established a general sense of basic bicycle friendliness. Car-oriented policies and planning philosophies still prevail when seminal decisions are at stake. These mental barriers of decision-makers surface in legislation, regulations, planning and design and play a decisive role for the emergence of factors that dissuade cycling. Although we have sketched a first draft of the mechanisms leading to and associated with mental barriers, there's a need for in-depth research on sociology of cycling marginalisation and the motives and drivers behind it – especially with a focus on decision-makers.

But why do such mental barriers arise in decision-makers from local (e.g. decisions for projects) to regional scale (e.g. law-making)? Some likely answers to our question in the introduction, why mental barriers exist, are:

- Decision-makers, especially politicians, have views on needs and wants of the general public that differ strongly from the general public's actual expectations.
- Decision-makers are "just normal people" in regard to cycling, they presumably ride even less for every-day purposes than average people do.
- Decision-makers predominantly have a car/chauffeur-dominated style of mobility and are therefore overcautious towards cycling as a transport mode.
- Decision-makers consider cycling in general as an inferior transport mode (Horton, 2007; Vigar, 2000) and therefore do not pay attention at all.

Mental barriers offer a wide research perspective. Sterman (2000) elaborated on the concept of double-loop learning by Argyris (1985) and a wide variety of barriers to learning on the individual and organisational level. Double-loop learning includes a second loop that represents the individual's mental models of the real world and the feedback decision rules. Our Horton-based control cycle appears promising to be adopted to this double-loop learning concept. Beyond these qualitative approaches to this phenomenon it appears necessary to obtain empirical data on mental barriers from project and implementation appraisal.

In our opinion not even very progressive minds are required as a first step for real bicycle promotion, we see just the consequent adoption of equal rights as a promising point to start from. Bikes have vehicular dynamics and need to be treated accordingly. Thus, for instance, if a decision-maker recognizes bicycles as what they are, vehicles, he/she can't possibly design bike infrastructures with right angle corners. Engineers need to learn from allegedly deviating behaviour of users on infrastructures that they designed. Badly designed structures lead to unfavourable behaviour and little satisfied users which then produce unsatisfactory mobility behaviour results. Therefore, cycling-literate engineers must make an extra effort to educate other transport engineers and politicians on issues of sustainable mobility, because politicians and engineers alike are responsible for the final results.

From a municipal point of view the allocation of funds on costly infrastructure that is only marginally accepted needs to be avoided to strengthen cycling's position as a cost-effective transport mode. For the sake of highest financial efficiency and maximised user satisfaction better designs need to be striven for by decision-makers that have left their mental barriers behind.

ACKNOWLEDGMENT

We thank Anna Mayerthaler, Peraphan Jittrapirom, Heidi Schmitt, Wolfgang Wehap and Peter Weiss for discussing selected items of the paper with us. A thank you belongs to John Parkin for suggestions on literature. Furthermore we thank two peer reviewers and the editors for their suggestions on a previous version of this paper. Declaration: In his spare time Tadej Brezina is a pro-bono board member of the non-profit Lower Austrian cycling advocacy group Radlobby NÖ. Funding sources: This work was done under the authors' base wage, no dedicated funding was provided. The article reflects the authors' materials collected and experience gained from their career in cycling projects and auxiliary cycling.

REFERENCES

Argyris, C. (1985). *Strategy, Change and Defensive Routines*. Boston, MA: Pitnam.

Ausserer, K., & Risser, R. (2001). *Cycling in Vienna - Differences in attitudes between cyclists and politicians*. Paper presented at the Velo-city 2001, Edinburgh, UK.

Banister, D. (1996). *Barriers to implementation of urban sustainability*. Paper presented at the European Regional Science Association - 36th European Congress, ETH Zürich, Switzerland.

Cole, R., Burke, M., Leslie, E., Donald, M., & Owen, N. (2010). Perceptions of representatives of public, private, and community sector institutions of the barriers and enablers for physically active transport. *Transport Policy*, *17*(6), 496–504. doi:10.1016/j.tranpol.2010.05.003

Daley, M., & Rissel, C. (2011). Perspectives and images of cycling as a barrier or facilitator of cycling. *Transport Policy*, *18*(1), 211–216. doi:10.1016/j.tranpol.2010.08.004

Davies, D., Brook-Carter, N., & Gardner, G. (2000). *Institutional and organisational attitudes to cycling (No. TRL Report 468)*. Crowthorne, UK: Transport Research Laboratory - TRL.

Envall, P. (2009). *What is hindering good cycle planning: how do city planners think?* Paper presented at the Velo-city 2009 - Re-cycling cities, Bruxelles, Belgium.

Envall, P. (2012). *What is the role of Bicycle Parking for Increased Cycling in Large Cites? A Literature Review*. Stockholm: CyCity.

Ferstl, K. (2013). *Verhandlungsschrift*. Baden: Bezirkshauptmannschaft Baden.

FSV. (2001a). *RVS 03.02.13: Nicht motorisierter Verkehr - Radverkehr*. Wien: FSV - Österreichische Forschungsgesellschaft Straße, Schiene, Verkehr.

FSV. (2001b). *RVS 03.04.12: Stadtstraßenquerschnitte; Querschnittsgestaltung von Innerortsstraßen*. Wien: FSV - Österreichische Forschungsgesellschaft Straße, Schiene, Verkehr.

FSV. (2011). *RVS 03.02.13: Nicht motorisierter Verkehr - Radverkehr*. Wien: FSV - Österreichische Forschungsgesellschaft Straße, Schiene, Verkehr.

Heinen, E., van Wee, B., & Maat, K. (2010). Commuting by Bicycle: An Overview of the Literature. *Transport Reviews*, *30*(1), 59–96. doi:10.1080/01441640903187001

Horton, D. (2007). Fear of Cycling. In D. Horton, P. Rosen, & P. Cox (Eds.), *Cycling and Society* (pp. 133–152). Aldershot, UK: Ashgate Publishing.

Jacobsen, P. L. (2003). Safety in numbers: More walkers and bicyclists, safer walking and bicycling. *Injury Prevention*, *9*(3), 205–209. doi:10.1136/ip.9.3.205 PMID:12966006

Jones, M. (2001). Promoting cycling in the U.K. – Problems experienced by the practitioners. *World Transport Policy and Practice*, *7*(3), 7–12.

Kleijnen, M., Lee, N., & Wetzels, M. (2009). An exploration of consumer resistance to innovation and its antecendents. *Journal of Economic Psychology*, *30*(3), 344–357. doi:10.1016/j.joep.2009.02.004

KTH, DTU, VTI, TOI, Austria Tech, & TSU. (2011). Deliverable 5: How to manage barriers to formation and implementation of policy packages in transport (OPTIC - Optimal Policies for Transport in Combination). Oslo: TOI.

Low, N., Gleeson, B., & Rush, E. (2003). Making Believe: Institutional and Discursive Barriers to Sustainable Transport in Two Australian Cites. *International Planning Studies*, *8*(2), 93–114. doi:10.1080/13563470305158

Rauh, W., & Maierbrugger, G. (2003). *Grenzen überwinden im Verkehr - Maßnahmen und Beispiele*. Wien: VCÖ.

Rosenkranz, B. (2009). *Resolution betreffend NÖ Bauordnung 1996 und NÖ Bautechnikverordnung 1997 bzgl. weiterer Erhöhung des Radverkehrsanteiles am Gesamtverkehrsaufkommen* (P. des Landtages von Niederösterreich, Ed.). St. Pölten.

Schmitt, H., & Landgraf, S. (2009). *Beobachtungsstudie: Einkaufsverhalten in Graz und Umgebung*. Graz: ARGUS Steiermark.

Sterman, J. (2000). *Business Dynamics - Systems Thinking and Modelling for a Complex World*. Boston: McGraw-Hill Higher Education.

Vigar, G. (2000). Local Barriers to Environmentally Sustainable Transport Planning. *Local Environment*, *5*(1), 19–32. doi:10.1080/135498300113246

Westhauser, C., & Hofecker, C. (2011). *Empfehlung über Fahrradabstellplätze bei verschiedenen Bauten*. Unpublished manuscript, St. Pölten.

Wiedmann, K.-P., Hennigs, N., Pankalla, L., Kassubek, M., & Seegebarth, B. (2011). Adoption barriers and resistance to sustainable solutions in the automotive sector. *Journal of Business Research*, *64*(11), 1201–1206. doi:10.1016/j.jbusres.2011.06.023

KEY TERMS AND DEFINITIONS

Cycling Barriers: Factors negatively influencing the level of cycling. The more and the bigger the barriers are, the less attractive is it to ride a bicycle.

Decision-Makers: Politicians, planners and technicians at national, regional and municipal levels that are responsible for designing transport infrastructures and systems.

Mental Barriers: The resistance of decision-makers to innovation adoption that appear to be the reason for inconsistencies, avoiding consequently a correct implementation of cycling policies.

Self-Experience: Decision makers experience the limitations of transport sub-systems due to either especially organized field trips or the effect that among an increasing number of cyclists, an increasing number of decision makers is likely.

Social Pressure: Pressure on decision-makers from an increasing number of people within society interested in a special issue.

ENDNOTES

1 Heinen, van Wee, & Maat (2010) originally considered a fifth category called "miscellaneous" that comprises of the following items: transport costs, safety and travel time and effort. Nevertheless, we have considered transport costs as a socio-economic aspect and safety, travel time and effort as a built environment aspect.

2 Figures published by Kuratorium für Verkehrssicherheit (Austrian Road Safety Board) on June 6th, 2012.

3 TEMS – The EPOMM Modal Split Tool, http://www.epomm.eu/tems/index.phtml.

Chapter 4
Comparative Analysis of Public Transport Users' Perception Targeting Sustainable Transportation

Yetis Sazi Murat
Pamukkale University, Turkey

Ziya Cakici
Pamukkale University, Turkey

ABSTRACT

Public transport is one of the main actors of sustainable transportation. To make an increase in the use of public transport systems and a decrease in private car usage are one of the main purposes of decision makers in many countries. Usage of public transport system has a strong relation with the quality of service. Service quality includes many parameters such as comfort, frequency, information system etc. Fare is another important parameter on demand for public transport systems. All of these parameters are varied by users' perceptions. This study focuses on public transport systems of Denizli, Turkey and investigates users' perceptions. Bus and paratransit (minibus) modes are regarded in the study. The public survey is performed for these modes. In the surveys, different questions have been asked to participants. Answer of each question is evaluated separately and distribution of each question for gender, ages and jobs are analyzed. Results obtained are summarized and discussed.

INTRODUCTION

In recent years, there has been a growing awareness of the need to use public transportation resources more efficiently. As a result, it has become very important for public transit systems to evaluate carefully all services so as to provide the most efficient and desirable transit services to the communities that they serve. Public transit customer surveys can play an important role in the evaluation of current and planned public transit services. When a public transit system decides to evaluate current or planned services

DOI: 10.4018/978-1-5225-2116-7.ch004

via the use of a customer survey, there are a number of important issues that needed to be addressed to facilitate the data collection process and to ensure that reliable and high quality data are collected, analyzed, and responsibly reported. In some cases, however, the collection of important information about public transit customers and the resulting evaluation has not been supported by comprehensive and methodologically valid surveying techniques. Despite the time and cost associated with such surveying efforts, the results obtained from surveys of public transit customers can be extremely useful to a public transit system's planning and operations functions, as well as to governmental boards, commissions, and councils. Therefore, it is in the best interest of public transit systems to conduct annual periodic surveys of its customers and to make sure that its surveying process is appropriate and correct to meet the desired information needs. The archiving of historical databases should be initiated and used for yearly comparisons of changes in customer demographics, travel patterns, and overall satisfaction with services provided, at a minimum.

Survey Elements

In order to fully identify the many aspects of public transit customer surveying, it is important to examine, review, and summarize the various types of survey instruments (questionnaires) and final reports from as many sources as possible.

The literature notes that all surveys have the following features in common, regardless of the group or topic being surveyed:

- Specific and clearly defined objectives,
- Method of data collection,
- Questionnaire construction,
- Sample size,
- Appropriate level(s) of analysis,
- Accurate and truthful reporting of results.

From this list of survey basics above, there are many interrelated steps involved in the design, planning, and administration of a survey of public transit customers. Before a survey can be conducted, important and sometimes subtle decisions must be made about the objectives and the purpose of the survey as well as its unique characteristics.

In this study, public transportation systems of Denizli, Turkey are investigated. Bus system and paratransit (minibus) modes are taken into account. Public survey method is utilized for the investigation. Two surveys are applied for the bus system and the minibus system. Results of the surveys are analyzed in details and both systems are compared. User satisfactions and perceptions about the public transportation systems are revealed. Studies including analysis of public transportation systems are summarized in the following section. Then application and some details of the surveys are reported. Results of the surveys and comparisons are given in the following sections.

BACKGROUND

User perceptions and user satisfaction levels for public transport systems are important issues in evaluating performance of public transport systems. Many studies relating to these issues have been conducted by public transport researchers from the entire world for many years. In the following, some important sample studies related to the issue are summarized.

Passenger's satisfaction level about public transport system is analyzed by Andreassen (1995). In this study, survey technique is used and various public transport systems such as tram, train, bus are considered separately and passenger's satisfaction levels about these different transport systems are analyzed statistically. Similar study is realized by Altan and Engin (2003). It is aimed to measure user satisfaction in a travel company making inter-urban transportation. Satisfaction degrees of the users are identified with survey studies. Survey studies include different important issues such as: cleanliness of vehicles, ticket selling service, driver's performance, driver's kindness etc. Data gathered from surveys are analyzed using SPSS program. Consequently, SPSS results are evaluated and interpreted in detailed. Relation of quality of public transport service and user satisfaction is examined by Morfoulaki et al. (2007). For this purpose, factor analysis and multinominal logistic regression model are applied on the gathered data. At the end of the study, it is revealed that user satisfaction can be improved by decreasing waiting time of users and improving frequency and terminal safety. Tyrinopoulos and Antoniou (2008) used two different statistical methods for assess quality implications of the variability of user's perceived satisfaction. In their study, five transit systems are investigated in Greece. The results showed that vehicle cleanliness and waiting time as well as service frequency and accessibility, play an important role on user satisfaction level and various corrective measures for transit systems should be taken into account considering these parameters. Fellesson and Friman (2008) compared perceived service satisfactions with public transport in: Stockholm, Barcelona, Copenhagen, Geneva, Helsinki, Vienna, Berlin, Manchester and Oslo cities with each other.

Totally 9542 public transport users are selected from these different cities and satisfaction surveys which include 17 items are applied on these users. At the end of the study, it is expressed that perceived satisfaction of public transport users in these nine European cities differs from each other. Rojo et al. (2013) proposed a new model to determine satisfaction of interurban bus service users. In this study, logit and probit models are used to determine important variables of user-perceived quality and most important variables are selected. At the end of the study, it is discussed that which features must be improved to increase the bus service quality (environmental protection, safety, ticket price, accesibility). Garrido et al. (2014) used Artificial Neural Network (ANN) approach for analyzing service quality in public transport. In this study, firstly, ANN approach is developed using data gathered from user satisfaction surveys in Granada/Spain bus metropolitan transit system. Then, three different methods are used for determining weights of different parameters which affects the service quality. The results showed that frequency is the most important parameter for determining the service quality in public transportation. On˜a et al. (2015) used decision trees and cluster analysis approach for determining transit service quality. In this study, when decision tree is used for determining most important service quality attributes affecting the overall user's satisfaction, different groups of the users are determined by cluster analysis.

In addition to user satisfaction, perceptions of users have a great importance on sustainable public transportation planning. Public transport systems can properly be organized considering users' attributes. For this purpose, many researches that focus on user perception have been realized. Urban transportation problems and passenger perceptions in Eskisehir, Turkey are determined with survey studies by Karacasu and Bilgic (1999).Various questions relating to service quality parameters affecting the public transport efficiency such asfrequency, ticket price, and distance between the stops, driver's kindness, and travel time, are asked to public transport users.

As a result, an overall evaluation for the public transport systems is made by researchers. Winder et al. (2005) presented the results of public perception surveys conducted on about 3000 people from EU member states. In the study, they investigated reliability, frequency/network coverage and personel security factors in public transportation.

Stradling et al. (2007) deal with analyzes of passenger perceptions on ideal urban bus journey. In the scope of this study, a survey which includes bus system's comfort, safety, cost and service quality parameters and personel information such as age, gender, income and frequency of bus usage, is applied on 2250 questionnaires.

Dell' Olio et al. (2010) studied on bus user perceptions for quality of public transport service. The overall service quality is analyzed using ordered probit models in the study. Consequently, the results showed that perception of quality changed with the category of users and different user profile. Belwal and Belwal (2010) aimed to determine public perceptions towards to public transport services in Al-Batinah region of Oman. At the end of the study, it is revealed that public transport services in Oman cannot be met the travel demand and these public transport systems are unsatisfactory. Besides, this study presents long-term strategies to policy makers in Oman for implementation of reliable and appropriate public transport services. Dell'Olio et al. (2011) suggested a new methodology to describe the quality of service desired by users of a public transport system. In their study, waiting time, comfort and cleanliness are determined as important public transport variables for users and the parameters such as vehicle occupancy, driver kindness etc. are determined as less important public transport variables for user. Eboli and Mazzulla (2012) studied on selecting of most appropriate performance indicators to evaluate public transport service quality. They investigated different attributes such as service availability, service reliabillity, comfort, cleanliness, safety and security, fare, information and etc. in detailed.

Das and Pandit (2013) focused on evaluating of level of service for bus transit for developing countries. They emphasized that user's perceptions about LOS for a transit system vary with level of development of a country. Belwal (2013) studied on people's perception of public transport services in sultanate of Oman. For this purpose, he used opinion survey approach. About 2000 surveys are applied on public transport users in the al batinah region of Oman. It is concluded that the public transport in Oman is still insufficient and the better and comfortable public transport service must be provided in Oman. Khalid et al. (2014) studied on public transport in Malaysia, including ktm komuter which is a heavy-rail public transport operator. They emphasized that ktm komuter cannot offer the high quality of service. User perceptions towards to ktm komuter are investigated in the study. So, a survey is applied on rail public transport users. It is concluded that ktm komuter must be improved. Maraglino et al. (2014) proposed to improve the quality of service of public transport systems. Karadeniz and Unver (2014) studied on customer's perception of service and satisfaction in maritime transport in Istanbul/Turkey. In this study, survey studies relating to customer's perception and satisfaction are applied on about 400 customers who use the Istanbul Maritime Transport system. And then, customer's perception and satisfaction for presented service are examined. Hernandez and Monzon (2016) aimed to identify key factors for defining efficient transport interchange. They considered user's perceptions and their experience in the study. For the appropriate transport interchanging, these parameters have a huge importance. In the scope of the study, traveler satisfaction surveys are prepared and these surveys are carried out in three European transport interchanges. In the evaluation stage, principal component analysis is used. As a result, it is concluded that reducing the waiting time at transport interchange and increasing the comfort level may be important factors for efficient urban transport.

SURVEY APPLICATION

A public survey is carried out to review available data about the patterns of use of existing public transport services in Denizli, Turkey. Surveys are applied separately for bus users and minibus users. Two main aims of these public transport surveys are to measure the quality of services of public transport systems and to reveal perceptions and satisfactions of users about the systems.

Sample Size Determination

Sample size is determined considering statistical methods. To determine appropriate sample size, confidence interval, confidence level and standart deviation values are considered. In this study, the confidence level value is selected as 99%, the confidence interval value is selected as 5% and standart deviation value is selected as 40% respectively. Using these values and the following formula the sample size is determined.

$$n = \frac{z \cdot s\left(1 - s\right)}{c} \tag{1}$$

where;

n: Sample size,
z: z score,
s: Standart deviation value,
c: Confidence interval.

Based on the formula the sample size is determined as 500 people for the bus survey and it is assigned as 882 people for the minibus survey.

Survey Location

Survey studies are applied on the users who prefer the minibus systems and bus systems in urban transport. The locations of the sample are are determined regarding the survey content. Some samples for bus stops and minibus stops in Denizli are selected as survey areas and the surveys are carried out on the public transport users in these areas. The stops located between the university campus and the city center and some others on the routes are determined randomly. Two main routes are considering in the survey. Figure 1 shows the public transport network of Denizli and the routes regarded in the survey. It should be noted that since bus system and minibus system have different properties in many aspects, these two systems are investigated and evaluated independent from each other.

Figure 1. The public transport network of Denizli and the routes considered in the survey

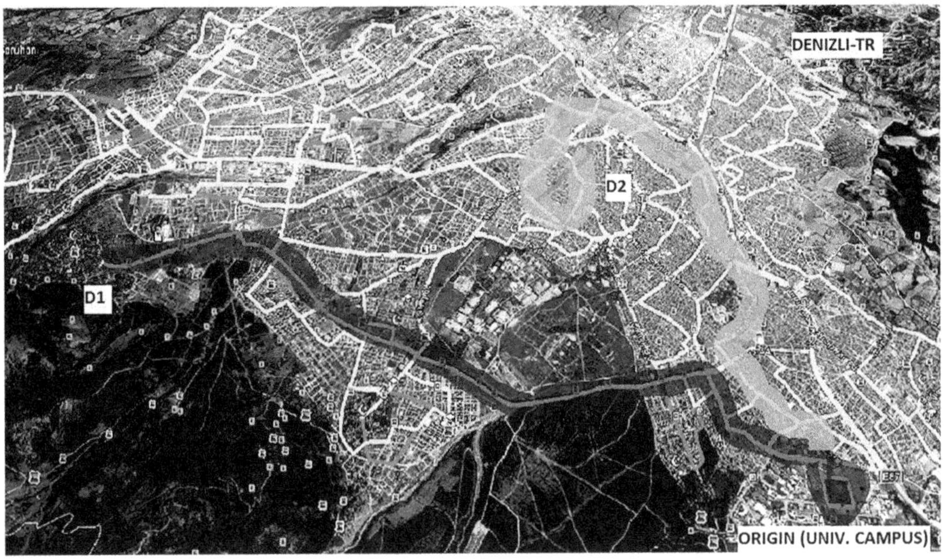

**For a more accurate representation of this figure, please see the electronic version.*

The Questionnare

In the questionnaire, satisfaction levels of participants and perceptions of users are investigated. Primarily, personal informations such as gender, age and profession are asked to participants in these surveys. Besides, these surveys include many questions relating to satisfaction of participants about the numbers of buses/minibuses, the numbers of bus stops/minibus stops, fare policy, bus/minibus routes, capacity, comfort issues, traffic safety etc.

The number of questions for survey on use of bus system and use of minibus system are also different from each other. While survey on the use of bus systems includes 29 questions, the survey on the use of minibus systems includes 25 questions. These questions are related to both perceptions of users and satisfactions of users. Details of the survey questionnares are given in appendices of this chapter of the book.

The sample questions relating to perceptions of users are following:

- What do you think about distances between bus stops in Denizli?
- Is the bus ride reliable for you?
- Are the bus arrival frequencies sufficient in the evening periods for you?
- Is the ticket price economic or not for you?

The sample questions relating to satisfactions of users are following:

- Are the minibuses comfortable enough for you?
- Are the numbers of bus/minibus stops sufficient for you?
- Are the numbers of bus routes satisfactory for you?
- Are the buses clean enough for you?

The classifications of the questions are shown in Table 1. Two main classes are determined as A-User Behaviors or Preferences and B-Operation respectively. The Operation class (B) is also taken into consideration by seven sub-divisions.

Answers of the questions and distributions of answers are determined by simple statistical approaches. In the following section survey results are defined in details.

SURVEY RESULTS

In this study, two surveys are applied for bus and paratransit (minibus) users. In the following, results of the surveys are summarized respectively.

Survey Results for Bus Users

The survey for bus users contains 29 questions. The survey is applied on 500 participants. The distribution of gender, ages and profession of participants are given in Figure 2.

The distribution rate of genders for the bus survey is very close to each other. But the figure is different for distribution of ages. The young participants are about nearly 82% of all. The results for distribution of jobs are so interesting that; about 67% of participants are student, 9% are civil servant, 8% are labor 7% are retired and rest of them are unemployed.

The bus survey results are shown in the following figures and tables respectively (Figure 3, Figure 4). Comments and analysis of each question is also given in the following.

Q1: In the first question of bus survey, the type of transportation system for daily usage is asked to participants. Based on the responses, it is clear that the use of bus system is higher than paratransit and others. The rate of bus system users is determined as 85% in the survey. It is determined that this rate is very high if the ages and occupation are considered.

Table 1. Classification of the survey questions

Class	Definition	Questions for Bus System	Questions for Minibus System
A	User Behaviors or Preferences	Q1, Q2, Q3	Q1, Q2, Q3
B	Operation		
B1	System Specifications	Q4, Q5, Q10, Q11, Q12, Q14, Q15, Q24, Q28,	Q4, Q5, Q9,Q10, Q11, Q20, Q24
B2	Information	Q8, Q9,	Q8
B3	Accessibility	Q6, Q13, Q16, Q17, Q18, Q19	Q6, Q12, Q13, Q14, Q15
B4	Comfort	Q20, Q29	Q16, Q25
B5	Capacity	Q21, Q22,	Q17, Q18
B6	Safety	Q23, Q25, Q26	Q19, Q21, Q22
B7	Reliability	Q7, Q27	Q7, Q23

Figure 2. Distributions of gender, age, and profession of participants for the bus survey

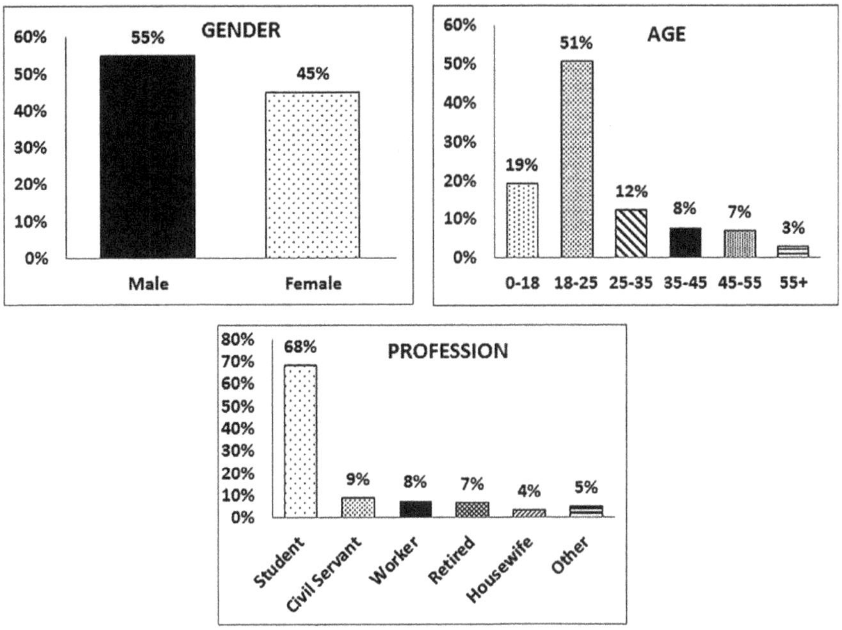

Figure 3. Bus survey results for the questions 1 to 6

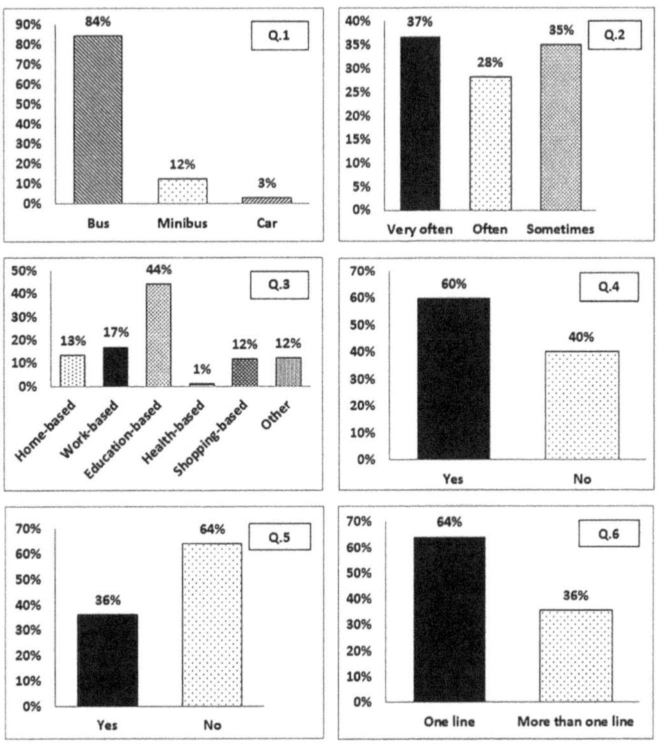

Figure 4. Bus survey results for the questions 19 to 24

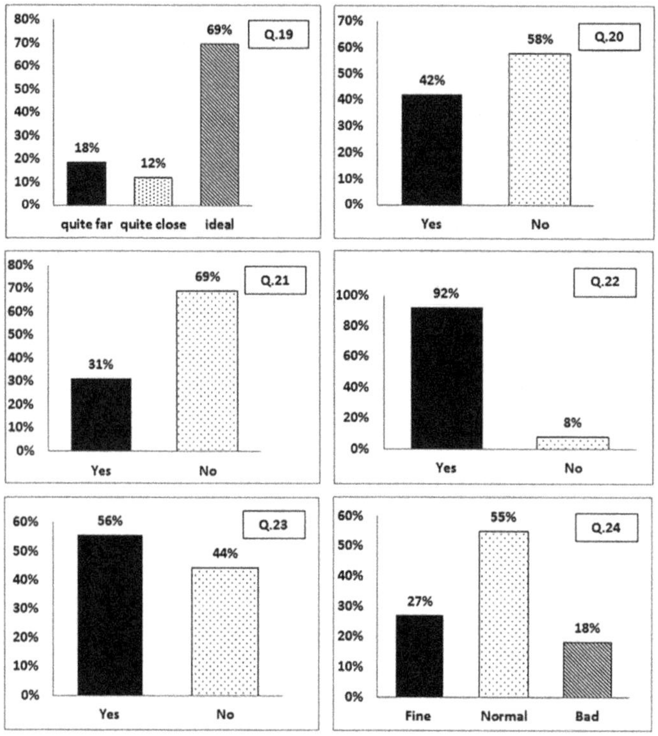

Q2: In the second question, the frequency of bus system usage is asked to participants. As seen on the figure, it is used very often by most of the participants.

Q3: The purpose of travel is asked to participants in the third question. Because of the most of the participants are students, the education based travels are the most preferred purpose. The work based travels and shopping based travels follow the education based travels.

Q4: In the fourth question of the survey sufficiency of bus stops are inquired. Results show that, although many participants agree about sufficiency of bus stops, it cannot be identical for all routes. On the other hand, results of this question can be varied by frequency of usage.

Q5: Sufficiency of the number of buses is inquired in the fifth question of the survey. In some routes, because of higher demand and capacity insufficiency, the numbers of buses are insufficient for the participants. Thus, these types of routes should be checked and the numbers of buses should be increased.

Q6: The number of bus lines used in the travel is asked to participants in the sixth question. The results show that 64% of participants use more than one line in the travel. This information can be used for re-organization of bus lines by decision makers.

Q7: The waiting times at the bus stop are asked to participants in this question. Results show that waiting times of participants can be changed during the day and it can be reached up to 30 minutes. This figure can be reduced by increasing the number of buses and frequencies.

Q8: Time table or schedule of bus systems is asked to participants in the eighth question. It is resulted that nearly half of the participants have not enough information about the schedule of bus system.

Therefore, waiting times of users increased and reliability of system decreased. These negative results can be handled by use of passenger information system and bus stop information system.

Q9: In the ninth question, the information systems for buses are asked. The result shows that 64% of participants cannot have enough information about bus systems. Therefore an information system should be supplied for bus users.

Q10: In this question regain rate of lost goods or belongings are asked. This rate has been determined as 50% regarding the survey results. It may be increased by establishing lost good office and a call center for users.

Q11: In this question, contact rate of users and operators for any problems encountered by users are asked. The responses show that 36% of participants have contacted with the operator.

Q12: Satisfaction level of smart city card usage is asked in the twelfth question. It is concluded that 83% of participants have been satisfied by use of smart city card in bus system fare collection.

Q13: Points of sales for smart city card are inquired in this question. 64% of participants' responses are negative for this question. Thus the number of POS should be increased.

Q14: The ticket prices are asked in the fourteenth question of survey. It is resulted that the prices are not economic for most of the participants. But it is economic for 41% of participants. In order to increase the number of users, it should be adjusted.

Q15: The fifteenth question is asked only for the participants whose answer is non-economic for the previous (fourteenth) question. The responses show that the ticket prices should be more affordable.

Q16: Another important issue for users is route of bus system. In the sixteenth question routes of bus system is asked to participants. The results show that the routes are not proper for 52% of participants. It should be improved and re-organized considering properties of catchment areas.

Q17: The accessibility of bus routes and shortest paths are asked to participants in the seventeenth question. The responses show that 53% of participants are not satisfied about the routes.

Q18: Average travel time of bus user is also inquired in the survey. It is resulted that average travel time (perceived) is about between 15 and 30 minutes.

The survey results for the questions 7 to 18 are given in Table 2.

Q19: Bus stops are another important issue in the system. Accessibility of bus stops can increase the number of users. In the nineteenth question the distances of bus stops are asked. It is resulted that the distance of bus stops are accessible for most of the users. On the other hand it is not proper for some locations or districts of city.

Q20: Quality of service is one of the important criteria for sustainable transportation. In this question comfort of bus systems is inquired. Results show that 58% of participants are not satisfied about the comfort of bus system. This issue should be investigated in details.

Q21: The comfort issue has a relation to capacity. Capacity of buses is asked to participants in the following question. It is inferred that most of participants are not satisfied about the issue. 69% of participants think that capacity of bus system is not enough for demand.

Q22: Overloading in bus system is also inquired in the survey. Results show that most of participants (92%) have disturbed about overloading. This issue is also associated about capacity of bus system. Bus system should be re-organized considering demand and capacity issues.

Q23: Safety is another important issue for users. In this question safety of bus system is asked to participants. It is resulted that 54% of participants are convinced about safety.

Table 2. Bus survey results for the questions 7 to 18

Question Number	Subject	Results
Q.7	Waiting times	0-15 minutes 44% 15-30 minutes 48% 30+ minutes 8%
Q.8	Time table	Yes 54% No 46%
Q.9	Information system	Enough 36% Not Enough 64%
Q.10	Lost Goods	Yes 12% No 10% I have no encountered 78%
Q.11	Communication	Yes 36% No 21% I have no encountered 43%
Q.12	Satisfaction about smart city card	Yes 83% No 17%
Q.13	Points of sales for smart city card	Enough 38% Not Enough 62%
Q.14	Ticket price (fare)	Economic 41% Not Economic 59%
Q.15	Recommendation for ticket price	10-25 cents 2% 25-50 cents 54% 50-75 cents 38% 75-100 cents 6%
Q.16	Routes	Enough 48% Not Enough 52%
Q.17	Shortest paths	Satisfied 47% Unsatisfied 53%
Q.18	Average travel time	0-15 minutes 7% 15-30 minutes 66% 30+ minutes 27%

Q24: Riding behavior of bus driver is also asked to participants. Most of participants agree that riding behaviors of bus drivers are good or normal. Only 18% of them have not agreed with the others.

Q25: In the twenty-fifth question traffic rules and driver compliance index is asked. It is concluded that bus drivers generally obey traffic rules due to 71% of participants.

Q26: Interaction with other vehicles in traffic and obstruction is also inquired in the survey. It is resulted that traffic is not obstructed by buses according to most of participants (61%).

Q27: The frequency of buses is asked to participants. Reults show that the frequency is not proper for the 78% of participants. This parameter is directly related to demand and satisfaction of users. Thus, it should be re-organized considering different time periods of day.

Q28: One of the main factors for bus users is cleanness. In this question cleanness of the bus is inquired. Results show that 59% of participants are satisfied by the cleanness of buses. This rate is not enough for a sustainable system. It should be increased by focusing on the system.

Q29: In the system, because of a direct relation with users, bus stops have a great importance. In this question, cleanness and comfort of bus stops are asked. Results of this question show that about half of the participants (51%) are not satisfied about the subject.

Table 3 depicts the bus survey results for the questions 25 to 29.

Survey Results for Paratransit (Minibus) Users

The survey for paratransit (minibus) system is applied on 882 participants. In the questionnaire 25 different subjects are asked. Figure 5 contains distributions of participants for the survey regarding gender, age and profession.

Table 3. Bus survey results for the questions 25 to 29

Question Number	Subject	Results
Q.25	Obeying rules (driver)	Yes 71% No 29%
Q.26	Effects on traffic	Yes 39% No 61%
Q.27	Frequency	Enough 22% Not Enough 78%
Q.28	Cleanness of vehicles	Enough 59% Not Enough 41%
Q.29	Design of stops	Enough 49% Not Enough 51%

Figure 5. Distribution of gender, age, and profession of participants for the paratransit survey

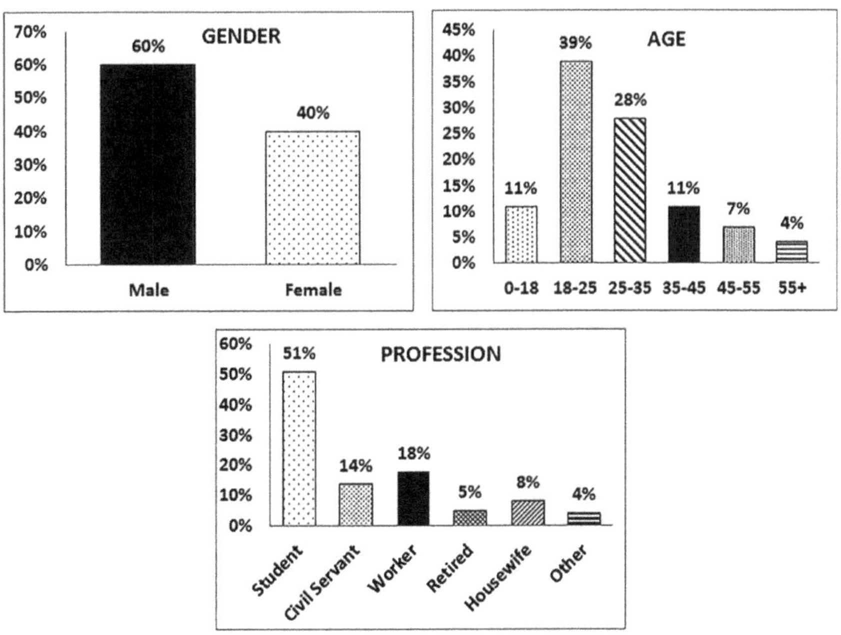

60% of participants are male and 40% are female. A significant amount of participants (61%) are young. Most of the participants are student (about 51%), 18% are worker, 14% are civil servant, 8% are housewife and 5% are retired. Survey results for each question are explained in the following. Sample results are given in Figure 6 and Figure 7.

Q1: Determination of modal split for urban transport is aimed with this question. As can be seen on the graph, the most preferred mode is minibus with a rate of 44%. Besides, the bus systems are preferred with a rate of 32%, the proportion of private car usage is about 19%. As a result, bus systems and minibuses have great importance in urban transport. In addition, it can be revealed that popularity of public transport in Denizli is quite high.

Q2: This question is asked for determining of minibus usage frequencies of public transport users. The graph shows that 41% of participants use minibus sometimes while 26% of participant use minibus everyday. Besides, it can be concluded that minibuses are used rarely in urban transport by 33% of participants.

Q3: The trip purpose of participants can be determined with this question. As can be seen on the graph, it is clear that the most of trips (with a rate of 35%) are educational-based. Rates of work-based trips and shopping-based trips are about 13% and 12% respectively. Health-based trips are quite low with a rate of 6%. It should be noted that, most of the participants are composed of teachers, academicians and students (about with a rate of %65) in this study. For this reason, rate of educational-based trips are quite high.

Figure 6. Minibus survey results for the questions 1 to 6

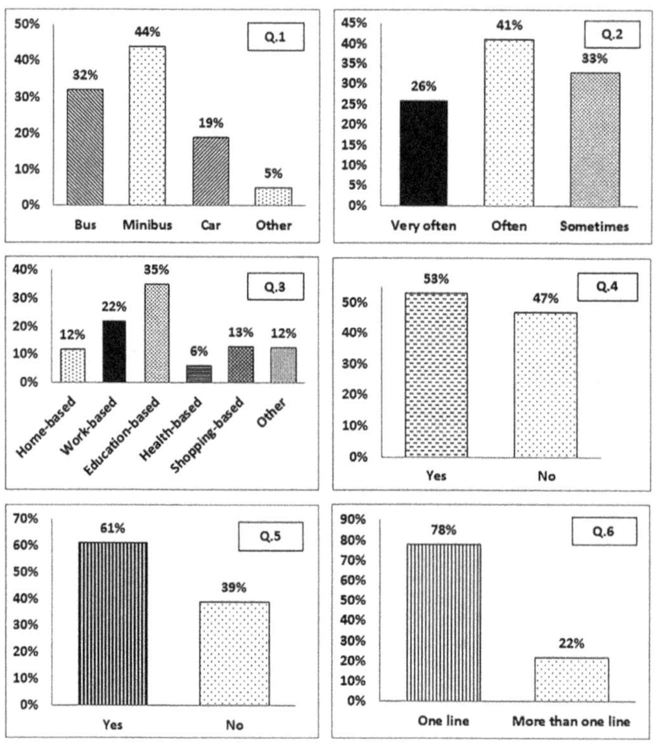

Figure 7. Minibus survey results for the questions 20 to 25

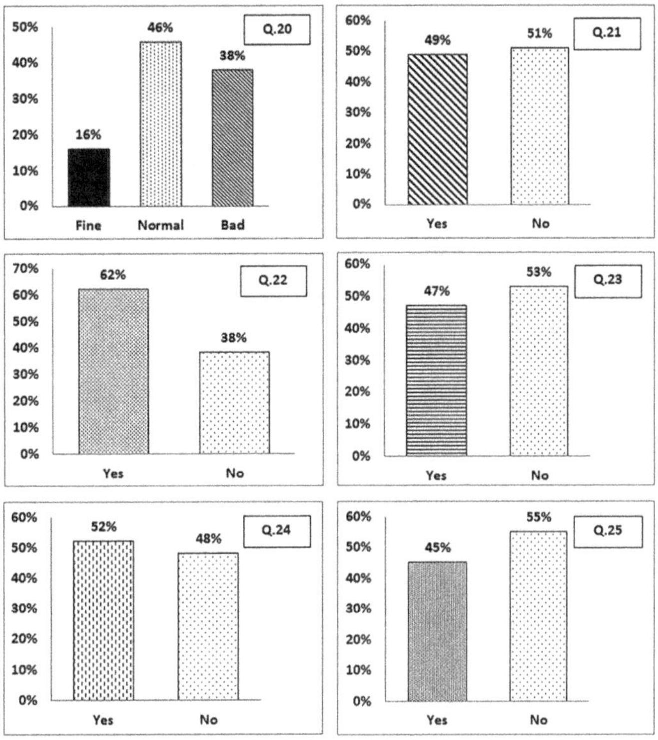

Q4: "Are the number of minibus stops sufficient for you?". This question aims to determine the adequacy of the number of minibus stops. As can be seen on the graph, answers consist of "yes" with a rate of 53% and "no" with a rate of 47% for this question. As a result, it is clear that the number of minibus stops should be increased by policy makers in Denizli city.

Q5: Determination of sufficiency of the number of minibuses is aimed with this question. The most of participants (about 61%) point out that the number of minibuses in Denizli is sufficient for urban transport.

Q6: Aim of this question is to determine the number of lines (transfer) which are used for arriving the destination points by the participants. As a result, it can be seen that the most of participants use one line for arriving the destination points.

Q7: Determination of waiting period of participants for arrival of a minibus to the minibus stop is aimed with this question. As can be seen on the graph, the most of participants have 2-4 minutes of waiting periods. Besides, very few of the participants have more than 6 minutes of waiting periods. As a result, it can be said that time schedule of minibus services is suitable and well-organized in Denizli.

Q8: Accesibility to informations relating to minibus lines is important issue for passangers in urban transport. Accesibility to informations relating to minibus lines is analyzed with this question. As can be seen on the graph, most of participants with a rate of 65% can not reach informations relating to minibus lines easily. So, policy makers must focus on this issue. Besides, new technological improvements should be taken into consideration by policy makers.

Figure 5 shows the minibus survey results for the questions 1 to 6. Table 4 summarizes the minibus survey results for the questions 7 to 19.

Q9: This question is related to belongings and goods which are lost in the minibuses. As can be seen on the graph, most of the participants with a rate of 56% have not lost their belongings and goods in minibuses yet. Although 23% of participants have lost their belongings and goods in the minibuses, they have not found them again. Besides, 21% of participants have not encountered with such an event yet. As a result, it is concluded that minibus drivers must be careful on this issue and they should help to the passengers.

Q10: When the passengers encounter with any problem relating to minibuses, it is needed that passangers should voice their problems to the authorities easily. This is a primary provision of reliable

Table 4. Minibus survey results for the questions 7 to 19

Question Number	Subject	Results
Q.7	Waiting times	0-2 minutes 10% 2-4 minutes 45% 4-6 minutes 25% 6-8 minutes 7% 8-10 minutes 8% 10+ minutes 5%
Q.8	Information system	Yes 35% No 65%
Q.9	Lost Goods	Yes 21% No 23% I have no encountered 56%
Q.10	Communication	Yes 45% No 55%
Q.11	Ticket price (fare)	Economic 37% Not Economic 63%
Q.12	Routes	Enough 70% Not Enough 30%
Q.13	Shortest paths	Satisfied 44% Unsatisfied 56%
Q.14	Average travel time	1-10 minutes 17% 10-20 minutes 62% 20+ minutes 21%
Q.15	Minibus stops	quite far 45% quite close 9% ideal 46%
Q.16	Comfort	Yes 64% No 36%
Q.17	Capacity	Enough 30% Not Enough 70%
Q.18	Overloading	Yes 82% No 18%
Q.19	Safety	Enough 37% Not Enough 63%

and comfortable trips. This question is related to communication facilities of passengers towards the authorities. As can be seen on the graph, 55% of passengers stated that they do not voice their problems to the authorities easily. According to this result, communication facilities must be improved by policy makers. For this purpose, passenger's satisfactions and passenger's complaints can be determined by the survey studies.

Q11: The aim of this question is to analyze minibus fare economically. The graph shows that the most of participants with a rate of 63% are unsatisfied about fare of minibus system in Denizli.

Q12: This question is related to the number of minibus routes. Sufficiency of the number of minibus routes is considered and then survey results are analyzed. As can be seen on the graph, the most of participants with a rate of 70% declared that the number of minibus routes is sufficient in Denizli. Besides, rest of participants point out that the number of minibus routes must be increased by the policy makers.

Q13: This question is related to whether the minibus routes comprise shortest paths or not. As can be seen on the graph, 44% of participants denote that the minibus routes compromise the shortest paths, the rest of participant point out opposite idea. As a result, 56% of participants argue that the lengths of minibus routes are quite long.

Q14: Determination of average trip time spent by the passengers in minibus system is aimed in this question. When the results are investigated carefully, it can be seen that 62% of the participants waste about 10-20 minutes in minibus system daily. As a result, it can be said that passengers who are using minibus system do not waste a lot of time in their trips.

Q15: This question aims to determine sufficiency of distance between minibus stops in Denizli. According to survey results, while 46% of participants state that distances between minibus stops are satisfactory, the rest of participants with a rate of 54% indicate that distances between minibus stops are not satisfactory. As a result, the distances between minibus stops should be reorganized by the policy makers.

Q16: Comfort is one of the important parameters in evaluation of public transport systems. This question is related to whether minibus systems are comfortable in Denizli or not. As can be seen from the survey results, the most of participants with a rate of 64% point out that minibus system in Denizli are quite comfortable. On the other hand, the rest of participants indicate that minibus systems in Denizli are not comfortable. As a result, comfort level of minibus systems should be improved by various policies. Renewal of vehicle fleet, vehicles maintanence on time etc. can be shown as sample for these policies.

Q17: Capacity parameter used for determining of public transport system performance is considered in this question. As can be seen on the graph, although 30% of participants state that the minibus capacities are sufficient, the most of participants (with a rate of 70%) think that the minibus capacities should be increased.

Q18: In Turkey, although it is banned, minibuses serve more passengers than their capacity sometimes. This situation adversely affects the quality of trip. So, the effects of this negative situation on passengers are analyzed in this question. As can be seen on the graph, the most of participants with a rate of 82% are not satisfied from the operation of over the capacity of minibuses. To avoid this unfavorable situation, the minibus drivers should not take more passengers than the number of seats of the minibus.

Q19: This question is related to safety parameter for minibus systems. As can be seen on the graph, the most of participants with a rate of 63% point out that minibus ride is not safe in Denizli. According to survey results, safety problems generally arise from drivers' failures. For this reason, minibus drivers should be careful and conscious in traffic. Besides, speed limits for urban roads in Turkey are also not exceeded by the minibus drivers.

Q20: Behaviors and attitudes of minibus drivers' are investigated in this question. While the most of participants (about a rate of 46%) point out that behaviors and attitudes of minibus drivers' are standard, 38% of participant are not satisfied from behaviors and attitudes of minibus drivers'. Besides, 16% of participants remark that behaviors and attitudes of minibus drivers' towards to passengers are satisfactory. As can be seen on the graph, it can be said that behaviors and attitudes of minibus drivers' are not satisfactory in general. As a result, it can be said that minibus drivers' should be more courteous towards to the passengers. In addition, level of education of minibus drivers must be improved with various education methods such as seminar, training and etc.

Q21: Traffic congestion and traffic accidents can be prevented by the traffic rules. So, vehicle drivers must obey the traffic rules to eliminate unfavorable impacts of traffic. This question is related to whether minibus drivers obey the traffic rules or not. As can be seen on the graph, the 51% of participant state that minibus drivers do not obey the traffic rules, the rest of participants claim that minibus drivers obey the traffic rules.

Q22: It sould be noted that traffic congestion is one of the major problems for developed and developing cities. Determination of the role of minibus systems in traffic congestion is aimed with this question. As can be seen on the graph the most of participants with a rate of 62% claim that minibuses cause traffic congestion in urban area, the rest of participants with a rate of 38% state that minibuses do not cause traffic congestion in urban area. It can be said that minibus drivers must be careful in traffic for uninterrupted traffic flow.

Q23: Arrival frequency of public transport vehicles is an important issue for evaluating public transport system performance. This question is related to whether arrival frequencies of minibuses to minibus stops are sufficient in the evening period or not. As can be seen on the graph, when 47% of participants claim that arrival frequencies of minibuses are sufficient in evening period, the rest of participants with a rate of 53% say the opposite of this. As a result, the arrival frequencies of minibuses should be rearranged in the evening periods.

Q24: This question is related to cleanness of minibuses. As can be seen on the graph, the most of participants with a rate of 55% state that minibuses are not clean enough. The rest of participants with a rate of 45% are satisfied from cleanness of minibuses. As a result, it can be said that more attention should be needed for cleanness of minibuses.

Q25: This question is related to comfort and cleanness of minibus stops. As can be seen on the graph, 52% of participants point out that they are satisfied from comfort and cleanness of minibus stops. Besides, the rest of participants with a rate of 48% are not satisfied from comfort and cleanness of minibus stops. As a result, it can be said that comfort level and cleanness of minibus stops should be improved by the authorities.

Figure 7 depicts the minibus survey results for the questions 20 to 25.

COMPARISONS AND DISCUSSIONS

The survey results for both systems are compared to each other considering the subjects used in the questionnaires. User and system specifications are considered in the comparisons. Table 5 shows the comparisons of the survey results regarding user specifications'.

As can be seen on Table 5, the average age of users who prefer minibus system is higher than the average age of users who prefer bus system. Namely, youngest PT users prefer bus system in urban transport. This situation can be arised from various reasons. Besides, it can be seen that students generally prefer bus system because of being more economical. In addition to this, it can be concluded that the purposes of travels resemble to each other for these two PT systems according to survey results. Table 6 shows the comparisons of the survey results regarding system specifications.

The results relating to user perceptions and user satisfactions about the bus and minibus PT systems are presented in Table 6. As can be seen on Table 6, the most of users of bus system point out that fleet size of bus system is not sufficient, the most of users of minibus system with a rate of 61% claim that fleet size of minibus system is sufficient. The number of stops is found enough by both PT users. When the criterion of fare is considered, similar results attract attention from both surveys. The fares for both bus system and minibus system are not found economical according to the results of surveys. Both survey results provide similar findings in terms of travel times. According to the results, it can be said that travel times using public transport systems are ranged from 10 to 30 minutes. One of the criteria of service quality of PT systems is also frequency parameter. As can be seen on Table 2, frequencies of both PT systems are not found as satisfactory by the users. The most of participants of both surveys with a rate of 70% point out that unit capacity of PT vehicles are not sufficient. Similarly, designs of stops are also fair for both PT system users. When the cleanness of vehicles is considered, although both PT systems are found clean enough by participants, the rates of participants are very low. Therefore it is commented that user expectations are not fulfilled about cleaning of the system. One of the most important parameters for PT users is also route. According to survey results, although the most of minibus system users state

Table 5. Comparisons of survey results for user specifications

Specification of PT Users	Bus Survey	Minibus Survey
Age (Young: under 25 years old users)	Young 70% Old 30%	Young 50% Old 50%
Gender	Male 55% Female 45%	Male 60% Female 40%
Employment	Student 68% Civil Servant 9% Worker 7% Retired 7% Housewife 4% Other 5%	Student 51% Civil Servant 14% Worker 18% Retired 5% Housewife 8% Other 4%
Purpose of Travel	Home-based 14% Work-based 17% Education-based 44% Health-based 1% Shopping-based 12% Other 12%	Home-based 12% Work-based 22% Education-based 35% Health-based 6% Shopping-based 13% Other 12%

Table 6. Comparisons of survey results for system specifications

Specification of PT System	Bus Survey	Minibus Survey
Fleet Size	Enough 36% Not enough 64%	Enough 61% Not enough 39%
Number of Stops	Enough 60% Not enough 40%	Enough 53% Not enough 47%
Waiting time	0-15 minutes 44% 15-30 minutes 48% 30+ minutes 8%	0-2 minutes 10% 2-4 minutes 45% 4-6 minutes 25% 6-8 minutes 7% 8-10 minutes 8% 10+ minutes 5%
Fare	Economic 41% Non-economic 59%	Economic 37% Non-economic 63%
Travel time	0-15 minutes 7% 15-30 minutes 66% 30+ minutes 27%	1-10 minutes 17% 10-20 minutes 62% 20+ minutes 21%
Frequency	Enough 22% Not Enough 78%	Enough 47% Not Enough 53%
Unit Capacity	Enough 31% Not Enough 69%	Enough 30% Not Enough 70%
Design of stops	Enough 49% Not Enough 51%	Enough 45% Not Enough 55%
Cleanness of vehicles	Enough 59% Not Enough 41%	Enough 52% Not Enough 48%
Routes	Enough 48% Not Enough 52%	Enough 73% Not Enough 30%
Driver Behavior	Fine 27% Normal 55% Bad 18%	Fine 16% Normal 46% Bad 38%
Obeying rules (driver)	Yes 71% No 29%	Yes 49% No 51%
Information system	Enough 36% Not Enough 64%	Enough 35% Not Enough 65%
Safety	Enough 56% Not Enough 44%	Enough 37% Not Enough 63%
Comfort	Enough 42% Not Enough 58%	Enough 64% Not Enough 36%

that the routes of minibus system in Denizli are enough, the most of bus system users have not agreed for the bus system. Although user perceptions for bus system and minibus system show similar trends in terms of driver behaviors and attitudes, minibus drivers are more criticized by participants. Besides, it can be said that both PT systems drivers obey the traffic rules. Information system is also one of the most important parameters for public transportation systems. As can be seen on the survey results, it can be concluded that both PT systems are not sufficient in terms of information system. The results of both surveys differ from each other in terms of safety and comfort. According to survey results, when the safety parameter is considered, it can be said that the bus system is safer than the minibus system.

Besides, when the comfort parameter is considered, it is concluded that the minibus system is comfortable than the bus system.

The results obtained may be used by the public transport authorities of Denizli city as a decision maker. The mayor of Denizli Metropolitan Municipality may have some implications from the results of this study. It may akso be used by the operator of minibus system. Based on the survey results some routes of both systems may be re-organized. The fares may be re-determined. Some incentives that support use of public transportation systems may be developed considering fare and transfer issues. Some personality development program may be organized for public transport drivers.

The survey questions and answers may be analyzed considering the classification shown in Table 1. For instance, the first three questions of these surveys were asked for determining public transport preferences and usage behaviors of the users. According to replies obtained from questions in this group, public transportation systems can be re-organized. In addition to this, a more stable and integrated public transportation system can be improved.

Problems related to system specifications for the operation of bus and minibus systems can be determined by the questions which are in the B1 class. For example, if the operation of public transport system is quite poor, the number of bus stops or buses can be increased, Thus, the inadequacy of system is prevented by decison makers. Similarly, the users are more satisfied from the economic and lower ticket prices.

The information (informing users) issue is one of the most important tasks in the operation of public transportation systems. The investigation of this issue was aimed with the questions which are in the B2 class. From the answers given for these questions, it can be concluded that the information system is insufficient and this system should be improved for increasing of the demand.

The accessibility issue were investigated with the questions which are in the B3 class. According to this, it is undersood that bus routes and accessibility of the system is insufficient. This problem can be prevented with the reorganization of the routes.

The questions which have been in B4 and B5 classes includes comfort and capacity matters. When the answers given to these questions were evaluated carefully, insufficiency of comfort and capacity were determined. Besides, it is understood that comfort and capacity should be improved by the decision makers. Standardisation of air conditioning systems and the improvement of vehicle fleets can be suggested for the prevention of these type of problems.

Safety and reliability matters are analyzed with the questions which have been in B6 and B7 classes. According to these, it is understood that the system is found insufficient by the users. Besides, it is concluded that system reliability which is an important operation parameter is also not found sufficient by the users. Traffic safety can be improved with the stimulating and controlling of vehicle drivers. Besides, the system reliability can be improved with controlling the time schedule plans in terms of compliance of plans and ensuring the regularity.

CONCLUSION

This study concerns with an investigation about the perceptions and satisfactions of public transport users in Denizli, Turkey. Two PT surveys are applied for the bus users and the minibus users.

In the bus survey, 500 users are participated and 29 different questions are inquired. Results obtained are classified for evaluation. Some findings obtained from the bus survey are given below.

- The numbers of buses and provided capacities are not sufficient for considerable rate of participants.
- Bus fares are not affordable especially for students.
- Bus trip is not comfortable for many users.
- Bus trip is defined as safer than others for many users.
- Smart commuter card is the best choice for many users but the problem is related to availability and re-loads.
- Buses are generally clean for many users but it is not fulfilled for most of the participants.

These are the general results for bus system. But it differs if the distributions for gender, ages and jobs are considered.

In the scope of this study, another survey is performed for minibus system. In this survey, 882 users are participated and 25 different subjects are asked. The results obtained are summarized in the following.

- The fleet size and unit capacity of minibuses are not met users' expectations for considerable rate of participants.
- Similar to the bus system, the minibus fares are not affordable for participants.
- Minibus system is defined more comfortable for many users.
- Minibus trip is not safer than others for many users. Drivers do not obey the speed limits and some other regulations
- Minibuses are defined as clean for 52% of participants.

The results obtained can be used for sustainable transportation planning purpose. One of the important finding that is related to both systems is reliability. According to the survey results, both of the systems are not reliable. Users have not enough information about time-tables, frequencies and also travel times. To increase reliability, passenger information system and bus stop information system should be designed. On the other hand, the frequencies of buses and paratransit should be adjusted regarding the survey results. Another important finding is about the comfort issue. The problems about comfort issue may be considered and handled by operator. The results pointed out that fare is also the common problematic subject for users. Regulation of fares including some policies (incentives etc.) can be helpful. Thus the satisfaction levels of users may be increased by this way. And the perceptions of users for public transport may be changed.

REFERENCES

Altan, M., & Engin, O. (2003). The measurement of user satisfaction in a travel company. *Journal of Social and Economical Researches of Selcuk University*, *6*, 84–98.

Andreassen, T. W. (1995). (Dis)satisfaction with public services: The case of public transportation. *Journal of Services Marketing*, *9*(5), 30–41. doi:10.1108/08876049510100290

Belwal, R. (2013). Peoples perception of public transport services in Oman. *Jurnal Teknologi*, *65*(3), 9–16. doi:10.11113/jt.v65.2140

Belwal, R., & Belwal, S. (2010). Public transportation services in Oman: A study of public perceptions. *Journal of Public Transportation*, *13*(4), 1–21. doi:10.5038/2375-0901.13.4.1

Das, S., & Pandit, D. (2013). Importance of user perception in evaluating level of service for bus transit for a developing country like India: A review. *Transport Reviews*, *33*(4), 402–420. doi:10.1080/01441647.2013.789571

De Ona, J., de Ona, R., & Lopez, G. (2015). Transit service quality analysis using cluster analysis and decision trees: A step forward to personalized marketing in public transportation. *Transportation*, 1–23. doi:10.1007/s11116-015-9615-0

DellOlio, L., Ibeas, A., & Cecın, P. (2010). Modelling user perception of bus transit quality. *Transport Policy*, *17*(6), 388–397. doi:10.1016/j.tranpol.2010.04.006

DellOlio, L., Ibeas, A., & Cecın, P. (2011). The quality of service desired by public transport users. *Transport Policy*, *18*(1), 217–227. doi:10.1016/j.tranpol.2010.08.005

Eboli, L., & Mazzulla, G. (2012). Performance indicators for an objective measure of public transport service quality. *European Transport*, *51*, 1–21.

Fellesson, M., & Friman, M. (2008). Perceived satisfaction with public transport service in nine European cities. *Journal of the Transportation Research Forum*, *47*(3), 93–103.

Garrido, C., de Ona, R., & de Ona, J. (2014). Neural networks for analyzing service quality in public transportation. *Expert Systems with Applications*, *41*(15), 6830–6838. doi:10.1016/j.eswa.2014.04.045

Hemedoglu, H. (2012). Measuring service quality in metro services: a comparison of perceived service quality and expected service quality. *Journal of Business Economics Instıtute of Business School of Istanbul University – Management*, *23*(72), 25-47.

Hernandez, S., & Monzon, A. (2016). Key factors for defining an efficient urban transport interchange: Users perceptions. *Cities (London, England)*, *50*, 158–167. doi:10.1016/j.cities.2015.09.009

Karacasu, M., & Bilgic, S. (1999). Survey study sample in Eskisehir/Turkey and Solution offers for transportation problems for Eskisehir city.*Proceedings of 2nd National Urban Infrastructure Symposium in Turkey*, 87-99.

Karadeniz, M., & Unver, N. M. (2014). Customer's perception of service and satisfaction in maritime transport in the city: An application. *Journal of Offering of Marmara University*, *11*(42), 43–68.

Khalid, U. A., Bachok, S., Osman, M. M., & Mansor, I. (2014). User perceptions of rail public transport services in Kuala Lumpur, Malaysia: Ktm komuter. Amer International Conference on Quality of Life. *Procedia: Social and Behavioral Sciences*, *153*, 566–573. doi:10.1016/j.sbspro.2014.10.089

Maraglino, V., dellOlio, L., Borri, D., & Portilla, A. I. (2014). dell Olio, L., Borri, D. & Portilla, A. I. (2014). Methodology for a study of the perceived quality of public transport in Santander. XI Congreso de Ingenieria del Transporte. *Procedia: Social and Behavioral Sciences*, *160*, 499–508. doi:10.1016/j.sbspro.2014.12.163

Morfoulaki, M., Tyrinopoulos, Y., & Aifadopoulou, G. (2007). Estimation of satisfied customers in public transport systems: A new methodological approach. *Journal of the Transportation Research Forum*, *46*(1), 63–72.

Rojo, M., dellOlio, L., Gonzalo-Orden, H., & Ibeas, A. (2013). Interurban bus service quality from the users viewpoint. *Transportation Planning and Technology*, *36*(7), 599–616. doi:10.1080/03081060.2013.845432

Stradling, S., Carreno, M., Rye, T., & Noble, A. (2007). Passenger perceptions and the ideal urban bus journey experience. *Transport Policy*, *14*(4), 283–292. doi:10.1016/j.tranpol.2007.02.003

Tyrinopoulos, Y., & Antoniou, C. (2008). Public transit user satisfaction: Variability and policy implications. *Transport Policy*, *15*(4), 260–272. doi:10.1016/j.tranpol.2008.06.002

Winder, A. D., May, A., & El-Araby, K. (2005). Urban transport performance and policy: results of a European public perception survey. *Association for European Transport and contributors 2005*. Retrieved from http://abstracts.aetransport.org/paper/index/id/2083/confid/11

Chapter 5
An Assessment Relationship between Social Exclusion and Transportation

Seher Ozkazanc
Yüzüncü Yıl University, Turkey

Nihan Ozdemir Sonmez
Ankara University, Turkey

ABSTRACT

Having been used extensively since the 1980s, the concept of social exclusion has given a new impetus to the discussions of poverty and disadvantageousness. The concept of social exclusion, which can be defined as the condition in which certain individuals or social groups cannot integrate into the society either socially or economically or politically, leads to social cohesion problems. This triggers segregation of the society, particularly in large cities, in both social and spatial terms. In the context of accessibility, "urban transport" appears as one of the most important factors determining level of social inclusion/ exclusion of the individuals or groups. In this study relation between social exclusion and transport has been evaluated as an attempt to identify socio-spatial segregation pattern of Ankara.

INTRODUCTION

Although social exclusion (seen as a new concept of social policy defined in France in the 1970's and subsequently spread all over the world, particularly in Europe), in essence, has existed throughout human history in various societies, it is felt in different dimensions and is greatly disadvantageous to individuals. Social exclusion, which in its simplest form is defined as not being involved in the community, isolation from social dealings, and inability to integrate into society, has become an interdisciplinary discussion matter with development policy, particularly that concerning poverty reduction, and social equity. While social exclusion encompasses issues such as long-term unemployment, poverty, inequalities in income distribution, lack of social protection facilities, it strongly affects individuals' states and the process of being deprived of rights and opportunities in economic, social, and political spheres. These deprivations

DOI: 10.4018/978-1-5225-2116-7.ch005

may arise from individual features, such as: poor skills, high crime environment, bad habits, and family problems, but lack of urban services can also restrict social inclusion.

Social exclusion that emerges in the form of deprivation from services offered (such as education, health, housing, nutrition and social security) is among the major social problems experienced in Turkey in recent years. Social exclusion, which has different manifestations in different geographic areas, is felt more particularly in the areas of the big cities that have received migrants, who are vulnerable and marginalized. Migration in Turkey from rural areas to the cities accelerated with the process of industrialization, which was actualized by individuals in order to improve their quality of life, reached its peak level in the 1950's. In this period, governments could not meet the accommodation needs of all newcomers to cities; thus, slums belts began to appear in the peripheries of urban areas. The morphological structure of the cities that continued to grow by receiving migration has been transformed with these marginal housing areas, and cities have entered a process of rapid expansion. Due to the policies addressing this, while housing plots in the central areas of the cities have intensified, socio-cultural and recreation areas have begun to shrink (Church et al., 2000). Loss of compact urban characteristics, expansions of the city limits, and changes in land use have brought various transport problems. Individuals that experience difficulties accessing various services, activities, and opportunities due to lack of transportation are isolated from society. However, in fulfilling the individual needs, the provision of social relations, and the prevention of social exclusion, "transportation" emerges as an effective tool.

In transport policy in developed countries (especially the United Kingdom and Canada), difficulties due to the relationship between social exclusion, social inclusion, and access to transport are considered significant problem areas. In fact, transport policies in essence accommodate access to equal and people-oriented services for all individuals, and main objective of these policies is to minimize the lack of access to vital activities such as health, education, employment by considering the variables based on socio-economic, demographic, physical, and cultural differences.

In the literature on Turkey, social exclusion appears to be discussed as an interdisciplinary concept that focuses on participation issues of poverty and labor force studies particularly. However, studies working to establish the connection between social exclusion and urban transport are scarce to non-existent. Based on this lack, this study seeks to demonstrate the relationship between social exclusion and urban transport.

From the perspective of social exclusion, we determine how accessibility and high mobility appear in different spatial embodiments of the city for understanding the factors and effects of transport and social exclusion approaches; this achieves a multidimensional, relational, and dynamic conception of exclusion. In other words, this study considers the demographic, economic, and socio-cultural constraints, as well as transportation elements, that could lead to exclusion.

In order to avoid unnecessary detailed examination of the literature in hand, the first part of the chapter is allocated to the discussions of social exclusion from only the transportation perspective. This is to provide a better understanding about to what extent transportation can affect social exclusion. In the second part of the chapter, a case study conducted in Ankara is examined. In this part, after a brief explanation of the methodology used in the field work, the relation between transportation and social exclusion is discussed within two separate contexts of "spatial patterns of social exclusion" and "spatial patterns of exclusion from transportation". Finally, in the third part of the chapter, findings of the case study are summarized and the relevance of these findings is discussed in relation to the hypothesis.

VIEW TO SOCIAL EXCLUSION FROM TRANSPORTATION PERSPECTIVE

Social exclusion is a relatively new social policy concept. Social exclusion in the context of social policy, as stated earlier in the study, incorporates poverty identification, measurement, and also understanding it. Based on this conception, Stanley and Vella-Brodrick (2009), when handling poverty as amount of income difference needed to sustain the life of individuals or households in their living environment in their studies, emphasize that social exclusion is a broader concept than poverty. Therefore, social exclusion is now focused on people's full participation in society and other obstacles that make it difficult for their participation. These obstacles include lack of employment, adequate housing conditions, education, health, and transportation. Because it directly affects the others, there is no doubt that transportation is the most significant of these barriers.

The relationships between social exclusion and transportation have been examined from various specific viewpoints. For example, urban geographers suggest that the position of the services and facilities provided by the public and the private sector and their urban space distribution levels are either advantageous or disadvantageous regardless of income of individual/households and other distinctive variables (Kirby, 1982). The point here is not disadvantages due to individuals' characteristics, it is disadvantage of place, and therefore it is the lack of transportation/access. Therefore, regarding the relationship between social exclusion and transportation, the concepts of accessibility and mobility commonly occur in the literature. The simplest definition of mobility (Hansen, 1959; Handy, 1994) covers actual mobility, potential mobility, and dynamism, which refers to individuals' ability to roam in a secure and independent way inside or outside the home, depending on their physical or mental abilities (Alsnih & Hensher, 2003). Metz (2000) and Burns (1999) define mobility as including one or more of the five dimensions that they place in order of (i) access to the desired location (e.g. socializing with family and friends), (ii) psychological benefits of social interaction and independence for individuals, (iii) the physical benefits of movement, (iv) social networks, and (v) trip potential.

In the transportation literature, the concepts of mobility and accessibility are often used interchangeably. This is due to the relationship between the two terms (Handy, 2002). However, the definition of accessibility is more complex than the definition of mobility. While Hansen (1959) defines accessibility as potential opportunities for interaction, Dalvi and Martin (1976) define it as ease of access to activity using a particular transport type, Burns (1979) as freedom of participation in various activities of individuals, and Bhat (2001) as reaching the facilities, products, and activities individuals want. Accessibility, in summary, refers to facilities for individuals accessing products, actions, and activities (Litman, 2011), and these facilities are generally associated with impedance of the state of the labor, time, and distance (Niemeier, 1997). Most researchers base their work on the previous definitions and seek to refine the accessibility concept by providing new contributions.

The concept of accessibility, as discussed, consists of several criteria, and in defining accessibility, four main determinants emerge: land use, transportation, and temporal and individual components related to these criteria (Geurs & Van Wee, 2004). *Land use* can be projected into accessibility criteria of activity type, nature, and spatial distribution, and desire for opportunities or supply-demand balance. According to the land use component, distance between activities, transportation density, and the level of traffic are important in terms of capacity and quality. The transport component projects the uselessness (standby, parking etc.) of start and end points of a particular type of transport in the journey in terms of time and difficulty (level of comfort, the risk of accidents, etc.). The time component projects temporal limitations – namely, the usability of facilities at different times of the day, and time needed to participate in a specific activity. Finally, the individual component is composed of the needs of people (depending on factors such as age, income, education level, household structure), their abilities (transport types that can be used depending on their physical condition), and opportunities related to their individual characteristics (income, transportation budget, education, etc.) (Geurs & Van Wee, 2004; Bocarejo & Oviedo, 2012).

The concept of the relationship of accessibility with notions such as poverty and income disadvantages has become a key element in social policy discourse. Indeed, social exclusion contains some disadvantages in spatial, political, cultural, social, and personal dimensions (including economic condition), and these disadvantages are exacerbated by poverty (Kenyon et al., 2002). This is because social exclusion causes decline in the participation of the parts of traditional social structure with normal processes of society as well as coexistence of group of social problems related to the increasing poverty/ deprivation among certain social groups (Burchardt, 1999; Witter, 2010).

Indeed, if an individual geographically lives in a society but is unable to attend its activities, he or she is socially excluded (Witter, 2010). The relationship between transport and social exclusion emerges fully in this centerline. In this context, social exclusion related to mobility/accessibility defines a deficiency of ability (or the lack of adequate tools for) for actualizing trips and, as a result, of accessibility to opportunities, which hinders participation of individuals in society's economic, politic, and social life (Kenyon et al., 2002). When it comes to people living in exclusion conditions as a result of urban layout (which was built on the concept of high mobility), trip selection and social integration are lessened due to lessened access to activities and services provided. These conditions are often reinforced with poor public transport services in the neighborhoods in which there is low private car ownership.

Church et al. (2000) state that intervention required for the development of mobility for fulfilling the social requirements of the links between exclusion and transport. It is essential to strengthen the role of transport for forming a more equitable and inclusive society. Rajé (2003), using a similar approach, has contributed significantly to demonstrating the relationship between social inclusion and mobility of transportation in the literature. She states that social exclusion is affected by accessibility to goods and services, and mobility and accessibility is an important concept for helping to understand how transportation contributes to the process of social exclusion.

Social exclusion issue and transportation disadvantage are significant problem areas in transport policy. This is because the relationship of transportation and social exclusion includes variables based on socio-economic, demographic, physical, and cultural differences. Hine and Grieco (2003), suggest that transportation plays an important role in the inclusion of individuals in society via better accessibility and mobility. Failure to meet the travel needs of social groups imposes negative externalities on these groups, in the areas of educational format and so on. On the other hand, these negative externalities include insufficient public transport, reduction in access to basic services, the negative impact of traffic, and the lack of mobility. The negative impact of the transport system stops individuals and groups from participating in the activities they want to participate in (for the reasons beyond their control), which leads to their social exclusion by means of this segregation in society (Bukhari et al., 2010).

REASONS FOR EXCLUSION FROM TRANSPORT AND DISADVANTAGED GROUPS AFFECTED BY SOCIAL EXCLUSION

Transport systems that cannot be equally provided to every individual, which substantially undermines social inclusion and social integration. Miller (2003) attributes inequitable distribution of the transport systems to two main causes. The first of these, poor transport, limits access to various services and opportunities for some parts of society but not for others. The second is negative effects of transport such as traffic congestion, accidents, air pollution, unsanitary conditions, noise levels being concentrated in areas in which there is a high social exclusion (SEU, 2002). These areas are generally poor neighborhoods that have low car ownership, where road links are not strong, and public transport is inadequate (Chakwizira et al., 2011).

Kenyon et al. (2002) assert that transportation and social exclusion are intricately connected. They define the reasons affecting exclusion from transportation in society as economic condition, space, and age of the individual. For instance, being very old will make it impossible to use motor vehicles, due to financial insufficiency and high transportation costs, and it will be difficult to participate in some events. In this context, space gains importance because, when considering the relationship of exclusion and transport, in order to avoid exclusion, either owning a car or residing in the central area (where the transport network is good) is necessary.

Cass et al. (2003) stressed encompassing a wide range of issues that cause exclusion from transportation. They identified the following reasons for exclusion: unemployment, deprivation and poverty, lack of education, disability or difficulty in mobility, lack of social participation, geographical isolation, difficult-to-reach groups, and self-exclusion with lack of access to services and information.

Reasons for exclusion associated with transportation are grouped under five headings in the report of the Social Exclusion Unit (SEU, 2003). These are listed as the spatial position of transportation services, transportation condition and physical accessibility, security, transportation costs, inadequate information, and individual limitations.

Studies on social exclusion from transportation are focused more on public transport systems. Thus, it is known that a good public transportation policy significantly redresses problems related to access to and mobility. Bukhari et al. (2010) examined the main causes of social exclusion linked to public transport under five headings:

- **Presence and Physical Accessibility of Public Transport:** Public transport network improvements, regulations on Infrastructure of transportation and changes in land use determine the direction of individual mobility.
- **The Cost of Transportation:** Rises in public transport costs negatively affect especially the low-income households.
- **Services and the Location of the Activity:** People prefer activity areas close to their living space that provide easy access to opportunities. The reason for this is a desire to avoid excessive material and temporal loss.
- **Security:** Reasons such as crowded public transport, experience of theft, risk of attack or harassment (risk increases in the trips at night), especially for women and children, and exposure to infectious diseases keep away individuals from public transport.
- **Journey Understanding:** Individual preferences may also cause the rejection of oneself in public transport use.

Kenyon et al. (2002) listed the reasons for exclusion from the urban public transportation as:

- Low service quality of vehicles and their accessibility,
- Security vulnerability,

- Public transport charges, and
- Being a relatively time consuming option.

Lucas (2012) in defining the relationship between social exclusion and transport disadvantages, expresses that when conditions of security, lack of information, high fare, lack of service, and inability to own private car combine with social exclusions, individuals/groups could be confronted by transportation deprivation. Similarly, Dodson et al. (2004) defined transport disadvantaged social groups as follows: low-income individuals, the unemployed, children and youth, women, the elderly, the disabled, those who live in the perimeters of city, and ethnic minorities. Individuals taking part in one/some of these groups it raises the risk of social exclusion from transportation.

Currie (2004), offering an alternative study for social groups excluded from the transportation, defines six groups: car-free adults, people over the age of 60, people who receive disability pension, low-income adults, adults who are unable to participate in the labor force, and students. On the other hand, Stanley & Stanley (2004) use the following groupings: youth, seniors, people with disabilities, low-income people, people living in semi-rural areas, and ethnic origins. Hurni (2006) defines the main factors affecting personal mobility as: physical ability, literacy, language, gender, age, family status, employment status, income, the ability to have access, and driver's license for motor vehicles:

- **Income:** It is commonly held that social exclusion is directly related to low income. Primarily, travel costs affect access (no matter what mode of transportation) for low-income people.
- **Car Ownership and Driving Ability:** Having a motor vehicle and being able to use it, if traffic congestion, parking problems, and transport policy are ignored, offers unlimited mobility opportunities to individuals.
- **Gender:** Gender is an important variable to establish the relationship of transport and social exclusion. It is seen that women are in a disadvantaged group. Hine and Mitchell (2001) emphasize that transportation exclusion experienced by women is similar to low-income groups, old people, the unemployed, and people with health problems.
- **Age:** For children and young people under the age of 18, getting a driver's license is not appropriate in terms of personal safety and the safety of other individuals in the society. However, often there can be conditions that limit opportunities of using public transport and walking for children and young people, and for these reasons they often require the help of parents or caregivers to meet travel needs. Other age-related disadvantaged groups are people 60 years old and over, because generally the mobility of people decreases with age. Pre-retirement people are so accustomed to using cars that they continue to focus on this transport mode in their retirement years in all of their activities, though they are often constrained or stopped by decrease in driving ability. The individuals in this age group are excluded from being pedestrians or bicycle use due to power loss, and are often excluded from public transport modes because of discomfort (Alsnih & Hensher, 2003).
- **Physical Disability:** It is very difficult to actualize basic activities such as:
  - Personal care,
  - Mobility,
  - Communication needs,
  - Education, or
  - Employment for most physically disabled people alone.

Generally, a large proportion of people with disabilities due to employment or difficulties commuting to work have low income, and thus their access to specially designed vehicles (motorized wheelchair, disabled cars, etc.) is limited.

- **Family Structure/Household Type:** The vast majority of families in the lowest income bracket are composed of households that live alone (divorced, widowed, single parents etc.). These individuals are in need of affordable transportation. On the other hand, families with very young children (0-4 years old) may also encounter a variety of obstacles in their mobility for the reasons of extra luggage or pushchairs.
- **Employment Status:** Unemployed people are more likely to have low incomes, and for this reason serious restrictions arise about mobility. However, it is important to search for employment or take vocational training for those people. Thus, they need more frequent journeys. Unemployment, particularly among young people, is quite high; individuals without any wealth in this group are riveted to their exclusion from transportation.
- **Migrants and Ethnic Origins:** Chances of finding work for the people in this group (because of problems of language, work permit, etc.) are very low; hence, they are located in the lower income bracket. Thus, immigrants and some individuals with ethnic origin are likely to be excluded from transportation more than others. In addition, their chances also to be excluded from transportation are increased as they tend to have more children.
- **Language and Literacy Knowledge:** Knowing the language of the country in which one lives and being literate emerge as two very important factors that determine communication ability and mobility. Lack of language skills and not being literate make it difficult to receive training and find employment, which makes it difficult to obtain a higher income. Additionally, mobility of individual is also hampered due to failure to read the transportation information and inability to follow a timeline.

CASE STUDY

Social exclusion, when representing the whole multidimensional process, brings together some problems of objective detection and measurement. Thus, a subjective point of view is required in the measurements of the exclusion that arises from the criteria such as: people's income level, their age, condition of disability, as well as gender orientation or ethnical originThe measurement of exclusion from transportation exacerbates existing measurement problems. Therefore, in the study, in determining social exclusion related to transportation, evaluation is related to both the trips and activities. In this way, both the effects of social exclusion from transportation and which socio-economic groups do or do not access activities in the city can be determined. This is because, in the study, homogeneous groups obtained by superposing exclusion from transportation and spatial patterns in which groups with inherent risk of social exclusion concentrate are scrutinized.

Case Selection

Turkey's capital, Ankara, was selected as the study area. Ankara has received heavy migration from rural areas and other cities since the 1950's. While this migration flow leads to rapid population growth, it has brought major changes in land use. It is seen that with the compact-growing urban macroform spreading by expanding further with the selection of housing space for workers in the western corridor in the 1980's, the city increased its topographic thresholds by developing in the northern and southern directions after the 1980's. Along with the preference of prestigious housing areas of upper-middle and upper income groups, the city continues to evolve along the axis of the road between Istanbul and Eskisehir (Eryaman, Umitkoy, Çayyolu, Konutkent, Koru, Yaşamkent, Baglica, and so forth.)

This rapid and uncontrolled expansion of the city has increased the dependence on motor vehicles in these years; in particular, middle- and high-income groups have shown a strong preference for private cars.

When the development of the transportation system is evaluated in the process of rapid change experience in the city of Ankara, unfortunately, it a consecutive integrated structure is not found. Even though the city has carried out several transportation surveys and planning studies, when the implementation stage is reached, most of the plans are shelved.

Studies use inadequate estimates of the future and have no detailed analysis, and implement irreversible arrangements fragmentarily (widening the existing roads, underpasses, interchange arrangements etc.), and this situation encourages more use of private cars in the city. In addition, many problems, such as the rail system projects that cannot be completed for many years in the city, show disregard for pedestrians and disadvantaged groups (the disabled, children, the elderly, etc.), including negligible level of bike paths, the issue of integration of public transport systems, and lack of pedestrian and traffic safety.

Determining the Limits of the Study Area

The administrative area of Ankara Metropolitan Municipality covered the most densely populated areas of the city between the years of 1984–2003. However, this structure has deteriorated with legal and administrative changes in 2004 and 2012. It is known that new borders that cover Akyurt, Ayas, Bala, Elmadağ, Kalecik, Kazan, and Cubuk are remote from center, and more remote still areKızılcahamam, Çamlıdere, Gudul, Beypazarı, Nallıhan, Haymana, Polatlı, Şereflikoçhisar, and Evren, which preserve their rural nature.

In this context, eight districts where urban transportation is seen frequently, have a dense population, and are in the center with urban characteristics (central city areas: Altındağ, Çankaya, Kecioren, Mamak, Yenimahalle, Sincan, Golbasi, and Etimesgut) are used as the study area (Figure 1).

Research Questions

It is possible to list the main questions looked for answers under the following five headings in this study that intend to determine the relational level of transport and social exclusion:

1. Does seeing oneself as socially excluded in society affect the individual's transportation behavior?
2. Is it possible to be excluded from transportation even if individuals do not see themselves as socially excluded?
3. What forms of transportation behavior vary between different socio-economic groups?
4. For individuals who see themselves as excluded from transportation, activities are they unable to participate in?
5. Can transportation be used as a means of minimizing social exclusion?

Figure 1. Location of case studies

The Hypotheses of the Research

"There is a relationship between dynamics of urban transportation and individuals' perceptions of social exclusion" is the main hypothesis of this study. On the other hand, other assumptions that will support this hypothesis are tested as follows:

H1: There is a relationship between social exclusion and exclusion from transportation (main hypothesis).
H2: As poverty of individuals/households increases based on income, sensation of being excluded from the transportation and social exclusion increases.
H3: Socio-demographic characteristics of individuals/households affect transport behavior and levels of social exclusion.
H4: Difficulties accessing urban activities lead to isolation from society, thereby causing social exclusion.

Sub-hypotheses generated in order to facilitate testing these four stated assumptions are shown in Table 1.

After the identification of research questions and hypotheses, research strategies are configured. However, exclusion from transport in the axis of social exclusion and accessibility/mobility are quite complex notions that are relational, and difficult to identify and measure.

In this context, in the study, factors affecting social exclusion, disadvantaged groups of social exclusion, and factors affecting exclusion from transportation and transportation disadvantages are evaluated by using different sets of data and criteria; to this end, spatial patterns of Ankara city were identified. By superposing these two sets of exclusion criteria, new spatial patterns are obtained that identify "problematic areas". In this way, the hypotheses envisaged make use of surveys conducted in the problematic area

Table 1. Sub-hypotheses of the study

Hypothesis	Sub-Hypothesis
H1 (relational)	*H1₁*. There is a negative correlated relationship between social exclusion from transport and neighborhood levels
H2 (economic)	$H2_1$. Participation to labor force reduces the risk of social exclusion and exclusion from transportation. $H2_2$. The increase in the income level reduces levels of social exclusion and exclusion from transportation. $H2_3$. The increase in the number of households increases the risk of social exclusion and exclusion from transportation.
H3 (Socio-demographic)	$H3_1$. Social exclusion and exclusion from transport by gender issues varies. $H3_2$. Age variable varies in rapid progress of social exclusion and exclusion from transportation. $H3_3$. Increasing the level of education reduces the risk of social exclusion and exclusion from transportation. $H3_4$. Social exclusion and exclusion from transport varies according to household type $H3_5$. Particularly, families with pre-school (6 years old) children differ in their social exclusion and exclusion from transportation issues compared to families without children.
H4 (activities, accessibility)	$H4_1$. The use of public transport in terms of exclusion from transport varies according to the use of private cars $H4_2$. Individuals having the driver's license will reduce the risk of exclusion from transportation. $H4_3$. Urban activities Made affect the modes of transportation used $H4_4$. The increase in the number of trips to be made affect the selection of mode of transport.

that carry risk at different levels in terms of exclusion of social and transportation; statistical analysis and in-depth interviews are used to test the hypotheses (Table 2).

Data Sets and Sample

In the process of obtaining data, *two separate data sets were used*. Firstly, data of the Statistical Institute of Turkey (TUIK) were obtained to put forth the spatial differentiation of excluded groups in order to be used in the clustering analysis. These sets of data contain information of general population census, socio-economic characteristics of the population, and building enumeration in the neighborhood (mahalle[1])

Table 2. In-depth interviews

Stage	District	Neighborhood (Mahalle)
1	Yenimahalle	Koru
1	Etimesgut	Eryaman
1	Yenimahalle	Çayyolu
2	Çankaya	Eti
2	Keçiören	Aşağı Eğlence
2	Keçiören	Ayvalı
3	Gölbaşı	Gaziosmanpaşa
3	Yenimahalle	Karşıyaka
3	Sincan	Tandoğan
4	Mamak	Araplar
4	Altındağ	Çalışkanlar
4	Altındağ	Fatih

level in the city of Ankara. On the other hand, the second data set, which is to be used in the clustering analysis in order to identify spatial differentiation of groups at risk of exclusion from transportation, is survey data prepared under the scope of "Transportation Master Plan for Ankara Metropolitan Area and Its Surroundings," which was undertaken as joint service project by Gazi University and Ankara Metropolitan Municipality. The main reason for the use of these data sets that they contain general information about households, personal information of household members, and detailed information on the trips made by individuals during the last 24 hours.

The Method Used For Data Analysis

In the first stage of fieldwork, cluster analysis was conducted, which enables the detection of spatial patterns, and non-hierarchical clustering method was used in the analysis. For the purpose of measuring the central tendency related to stages obtained as a result of comparison of cluster analyses, the frequency analysis, mean, weighted mean, median, and mode values were used. For trends as a result of the comparison of cluster analysis, mean, weighted mean, median, and mode values were used; for the purpose of measuring the relationship between variables, the chi-square test and correlation were used. In addition, while assessing the research findings, descriptive and inferential statistics were used for compiling, summarizing, analyzing the numerical data we used; the obtained survey data were evaluated with the help of IBM SPSS Statistics 20.

Acceptance of the Study

The present analysis excludes data for individuals who exclude themselves willingly from society or those who made 5 or more trips made during the day (as these groups do not exceed 5% of the sample).

Social Exclusion and Spatial Patterns of Exclusion from Transportation

As spatial patterns were used, non-hierarchical cluster analysis was utilized. Non-hierarchical clustering, due to having a more powerful theoretical basis, is more effective according to hierarchical (Yilmaz & Temurlenk, 2005). Cluster analysis was carried out in four stages. In the first phase, the data matrix was generated; in the second phase similarity was measured between objects observed; in the third phase, by selecting suitable clustering techniques, variables was divided into clusters according to the similarity matrix. The last phase is the interpretation of obtained clusters according to their characteristics.

Spatial Patterns of Social Exclusion

For the purpose of determining spatial differentiation of social exclusion in the cluster analyses, TUIK data was used. Sixteen variables related to housing, fifteen variables related to employment, and twenty variables related to population, education, households were used; a total of seven different clusters were obtained (Table 3).

By conducting an ANOVA test for the purpose of examining the variables of cluster analysis, H_0 and H_1 were tested (H_0: There is no significant difference between clusters based on variables, H_1: There are significant differences between clusters based on variables). Accordingly, it is seen that, except disability condition, p-values indicate statistically significant difference.

Table 3. Main variables used

Variables Related to Residence	• Residence age • Housing purpose • Residential infrastructure, waste water installations • Number of rooms in recidence • Recidential heating system • Recidence/land ownership status
Variables Related to Employment	• Employment participation levels • The sectoral distribution of the workforce • Employment status
Variables Related to Population, Education and Households	• Number of households • Level of education • Disabled population • Gender • Age • Marital status

The resulting exclusion risk sequence of the seven clusters is in the form of clusters of 1 <6 <5 <4 <3 <2 <7. Thus, Cluster 1 consists of advantaged individuals with regard to income, education level, and social status and who therefore have the lowest risk in terms of social exclusion. Clusters 6 and 5 consist of high concentrations of employers, entrepreneurs, directors, and senior managers. However, Cluster 5 consists of disadvantaged groups with regard to number of women, broken families, and population aged 65 and over; therefore, it could be described as the upper middle cluster. In Cluster 4, which can be regarded as the transition zone, trading and sales staff, high school graduates, and women not in the labor force are present in high proportions. Generally, all variables have a mean value in this cluster. There are the highest proportions of women and unpaid family workers in Cluster 3, and in the sequence of risk it comes after Cluster 2 and Cluster 7. In Cluster 2, tenants with one- or two-bedroom houses and illiterate that are in the highest concentration and, compared to other clusters, it can be said to consist of people that have high potential of social exclusion. Cluster 7 is a cluster that has the highest concentration of unemployed population, employees in the manufacturing industry, households that have 5 or more people, and primary school graduates. Due to most of the variables taking place at high rates that would increase the risk of social exclusion, Cluster 7 is the most disadvantaged group from the point of view of exclusion.

When spatial distribution of clusters of 7, 2 and 3 with high risk of social exclusion examined (Figure 2), these areas are away from the center, generally located on the north-northeast of the city. In particular, it is remarkable that there are multitudes of high risk of social exclusion neighborhoods in Altindag and Mamak districts. Mainly are clusters 1, 6 and 5 located, which have low risk of exclusion; the transition zone of Cluster 4 is between the city center and the periphery. The risk of exclusion increases gradually from the center to the periphery in the city of Ankara, though it is seen that transitions between clusters are not sharp.

Spatial Patterns of Exclusion from Transportation

In the clustering analysis conducted to be able to discern the spatial differentiation of groups at risk of exclusion from transport, data of the Ankara Transportation Survey were used. In analysis, nineteen variables related to population, education, and households; ten variables related to employment and

Figure 2. The spatial distribution of social exclusion clusters

**For a more accurate representation of this figure, please see the electronic version.*

income levels; six variables related to housing condition; and eight variables related to transportation were used. While determining these 43 variables, the studies in the literature considered women, youth, the elderly, sick and disabled, crowded households, individuals with low levels of education, those who could not participate in employment, those with low-income ones, non-homeowners, and those who do not have a driver's license and own a car (Table 4).

By conducting an ANOVA test for the purpose of examining the variables of cluster analysis H_0 and H_1 hypothesis is tested. Accordingly, it is seen that p values of gender, population of young people between the ages of 0–5 and 13–18 years old, the disabled and hospitalized, illiterates, and university graduates are greater than 0.05. Therefore, the cluster analysis of these variables shows no significant difference.

According to cluster analysis conducted to determine the spatial patterns of exclusion from the point of view of transportation, seven clusters give the best results in the city of Ankara. The resulting risk sequences of these clusters is in the form of clusters of 5 <7 <6 <4 <2 <1 <3. Clusters 5 and 7, with the

Table 4. Main variables used

Data of Variables Related to Population, Education and Households	• Gender • Disabled and chronically ill • Number of Households • Education level • Age
Variables Related to Employment and Income	• Employment status • Income rate
Variables Related to Housing	• Residence ownership • Residential rental levels
Variables Related to the Transportation	• Driver's license ownership • Private car ownership • Purpose of trip

least risk of being excluded from the transportation, have the highest concentration of individuals that own second and third private cars, have income of 5000 TL and over, and have postgraduate degrees; and again they are the clusters that have the lowest concentration of individuals that are unemployed, with low education, and with low income. In this aspect, it could be said that they carry very low risk in terms of social exclusion and exclusion from transportation. Cluster 6 is a transition cluster that shows the lowest concentrations of those aged over 65 and those who take trips on foot, and it could be said that it carries an average level of risk in terms of exclusion. Cluster 4 also has an average value in terms of risk of exclusion from transportation. When the clusters are evaluated according to the risk of social exclusion from transport, it is seen that the most disadvantaged cluster is number 3. Other clusters that accommodate exclusion potential from transportation are clusters number 1 and 2. The risk of exclusion is higher in these clusters than the others, as the number of households, population over 65 years old, and people who cannot participate in employment is high in these clusters, while the level of income, the ownership of housing, and the level of education is low. In addition, driver's license and private car ownership rate is even lower in these areas than other parts of the city. When examining (see Figure 3) the spatial distribution of Clusters 3, 1, and 2, which carry high risk of being excluded from the transportation at the neighborhood scale, it is seen that these areas are generally located in the periphery of city. Again, when examining the spatial distribution of Clusters 5 and 6 (that carry lower risk), it is seen that they concentrate in the center of city and its surroundings and along the course of Eskisehir road.

Figure 3. The spatial distribution of cluster exclusion from transport

**For a more accurate representation of this figure, please see the electronic version.*

While minimum income is TL 803 in Turkey, according to research conducted by Confederation of Turkish Trade Unions, the hunger limit is TL 1020, and the poverty limit has been identified as TL 3322 (Turk-Is, 2015).

A Comparison of Clusters of Social Exclusion and Exclusion from Transport

While TUIK data aims at social exclusion, generated clustering from data of Ankara Transport Survey aims at exclusion from transportation. On this basis, results of two clustering were compared. As a result of this comparison, it was found that Clusters number 1 and 6 in the clustering of TUIK show similar characteristics with Clusters 5 and 7 of the other survey. Likewise, it is seen that Clusters 5 and 4 in the TUIK match 4 and 6 in the other data, and 3 and 2, respectively, 2 and 7, respectively, and 1 and 3, respctively, show similarities. These similar clusters were spatially overlapped (See Figure 4) with the help of geographic information systems; by obtaining spatial distributions of similar clusters, a new classification has become possible.

According to the new classification in the study, four stages have been identified (Table 5). These stages are listed again according to the risk of social exclusion and exclusion from transportation.

While risk is in the lowest level in stage 1, it increases more in stages 2 and 3, and it reaches the highest level in stage 4.

Research Findings

Restrictions that prevent the participation of individuals in society, inadequate transport and effects of private car-dependent structure of city of Ankara on social exclusion are reconsidered especially through the eyes of the disadvantaged of society in this study. Research findings are evaluated in the light of hypotheses generated.

Figure 4. Problematic area: the spatial distribution of similar clusters result of overlapping

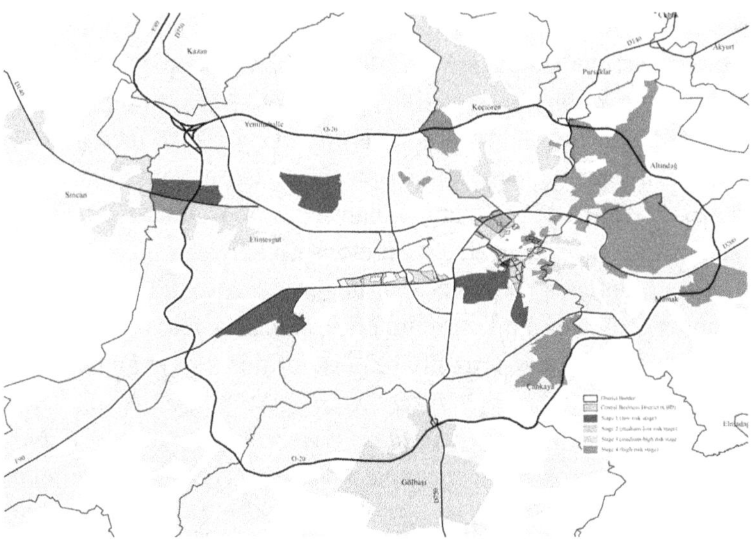

**For a more accurate representation of this figure, please see the electronic version.*

Table 5. Information related to stages of neighborhood (mahalle)

Stage Number	Stage Name	Number of Neighborhood (Mahalle)
Stage 1	Low risk stage	14
Stage 2	Medium-low risk stage	5
Stage 3	Medium-high risk stage	38
Stage 4	High-risk stage	38

In 95 neighborhoods of problematic areas, 27,297 people were surveyed, divided into two sub-headings. In the first, socio-demographic structure of neighborhood stages were assessed and exclusion levels of individuals determined. Again, for the purpose of determining transport-induced exclusion particularly, car ownership and trip information are evaluated. In addition, level of participation in urban activities which are important measurements in providing social integration are discussed in detail under the first sub-heading.

Socio-Demographic Structure of the Neighborhood Stages

Social and demographic structure of the neighborhood stages were assessed using statistical analysis of survey data as well as from in-depth interviews[2]. Firstly, 52.2% of the sample is male and 47.8% of them are female. Gender is an important variable to establish the relationship of transport and social exclusion, and it is known that women have inherent disadvantages. Most of the women who participated in interview pointed out that they experience social exclusion due to both faith-based and moral discourse by most of society over women. The average age of those who participated in the survey is 36.6.

The rate of vulnerable age groups (0–5 years, 6–17 years, and 65 years and over) that have potential of exclusion from transportation is 24.7% in the sample. In the interviews conducted with young people with varying ages between fourteen and seventeen, it was determined that none of them see themselves as isolated from the community or socially excluded. However, the situation is a bit different in the elderly population. That is, older people think that they are not socially excluded from the society, but their social relations are disrupted for physical reasons. The average household size is 3.7 in the sample. When correlation analysis was conducted between the number of people living in the household and the stages, it is seen that there is a significant positive correlation. Thus, as the number of people living in households increases, risk of exclusion at the neighborhood stage also increases. These risks are often caused by splitting household income between people living in the households (Table 6).

According to the and the result of correlation analysis of household type and neighborhood stages, where there is a negative significant correlated relationship between them, it was determined that the broken families are more common at the lower risk stages.

The type of households generates a pressure in the community, and the interviewees stated that they also sometimes experience social exclusion only because of this. For example:

(...) Most of the time, even though I don't think there is a deliberate exclusion but I see that there are many decomposition such as married women, single women, a widow, women with children and etc. in

Table 6. Household type according to neighborhood stage

		Completed Family	Childless Family	Single-Parent Family	Living Alone Man	Living Alone Woman	Total
Stage 1	Number	2776	896	466	69	68	4275
	%	64.9	21.0	10.9	1.6	1.6	100
Stage 2	Number	1905	496	161	54	33	2649
	%	71,9	18,7	6,1	2,0	1,2	100
Stage 3	Number	11805	2152	1477	204	129	15767
	%	74,9	13,6	9,4	1,3	0,8	100
Stage 4	Number	3376	748	397	56	52	4629
	%	72,9	16,2	8,6	1,2	1,1	100
Total	Number	19862	4292	2501	383	282	27320
	%	72,7	15,7	9,2	1,4	1,0	100

the community. I see that there are so much decomposition. Just not to be physically or verbally abused, women who had separated from her spouse, doesn't say that in my work place ...

According to the correlation analysis conducted between the neighborhood stages and education variables, there is a negative correlation. As the risk of the neighborhood exclusion increases, there is a decline in the levels of education. Increase in the level of education increases the possibility of finding jobs for individuals. Indeed, when working condition data are examined according to the neighborhood stages, in stage 1, 1736 people (40.7%), in stage 2, 870 people (33.0%), in stage 3, 5482 people (34.9%), and in stage 4, 9566 people (35.1%) have participated in employment, and it is seen that men have more job opportunities than women (Figure 5).

Figure 5. The level of education of the individual in the neighborhood stages

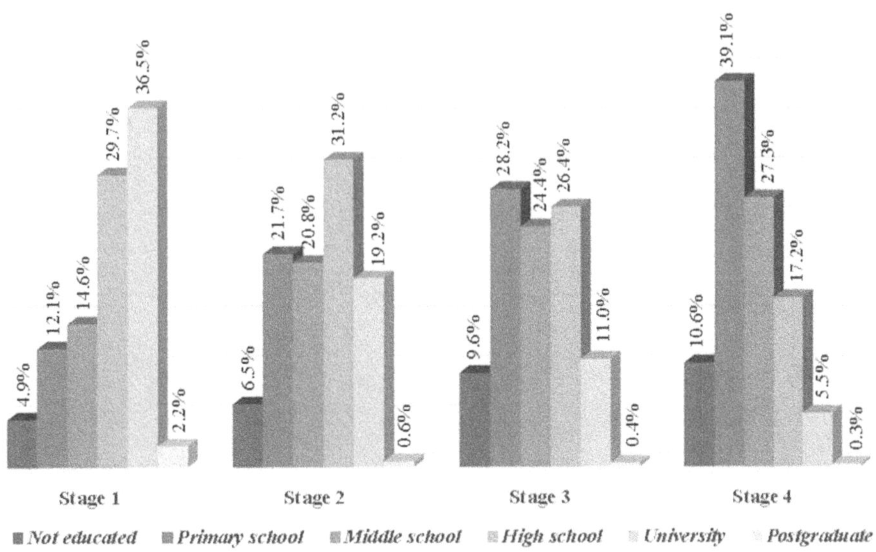

**For a more accurate representation of this figure, please see the electronic version.*

Car Ownership and Travel Information in the Neighborhood Stages

Car ownership emerges as an important variable in terms of increasing the mobility of the individual. Thus, car ownership increases the level of participation in urban activities as well as it being effective in diversifying the activities carried out during the day. When the public transportation is inadequate, in some cases individuals who could not have private cars are excluded indirectly from transport.

Although there have been fluctuations (see Figure 6) over the years in Ankara, the increase in car ownership rate shows a linear cumulative increase (see Figure 7). There are 9748 households in the sample, of which4686 (48.1%) own at least one vehicle.

When private car ownership information is analyzed by sample stages, it is seen that the most advantaged one is Stage 1 and the most disadvantaged is Stage 4.

It was determined that the variables of income and the household type significantly affect car ownership in the sample. Indeed, when examining Figure 8, it is seen that the income level in Stage 1 is quite high compared to those in other stages.

In addition, income as well as household typology affect status of car ownership income and, therefore, also affect the individual's transportation behavior (Figure 9).

Figure 6. The rate of car ownership changes over the years

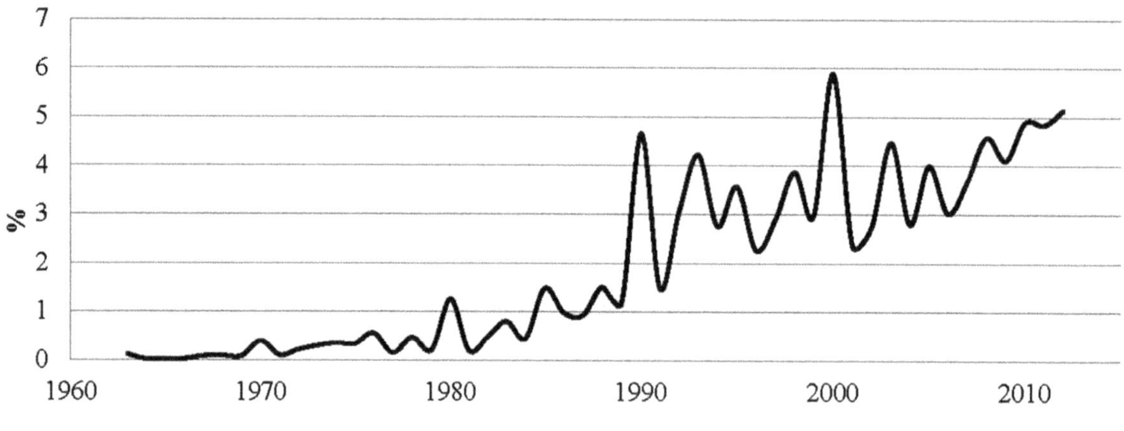

Figure 7. Cumulative increase of car ownership by years

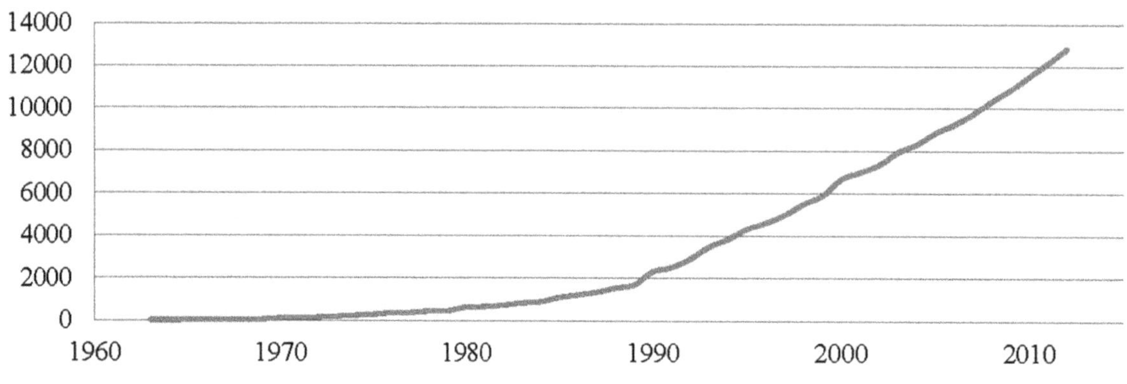

Figure 8. Car ownership in the neighborhood stages

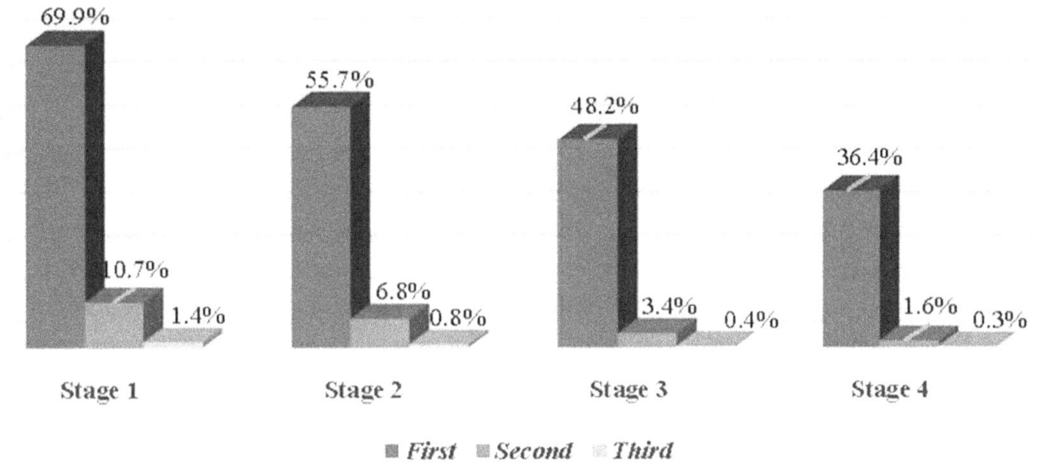

Figure 9. Income distribution in the neighborhood stages

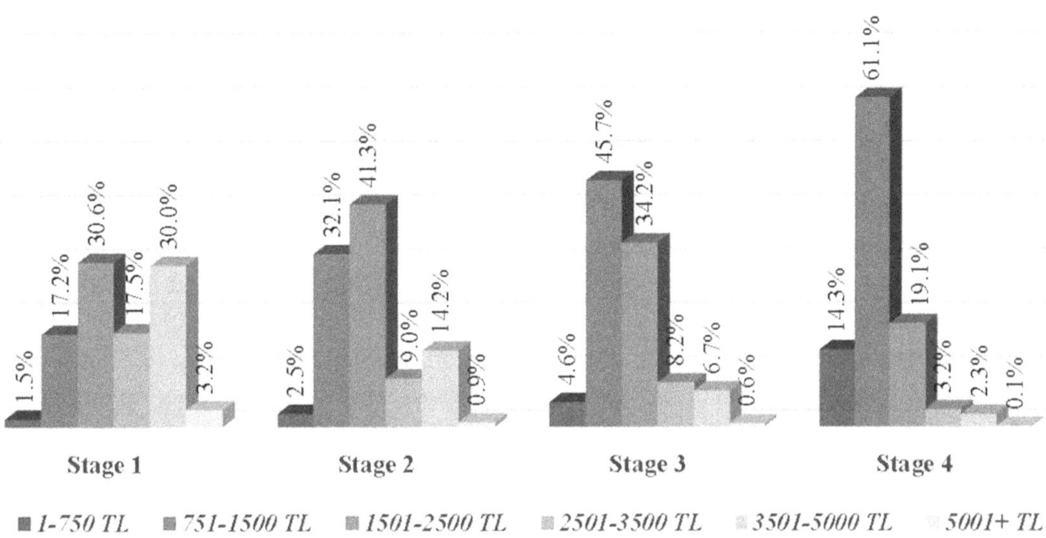

**For a more accurate representation of this figure, please see the electronic version.*

In this context, transportation patterns of families with children and childless families, single-parent families and families with a single person, women living alone, and men living alone may substantially differ from each other. When the data of car ownership distribution is examined by the type of households, 53.9% of completed families, 47.3% of one-parent families, 38.6% of families without children, 37.1% of men living alone, and 17.0% of women living alone have private cars. Thus, it emerges that families having children is an important criterion in the family's stance on the ownership of the vehicle. In fact, 45.6% of households that own private cars consist of families without children, while 6% of the families with children own private cars. This case can be explained by changing their transportation patterns/behavior in the case of people having children. For instance, even conditions such as dropping off and picking up children to/from educational facilities (especially pre-school period), hospital visits for health checks, and so on mean parents make extra trips. Due to the increase in the number of trips as well as in terms of ensuring road safety for children, most parents may prefer to use private cars. One of the interviewers who have children quotes this situation as follows:

I lived and am living exclusion from transportation in several ways. Because it is different in the case of being pedestrian, different in the case of being female driver, if women have a child but no car and then even more different. (...) Impossible reaching to some places as pedestrian with child, for instance. Pedestrians, elderly, disabled, pregnant women or woman with have priority on passing in all cultures, unfortunately, such a condition is not the case with us here. In particular, drivers drive their car over you so that you pass quickly. It is not easy climb onto half a meter sidewalk with a child or stroller on the lap. On the go somewhere, baby is going to need a lot of things, you are walking in your hand with handbag looks like luggage. Especially the ride on public transport alone with bags and pushchairs is much more difficult. Moreover, because drivers of public transport use the vehicle ignoring the presence of child in the vehicle, it turns out a serious security problem. (...) We have got our car just after the birth of our child. System somehow pushes you to be having a private car as I told you ...

Two other determinants of exclusion from transport are undoubtedly the number of trips and purpose of trips of individuals made in a day. When looking at the average number of trips of participants made in a day in the different stages, in Stage 1, 1.46, in Stage 2, 1.14, in Stage 3, 1.28 and in Stage 4, 1.11 trips are seen on average. Considering that the average trip number of the sample is 1.26, it is understood that number of trips in Stages of 1 and 3 are higher than average. On the other hand, the average trip time in Stage 1 is 34.5, in Stage 2, 30.9, in Stage 3, 35.0, and in Stage 4, 37.2 minutes. More time is spent on trips in stage 4, despite making fewer trips. This could be explained by the type of transportation used.

When the distribution data of the transport vehicles used in first trips of the sample is examined, it is seen that the rate of trips made on foot or bicycle in Stage 1 is 20.3%, in Stage 2 is 31.0%, in Stage 3 is 33.6%, and in Stage 4 is 39.4%.

The rate of private car and taxi trips in Stage 1 is 33.6%, in Stage 2 is 24.7%, in Stage 3 is 21.7%, and in Stage 4 is 14.4%. Pedestrian or bicycle use is highest in Stage 4, and the use of private cars is the lowest. Again, it is seen that the rate of using public transport in Stage 1 is 34.0%, in Stage 2 is 234.7%, in Stage 3 is 34.1%, and in Stage 4 is 28.8%.

The reason for the low rate seen in the use of public transport in Stage 4 is actually due to the high pedestrian travel in this neighborhood. As Stage 4 consist of high risk group neighborhoods, where disadvantaged groups (housewives, the unemployed population, the sick and the disabled, low-income earners) concentrate, it is expected to have more share in trips on-foot. Indeed, an interviewee who lives in Stage 4 explains the reason for preferring more pedestrian trips as follows:

Riding a bus/dolmus with children is both difficult and costly. If you go and return to/from Kizilay (business area) for shopping with the family, whichever way you look at it you spend 20 TL. This is too much for us. Because we have a lot of expenses other than transportation. That is why we generally prefer places that we go on foot.

Another interviewee working as a cleaner pointed out that there are some economic dimensions as well as social reasons behind the preference of pedestrian transport instead of public transport.

Most of the time I am walking to work, so I do not ride the bus much. Since I don't have reading or writing skills, I am ashamed of asking where the vehicle goes because either the ones in the bus stop or driver get angry with me... so I do not ride much as I have already said, may be two times a year in total...

On the other hand, when the distribution of trips is analyzed according to motor vehicle use by excluding pedestrian trips, cases differ widely. Use of private cars and taxis in Stage 1 it is 42.2%, in Stage 2, 35.8%, in Stage 3, 32.8%, and in Stage 4, 25.5%. Transport use in stages reach to level 42.6% in Stage 1, 50.2% in Stage 2, 51.3% in Stage 3, and 66.0% in Stage 4.

Many factors can influence the preferences of the individuals related to selection of their transportation type. These factors can include household type, age, income, having children, disability or chronic illness status, having private car and driver's license, employment status, education level, and fares. In addition, the level of public transport in the city, which is one of the indicators of a healthy transportation system, and some problems in this regard also exclude individuals from transportation. Participants list public transport problems in the forms of non-compliance timetable of vehicles, inadequate number of vehicle service, inadequate number of lines, expensive fare, insufficient ticket offices, crowded or dirty vehicles, difficulty getting information about services, and inadequate arrangements for the disabled.

A disabled female interviewee expresses her transportation type selection rationale and the exclusion from public transport process as follows:

If you live in Turkey and disabled you are in prison in the house. Maybe I am one of those lucky disabled person in terms of transportation. Because my father and my sister are taking me to place I go by car but it is not possible to use public transport and taxis. There are scarcely any bus for disabled, existing ones cannot be ridden with the wheelchair because slope of the sidewalks and the ramps are not appropriate. Disability is more difficult for those who have economical hardship. Depending on the type of wheelchair their costs and ergonomics are changing. Taxi drivers do not allow heavy foldable chairs (which can be disassembled into small pieces like Lego) in their taxis with the excuses of such as trunk is full, stains rear seat by mud and so on. Use of metro is impossible because I did not find any working elevator in the stations as in the overpasses...

Another interviewee, who is chronic kidney failure patient, describes the reason of his/her inability to use the public transport by associating with his/her illness.

I'm going to dialysis four days a week. The center I go, pick me up from my home with service vehicle and again drop me off to my home. My body is quite exhausting after dialysis, do not drive, I do not ride bus that I do not know what time it will come, in the crowd, I am not likely to travel by standing. Also my close ones occupied with their own work, they couldn't wait me at least three hours! (...) I cannot use public transport outside dialysis also. There is a serious sterilization problem, also my immune system is very weak...

Reduced service of the public transport system (due to optimization works in the city) increases peoples' tendency to avoid public transportation systems. Indeed, due to the reduction of the number of services in the transport route, many people state that they started to use private car.

Another restriction in public transport is the security problem. In this aspect, gender-related differentiation emerges in the conducted interviews, and safety concerns are felt more particularly in the younger age groups. For example, a16-year-old (male) interviewee expresses his concerns about physical violence as follows:

We meet frequently with my friends in Kizilay. I can ride the special bus or municipality buses discounted for being a student. I have never stayed there after ten o'clock. Bus services going to our neighborhood are decreasing as well as those who use drugs, people who can do extortion and can injure other people increase at that hour. I'm refraining, so do not delay too much.

Another 16-year-old (female) interviewee point out that she try not to stay alone because of harassment and sexual violence in transport vehicles, if she will make trip alone, she returns home in the early hours. The elderly also have different concerns over the use of public transport. For example, a 78-year-old participant describes the process of exclusion from public transport as follows:

I do not get on the bus or minibus/dolmus as before because I'm afraid of falling. I used to prefer minibuses with low entry instead of buses. Taxi or car is more suitable for me. I do not have car and driver's license, if I got car, I would have difficulty for driving in this age. Therefore, if I need to go somewhere by car, I expect either my children or my grandchildren to take me.

Another 68-years old participant stated that "in Ankara transportation/public transport system is quite inadequate and therefore, there is no possibility to remain outside of non-existent system."

While municipality buses have the highest utilization rate among public transport vehicles in stages, the much diminished preference for the rail systems stems from inadequate service and infrastructure in Ankara. In fact, Ankara has a quite weak rail system network compared to other capital cities of Europe.

Increase in the number of trips of people (in all stages) causes the vehicle type used slide from public transport to private cars. In fact, while in the second trips private car usage rate is 41.2% in stage 1, %34.9 in stage 2, 32.2% in stage 3, %25.0 in stage 4, public transport usage rate is 41.8% in Stage 1, 60.7% in Stage 2, 51.8% in Stage 3, and 66.5% in Stage 4.

In the third trips, private car utilization rate is 51.8% in Stage 1, 51.4% in Stage 2, 48.4% in Stage 3, and 33.7% in Stage 4, while public transport usage rate is 41.7% in Stage 1, 42.9% in Stage 2, 47.9% in Stage 3, and 64.1% in Stage 4. When evaluating the fourth trips, private car usage rate is 63.3% in Stage 1, 58.3% in Stage 2, 45.9 in Stage 3, and 45.0% in Stage 4, while public transport utilization rate is 26.5% in Stage 1, 41.7% in Stage 2, 44.8% in Stage 3, and 55.0% in Stage 4. In this case, it will not be wrong to say that people who make three or more trips prefer private cars to public transportation.

In order to determine urban activities that people do in a day in the sample, the data related to the trip purpose was used. However, due to friend visits accounting for less than 5% in the sample, it is not classified as a separate sample but it has been included in the social and cultural trips. In this context, according to results of conducted analysis, in all 4 stages, the first line trips have the purpose purpose of home, work, and work-related, in the second line school trips take place.

Shopping trips account for 11.5% in Stage 1, 6.9% in Stage 2, 7.3% in Stage 3, and 5.5% in Stage 4; the health trips are 3.8% in Stage 1, 2.8% in Stage 2, 3.8% in Stage 3, and 5.7% in Stage 4. It is determined that social and cultural trips account for 15.2% in Stage 1, 14.0% in Stage 2, 16.9% in Stage 3, and 13.4% in Stage 4.

When trips made mandatorily (work, education, and health) are ignored, it is remarkable that visits of relatives/friends of individuals and the participation level to socio-cultural activities is very low.

However, among the most important determinants of this case are income and time constraints. Indeed, while another interviewer, who is a minimum-wage worker, links the activities he wants to participate in but cannot with his low income, another interviewer indicates that transportation plays a very crucial role in participation in urban activities.

There are no big parking lots around my near surrounding. There is no public transport goes from here to Eymir or Mogan [large green areas] as far as I know, however, I can get with the cars of my friends. Metro going to big shopping centers are nice, but this time again you are caged indoor space out of the house. I am having problems with the participation of particular cultural activity. Ankara is already a bit more limited in terms of these activities. You have to go to weekend hours during the day for a show in State Theatres or State Opera Ballet. Because weekdays show starts at eight in the evening and ends somewhere around ten or eleven, getting home after the show is rather difficult. I want to go to the library for studying, there are libraries open for24 hours, but there are no open transport for 24 hours...

Most of the people living in Stages 1 and 2 state that they have difficulty accessing urban activities due to transportation. However, individuals living in Stage 3 and 4 did not complain about accessing urban activities as much as other interviewers did. Indeed, an interviewee reported carrying out his/her visits for shopping, health needs, education of children, and relatives and friends in the same neighborhood, in addition, he/she stated that because of the closeness of his/her home to his/her work place, he/she does not frequently use public transportation. Due to there being more similar response in-depth interviews conducted in high exclusion-risk groups, "what do you describe in urban activities?" was asked of these participants. However, no activity outside of visiting for shopping, health, education, and relatives and friends was reported. This situation reveals that many individuals are deprived of information about what kind of activities there are in the city, which indicates that they are indirectly excluded from urban life and transportation.

Finally, the transport individuals who participate in activities that take place in the city are analyzed. According to trip purpose, mode choice in the first trip was examined, revealing variations between the stages. Thus, while private car or taxi are preferred in the trips for the purpose of home, work, and work-related in Stages 1, 2, and 3, pedestrian or bicycle use are preferred in Stage 4. In school trips, pedestrian or bicycle use in Stage 1 is 35.8%, in Stage 2 is 56.3%, in Stage 3 is 65.3%, and in Stage 4 is 67.3%. However, for students that go to school with service vehicle with a fee, the opposite situation arises: while students in stage 1 use a service vehicle as secondary, in other stages, use of municipality/city bus is common.

While in shopping trips, in Stage 1 use of private car or taxi (39.5%), in Stage 2 municipal/city bus (32.7%), in Stage 3 (44.6%), and in Stage 4 pedestrian or use of bicycle (44.4%) take first place in terms or preference; second place in stage 1 (30.1%) and in Stage 2 (31.7%) is pedestrian or bicycle, private car or taxi in Stage 3 (20.3%), and minibus/dolmus use in Stage 4 (17.0%). While private car or taxi ride is preferred for health trips in the first three stages, in Stage 4, pedestrian or bicycle is preferred. Second place in Stages 1, 2, and 3 is minibuses/dolmus, and municipality/city buses are preferred in stage 4.

Again, in social and cultural activities and visiting friends, which are non-compulsory (recreation) trips, it is seen that, in Stage 1, private car or taxi use is the top preference. In visiting friends pedestrian or bicycle is the secondary preference, while in social and cultural activities it is municipality/city bus. In Stage 2 for both activities, pedestrian or bicycle is the first and private car or taxi the second preference. In Stage 3 in social and cultural activities pedestrian or bicycle is preferred, and in visiting friends, private car, while in Stage 4, in both activities, pedestrian or bicycle are the highest preferences. As shown, although mode choice varies by type of activity, in Stage 1 there are more private cars, and in Stage, 4 pedestrian and bicycle use is widespread.

CONCLUSION

With a direct impact on raising the level of prosperity and improving the quality of urban life, social exclusion has become a major problem, especially in big cities. For that reason, while social exclusion studies conducted in the past addressed issues focused on individuals or specific groups such as income, language, religion, and ethnic origin, the concept of space is included in contemporary discussions. Indeed, migration into big cities and rapid urbanization taking place along with it have revealed both individual and space poverty. While the excessive growth in cities led to the spread of living spaces out from the central areas, these changes in the urban morphology lead to increases in the use of private cars. While high-income groups with a high rate of private car ownership settle in areas outside the center in an effort to escape the negative aspects of the city, low income groups suffering financial difficulties prefer the suburbs, where land prices (rents) are lower. However, several studies have shown that social exclusion has been triggered, and that access of individuals to work, education, healthcare services and sociocultural activities has become more difficult in these areas with inadequate public transport facilities, which makes private car ownership a must. This aspect of transportation has turned out to be the most important factor in triggering social exclusion.

Understanding social exclusion correctly, including all of its other parameters, requires a thoroughly analysis of its urban space dimension, and therefore, transportation. On this basis, the study primarily attempts to concretize the spatial patterns of exclusion on the axes of both social exclusion and exclusion from transportation. The purpose of determining the levels of groups at transport and social exclusion risks is to enable the identification of social exclusion and transport relations. It has thus been possible to specifically address transport behaviors of individuals working in marginal sectors and living in segregated areas, where the proportion of disadvantaged groups intensifies. The transportation and exclusion problems that differentiate from or are similar to the other sections of the society (i.e., of those with strong access to transportation resources and urban services, who live in areas with low risk of exclusion in socio-economic and cultural dimensions) have been identified more easily in the present study. In addition, the obtained homogeneous regions (levels) also increase the reliability of the research.

In transport-related social exclusion studies, less educated and lower income individuals who are unable use public transport systems and who do not own private cars are often seem to be defined as disadvantaged groups. Upon examining the findings on the educational level of the sample, it was determined that the literacy rate of 95.6% of the sample fell behind that of the Ankara province average of 97.7%, and that women formed the majority among the illiterate (about 3 times that of men). Only 58.5% of the sample had received education up to the high school level (8.7% did not finish any school, whereas 26.8% are primary school and 23.0% are secondary school graduates). An evaluation of participation in work life and income level findings showed that 65.4% of the sample did not work and that the income of 42.6% of those who work was concentrated between TL 751–1500. Considering that the survey data were collected in 2013 and the minimum income calculated that year was TL 803 and the poverty line was TL 3322, it can be concluded that 83.3% of the sample members sustained their lives between the minimum income and poverty line. While the income of 5.6% of the sample is less than the minimum wage, the income of only 11.1% is above the poverty line. In this context, the often focused on poverty and deprivation findings (inequality in income distribution, non-participation in the labor market and low educational level) in the literature raise the possibility of exclusion of the sample may be higher than expected. Indeed, a vast majority of individuals with low income articulate that they are unable to use the transportation system because of their financial problems.

It has been found with this study that not only the poor people but also a large part of the population is in fact directly or indirectly excluded from the transport systems in Ankara. As stated in the previous sections of the study, Ankara is actually a car-dependent city, which is very worrying in terms of exclusion. The reason of this concern is that the car, which is a non-sustainable mode of transport, is criticized for isolating people from a certain section of the society, isolating people, and depriving them of the ability to empathize. An examination of the car ownership data of the sample revealed a cumulatively linear increase in the number of cars, and 48.1% of households are identified to be private car owners. Indeed, per capita car ownership rate in Ankara is 184‰ (AUAP, 2014). The demand for private cars in the city of Ankara is actually caused by limitations in the city's public transport. Indeed, while the city's sprawling macroform mandates the use of private cars, supply-related problems, such as the absence of roads allocated to public transport, private car travel time's being about 10 to 20 minutes shorter compared to public transport, and difficulties associated with using public transport in peak hours, lead individuals to private cars.

The continuation of the tendency to obtain private cars increasingly will in the long run cause a decrease in the number of people using public transport, restriction of scheduled shuttles, and a reduction in the quality level of public transport services. Such difficulties encountered in public transport will in turn lead to an increase in the tendency of individuals to obtain private cars and, with the increasing car numbers, the disadvantaged individuals who cannot afford private cars will be obliged to use a public transport system with a reduced number of scheduled shuttles and higher ticket prices in traffic that has turned into chaos. Unfortunately, it is clear that both private car users and public transport passengers will be excluded due to this vicious cycle over time.

The study shows that private car ownership is significantly determined by income level. Car prices are very high in Turkey due to high taxes, and as a natural consequence of this situation, a positive correlation between car ownership and income has been identified. Therefore, a car ownership rate of over 80% of households with an income of TL 3500 and above is expected. Hence, it has been established that the tendency to have cars increases as income rises in the sample, and the first-car ownership rate was found to be 69.9% at the 1st level, 55.7% at the 2nd level, 48.2% at the 3rd level, and 36.4% at the 4th level. Although the private car ownership rate changes at different levels, the car ownership rate of 36.4% of households at the 4th level (high risk group) is remarkable. This can be explained by orientation of individuals to inexpensive old cars due to the absence of vehicle age quotas in Turkey. The reason is that due to social norms, private cars are also considered a status symbol. While newer, higher-segment cars are preferred in the upper income group, largely old, domestic cars are used by the lower income groups. However, the difficulties in meeting the cost of using cars are manifested at different levels of groups. In this context, the proportion of individuals who cannot use their cars although they own one in the sample is 54.5% at the 1st level, 58.4% at the 2nd level, 59.1% at the 3rd level, and 62.4% at the 4th level. Due to high vehicle insurance, motor insurance, and fuel prices and taxes, using private cars brings an additional cost burden on individuals. On the other hand, efforts to reduce costs (such as refraining from having vehicle insurance, transformation of non-compliant vehicles for LPG fuel consumption, lack of suitable tires for the season) reduce traffic safety.

The study also reveals that the type of household also affects ownership of private cars. The results of the correlation analysis conducted with car ownership and household type variables show that it is possible to speak of a negative relationship between the two variables. In this context, the household type that is nearest to car ownership is families with children, while the farthest ones are women living alone. Indeed, 53.9% of completed families, 47.3% of single-parent families, 38.6% of families without children, 37.1% of men living alone, and 17.0% of women living alone have a privately owned car. The fact that private car ownership of men living alone is higher compared to women living alone seems to be associated with income. Men have advantages in terms of participation in employment and obtaining high-income compared to women, and the almost-three-times-higher private car ownership ratio of men to women in Ankara is a reflection of these differences. However, women often state that they are sexually or otherwise abused while they are pedestrians or using public transport, especially when it gets dark, and that they would like to obtain private cars even if for safety reasons only.

Families' having children pushes them out of public transport while increasing their need for private cars (86.6% of families with children have private cars). This can be attributed to changes in travel behaviors and extension of travel chains of families with children. Children's education, health, sports, and socio-cultural activities necessitate extra trips for parents. With the increase in trip chains, most parents may prefer to use private cars. An examination of the overall sample shows that while the rate of use of a private car is 51.9% with one trip and 51.1% with two trips, it rises to 74.2% when more than two trips are performed. Based on these findings, it would not be wrong to argue that as mobility during the day increases, the tendency towards private cars increases. Along with the increase in the number of trips, the difficulties in using public transport when accompanied by pre-school children and failure to provide security increase the preference for private cars. In addition, some of the interviewed parents articulate that time becomes even more precious with the rush of reaching urban activities with children, and waiting periods at the public transport stops lead to serious loss of time.

This study identified that, along with individuals that are unemployed, with low education, poor, women, elderly, and young, the group that constitutes the other disadvantaged section of the society, consisting of individuals who are disabled or with chronic illness, also face many problems in the city. However, these problems seem to focus around lack of access to transportation rather than social exclusion. The interviews conducted showed that these individuals do not feel socially excluded, thanks to their strong family and friendship ties. Unfortunately, it is difficult to assert that individuals with disabilities or chronic diseases are as lucky when it comes to urban transport. While obstacles such as rough pedestrian paths, narrow and high sidewalks lacking ramps, invasion of sidewalks by vehicles, lack of audible signaling at crosswalks, and lack of elevators at places with high-rise stairs limit their navigation throughout the city as pedestrians (or on wheelchairs), public transport systems ignoring the disabled oblige them to use private cars or ask their relatives or friends to drive them.

The concept of exclusion from transportation becomes well evident in the mode choice of individuals for their trips. Indeed, those individuals who cannot afford the costs of private car use tend to use public transport and those who cannot afford even those of public transport tend to be pedestrians or use bicycles. The fact that public transport use is 34.0% at the 2nd level, 34.7% at the 2nd level, 34.1% at the 3rd level, and 28.8% at the 4th level, whereas private car use and taxi rides are highest at the 1st level and pedestrian trip or bicycle use is highest at the 4th level proves this. The higher share of the 4th level in pedestrian trips may be associated with the fact that this is a cluster consisting of more disadvantaged neighborhoods where housewives, the unemployed, the sick and the disabled and low-income groups are concentrated. In the interviews made with members of the 4th level cluster, it has been understood that individuals have difficulty in meeting the transport costs, and therefore, are obliged to perform pedestrian trips that last more than an hour.

The preference of private cars or pedestrian-bicycle trips to public transport is due to the deficiencies in the service quality of public transport modes, services not complying with scheduled shuttle times, inadequate number of scheduled shuttles, inadequate number of transport lines, expensive fares, inadequate number of ticket sales offices, crowded or dirty vehicles, difficulty in obtaining information about the scheduled shuttles, inadequacy of the regulations for the disabled, security issues, and public transport's being a time-consuming option compared to other types of transportation. Yet, public transport is known to include social integration, providing notions such as people's showing respect and understanding of each other and enabling communication among individuals from all sections of the society. This is why a well-constructed, well-run, low-cost, easy-to-access transportation system is of great importance in the prevention of exclusion.

Lastly, when the aim of trip of individuals findings are examined from a transport-related social exclusion perspective, it is seen that house-, work- and work-related trips rank in the first place followed by school-related trips at all the four levels. Trips for shopping and sociocultural objectives are the highest in the 1st level. When mandatory trips such as business-, education-, shopping-, and health-related ones are neglected, the participation levels of the sample in relative/friend visits and sociocultural activities appears to be quite low. It is identified from the interviews that issues such as relatives/friends residing in the city's peripheries; cinemas moving away from the city center and jamming inside the shopping malls; late ending hours of theater, opera, or ballet performances; and difficulties of access to recreation areas make private car or taxi use compulsory. However, individuals who cannot afford these costs are excluded from urban activities. In this context, the biggest obstacles that prevent individuals from participating in urban activities emerge to be poverty and deprivation from transport.

Analyzing transportation by associating it with a social exclusion approach, this study illustrates once again that there is a requirement for fairer, people-oriented transport services that will ensure social inclusion in the city of Ankara. Exclusion from transportation in the city indeed manifests itself in economic, time-based, fear-based, and exclusion from opportunities dimensions.

REFERENCES

Alsnih, R., & Hensher, D. A. (2003). The mobility and accessibility expectations of seniors in an aging population. *Transportation Research Part A, Policy and Practice*, *37*(10), 903–916. doi:10.1016/S0965-8564(03)00073-9

AUAP. (2014). Ankara Metropoliten Alanı ve Yakın Çevresi Ulaşım Ana Planı (2013-2038). *Gazi Üniversitesi Ulaşım Ana Planı Ofisi*, 1-27, 34-41, 70.

Bhat, C., Handy, S., Kockelman, K., Mahmassani, H., Chen, Q., Srour, I., & Weston, L. (2001). *Assessment of accessibility measure* (Research Report No. 7-4938). Texas Department of Transportation.

Bocarejo, S. J. P., & Oviedo, H. D. R. (2012). Transport accessibility and social inequities: A tool for identification of mobility needs and evaluation of transport investments. *Journal of Transport Geography*, *24*, 142–154. doi:10.1016/j.jtrangeo.2011.12.004

Bukhari, S. M. A., Hine, J., Gunay, B., & Blair, N. (2010, October). *Transport disadvantage and public transport network change: a case study of Belfast City*. In *12th World Conference on Transport Research*, Lisbon, Portugal.

Burchardt, T. (1999). Participation not consumption: The need for new participatory practices to address transport and social exclusion. *Transport Policy*, *10*, 272.

Burns, L. D. (1979). *Transportation, temporal and spatial components of accessibility*. Toronto: Lexington Books.

Burns, P. C. (1999). Navigation and the mobility of older drivers. *The Journals of Gerontology. Series B, Psychological Sciences and Social Sciences*, *54*(1), S49–S55. doi:10.1093/geronb/54B.1.S49 PMID:9934402

Cass, N., Shove, E., & Urry, J. (2003). *Changing infrastructures, measuring socio-spatial inclusion/exclusion*. Final Report to The Department for Transport.

Chakwizira, J., Bikam, P., Dayomi, M. A., & Adeboyejo, T. A. (2011). Some missing dimensions of urban public transport in Africa: Insights and perspectives from South Africa. *The Built & Human Environment Review*, *4*(2), 56–84.

Church, A., Frost, M., & Sullivan, K. (2000). Transport and social exclusion in London. *Transport Policy*, *7*(3), 195–205. doi:10.1016/S0967-070X(00)00024-X

Currie, G. (2004). Gap analysis of public transport needs: measuring spatial distribution of public transport needs and identifying gaps in the quality of public transport provision. *Transportation Research Record: Journal of the Transportation Research Board*, (1895), 137-146.

Dalvi, M. Q., & Martin, K. M. (1976). The measurement of accessibility: Some preliminary results. *Transportation*, *5*(1), 17–42. doi:10.1007/BF00165245

Dodson, J., Gleeson, B., & Sipe, N. G. (2004). *Transport disavantage and social status: a review of literature and methods*. Brisbane: Urban Policy Program, Griffith University.

Geurs, K. T., & Van Wee, B. (2004). Accessibility evaluation of land-use and transport strategies: Review and research directions. *Journal of Transport Geography*, *12*(2), 127–140. doi:10.1016/j.jtrangeo.2003.10.005

Handy, S. (1994). Highway blues: Nothing a little accessibility can't cure. *ACCESS Magazine*, *1*(5), 3–7.

Handy, S. L. (2002). Accessibility-vs. mobility-enhancing strategies for addressing automobile dependence in the US. Institute of Transportation Studies.

Hansen, W. G. (1959). How accessiblity shapes land use. *Journal of the American Planning Institute*, *25*(2), 73–76. doi:10.1080/01944365908978307

Hine, J., & Grieco, M. (2003). Scatters and clusters in time and space: Implications for delivering integrated and inclusive transport. *Transport Policy*, *10*(4), 299–306. doi:10.1016/S0967-070X(03)00055-6

Hine, J., & Mitchell, F. (2001). The role of transport in social exclusion in urban Scotland. *Scottish Executive*, 11-42.

Hurni, A. (2006). Transport and social disadvantage in Western Sydney: a partnership research project. *Western Sydney Community Forum*, 11-93.

Kenyon, S., Lyons, G., & Rafferty, J. (2002). Transport and social exclusion: Investigating the possibility of promoting inclusion through virtual mobility. *Journal of Transport Geography*, *10*(3), 207–219. doi:10.1016/S0966-6923(02)00012-1

Kirby, A. (1982). *The Politics of Location*. London: Methuen.

Litman, T. (2011). *Evaluating accessibility for transportation planning: measuring people's ability to reach desired goods and activities*. London: Victoria Transport Policy Institute.

Lucas, K. (2012). Transport and social exclusion: Where are we now? *Transport Policy*, *20*, 105–113. doi:10.1016/j.tranpol.2012.01.013

Metz, D. H. (2000). Mobility of older people and their quality of life. *Transport Policy*, *7*(2), 149–152. doi:10.1016/S0967-070X(00)00004-4

Miller, H. J. (2003). Travel chances and social exclusion.*10th International Conference on Travel Behavior Research*, 1-38.

Niemeier, D. A. (1997). Accessibility: An evaluation using consumer welfare. *Transportation*, *24*(4), 377–396. doi:10.1023/A:1004914803019

Rajé, F. (2003). The impact of transport on social exclusion processes with specific emphasis on road user charging. *Transport Policy*, *10*(4), 321–338. doi:10.1016/S0967-070X(03)00038-6

SEU. (2002). *Making the connections: final report on transport and social exclusion*. London: Social Exclusion Unit.

SEU. (2003). *The future of social exclusion: drivers, patterns and policy challenges*. London: Social Exclusion Unit.

Stanley, J., & Stanley, J. (2004). Improving public transport to meet community needs: A Warrnambool case-study. Report for Bus Association Victoria and Warrnambool Bus Lines, Melbourne, Australia.

TÜİK. (2011b). *2000 Bina sayımı veri seti: Ankara kenti*. Ankara: Türkiye İstatistik Kurumu.

Türk-İş. (2015). Retrieved from http://www.webcitation.org/query?url=http%3A%2F%2Fwww.turkis.org.tr%2Fdosya%2F8BPrv9CducUP.pdf+&date=2016-06-30

Witter, R. (2010). Public urban transport, travel behaviour and social exclusion–the case of Santiago de Chile.*XII World Conference on Transportation Research, Lisbon*, 1-23.

Yılmaz, Ö., & Temurlenk, M. S. (2005). Türkiye'deki Istatistik Bölgelerin Kişi Başina Düşen Gelir Açisindan Hiyerarşik ve Hiyerarşik Olmayan Kümeleme Analizi Ile Değerlendirilmesi: 1965-2001. *Atatürk Üniversitesi İktisadi ve İdari Bilimler Dergisi*, *19*(2), 75–92.

KEY TERMS AND DEFINITIONS

Cluster Analysis: One of the social analysis method in which unclassified observations are put together according to their similarities.

Disadvantaged Groups: The group of people who are either lack of economic, cultural and social resources or left alone in need of such resources.

Exclusion from Transportation: The condition in which certain individuals or social groups cannot use or access to the transportation services, due to the individual, economic or physical constraints.

Individual Limitations: The condition in which individuals have some restrictions or difficulties in terms of physical or mental abilities.

Mahalle: The Turkish name of the smallest local administrative unit which also has an elected headman (muhtar in Turkish).

Social Exclusion: The condition in which certain individuals or social groups cannot integrate into the other parts of the society either socially or economically or politically.

Social Inclusion: The state of being and feeling as part of the society.

Socio-Spatial Segregation: The spatial separation of people living in urban areas parallel to their social, economic or cultural differences.

Spatial Pattern: A socio-spatial term which refers to the formation of the physical structure of an urban area by the social, economic and political constraints.

ENDNOTES

1. Mahalle is the Turkish name of the smallest local administrative unit which also has an elected headman (muhtar in Turkish).
2. In the research, out of 4 neighborhood stages, neighborhoods hosting most of its own characteristics of the stage that has been tried to be selected. To do this, it has benefited from the analysis of data envelopment. In-depth interviews were conducted in the first three neighborhoods showing the most similarity to the stage according to data envelopment analysis.

Chapter 6
Equidistance:
Evidence of the Influence of Parking Organization on Mode Choice

Guenter Emberger
Technische Universität Wien, Austria

Paul Pfaffenbichler
Technische Universität Wien, Austria

ABSTRACT

The chapter will provide empirical evidence how parking organization (location, number and pricing policies of parking spaces in urban areas) influences the mode choice of transport system users. Following from that, the presently applied practice of parking space modelling within existing transport models will be presented and discussed critically. It will be shown which influence on transport demand the different parking space instruments have in existing 4-stage models (FSM) and where flaws and errors in the transport demand and mode choice estimation can occur. Finally, potential improvements how to overcome these afore mentioned shortcomings are presented and the impacts on future transport planning and transport policy formulation towards sustainability will be introduced.

INTRODUCTION

Transport related problems in urban agglomeration are of growing concern all around the world. Cities, settlements and agglomerations apply a wide range of transport policies such as the improvements of public transport systems, improvements of road capacity through new construction (where it is still possible) or through the application of ICT (e.g. demand responsive traffic signaling), increase of parking space capacity, introduction of parking space management instruments such as electronic parking place management systems, implementation of parking fees, dedicated bus lanes, cycle lanes, pedestrian zones, shared space, e-mobility, and many more instruments – to mitigate the negative impacts of urban traffic issues.

DOI: 10.4018/978-1-5225-2116-7.ch006

Political declarations, international (United Nations, 1998, 2015), EU-wide (Commission of the European Communities, 2001; Department for Transport (DfT), 2000; European Commission Directorate-General Energy and Transport, 2011) and national (BMVIT, 2002, 2013; Kovacic, 1999; Magistrat der Stadt Wien, 2003; Stadt Wien - Stadtentwicklung, 2002; Steininger, 2008), provide visions and targets which should result in the near and much more often, in the far future in a sustainable, green, and "smart" transport system, where every person and good will be able to travel whenever he/she/it wants, paying just a low price, with high speed at every time of the day and night, and between any origin and any destination.

The vision of transport and traffic engineers as well as politicians is to create this "ideal" transport system, which is, when taking laws from physics into consideration, per definition not possible.

As Knoflacher shows in his chapter of this book there exists no growth in (physical) mobility, so an increase of the attractiveness of motorized (and mechanized) means of transport will always result in a decrease of walking trips within the population.

By looking at the afore mentioned transport related policies nearly all of them increase the attractiveness of motorized means of transport and support so the increase of car use in and within settlements.

When looking from a general perspective on a dynamic system (and the land use – transport system is a dynamic system) then it can be said, that as long there is a growth, for example in the distance travelled by cars or the number of trips made by cars, there is no dynamic equilibrium reached. The transport system is continuously looking for such a dynamic equilibrium, where balancing feedbacks will become strong enough to keep the level of travelled distance or number of trips by car constant.

As Knoflacher explored in very detail in his books (Knoflacher, 1997, 2001) road congestion, for example, is such an important balancing feedback, which is an important instrument to limit motorized growth in the transport system. Interesting is here, that traditional educated transport engineers as well as politicians do not understand the importance of road congestion as a transport policy instrument and its adequate application in modern transport planning. Congestion can be used to steer the overall land use transport system towards the desired, sustainable direction - on the contrary, traditional educated transport engineers see road congestion as one of the major issues, which has to be eliminated to reach the above mentioned unlimited motorized mobility of persons and goods.

WHY IS PARKING ORGANIZATION SO IMPORTANT FOR THE MODE CHOICE OF INDIVIDUALS?

Another important steering instruments besides congestion is the organization and much more important the location of parking spaces for cars. It is common sense in transport planning that every trip either made by public transport or by car starts and ends with a walking trip. Therefore the distance to (origin) and from (destination) the parking space (car trip) or to/from the public transport stop plays a dominant role within the intra personal mode choice.

In other words, it could be said from a systemic point of view there is no difference between a public transport stop and a car/(motor-)bike parking place. Both are the interface between walking and the motorized (mechanized) means of transport. Interesting is, that although a public transport stop and a parking place are from a systemic point of view absolute the same, the treatment in transport planning science was/is completely different.

E.g. by looking at the past developments of settlement structures it can be seen, that nearly in all cultures and societies car parking places are located as close as possible to the origins (housing) and destinations (shopping centers, office buildings, factories, etc.). This is due the fact that in nearly all countries exist guidelines which provide regulations for the provision, number and location of parking places (Asian Development Bank, 2008; Attoney General's Chambers, 2010; Beaumont, 2002; City of Vienna, 2008; Davidge, 1924; Development & Building Control Division Roads Group, 2005; Greater London Authority, 2008; Hong Kong Transport Department, 2002; Morikawa, 2010; Planning Department of Hong Kong, 2006; Stadtentwicklung Wien, 2003; Westminster City Council, 2010)[1].

These guidelines were developed to avoid car parking in public space but had the unintended side effect that the distance between car parking places and public transport stops developed in different directions. Car parking places tended to be located as close to the origins and destinations as possible, whereas the public transport stops were allowed, depending on the type of public transport, to be between 300m (bus, tram) to 2000m (light rail) away from the origins respective destinations of the trips.

(Knoflacher, 1980, 1981) described already in the early 1980ies that there must exist a relation between parking location and individual mode choice. Based on this research (Peperna, 1982) investigated 1982 the influence of the walking distance to/from public transport stops and mode choice and found out that with increasing distance to a public transport stop the likelihood to use public transport diminishes following an negative exponential-function (see Figure 1).

On the y-axis of Figure 1 one can see the acceptance of travelers to walk and on the x-axis there is the distance to the destination point. Is the destination point less than 100m away from the starting point everyone walks to the destination point (shop, workplace, etc.). Increasing the distance between

Figure 1. Relation built structure and behaviour, "Die Einzugsbereiche von Haltestellen öffentlicher Nahverkehrsmittel im Straßenbahn- und Busverkehr," Master Thesis, Technische Universität Wien, Supervisor Prof. H. Knoflacher
Source: Peperna, 1982.

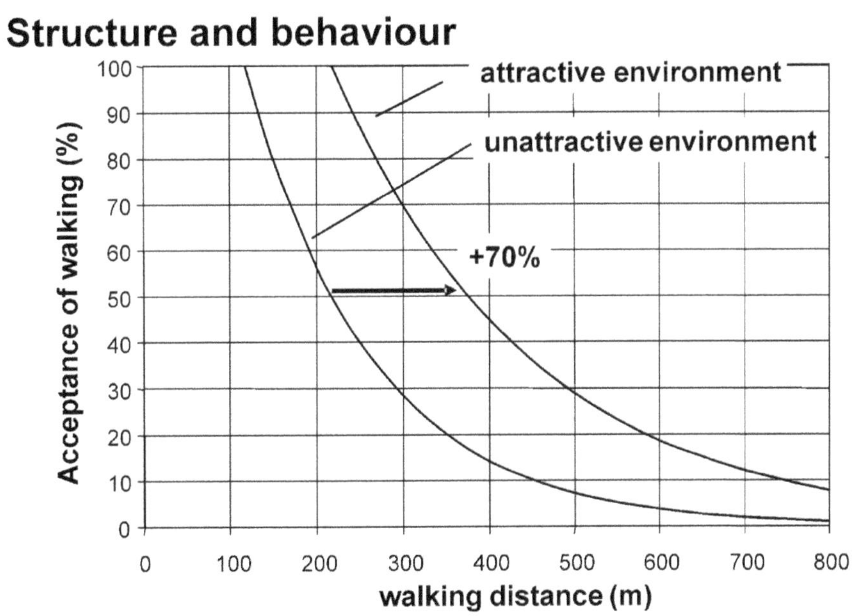

origin and destination, say to 200m, the acceptance to walk is reduced by 50%, people tend to use other means of transport (car, cycle or public transport). If the destination point is more than 450m away, the acceptance to walk drops to a mere of 10%. What can be learnt from this? It is important to keep the distances between origin and destination short – in other words it is necessary to keep the density and the mixture of the land use high. And this can only be realized when space consuming means of transport are not allowed in this kind of settlement structures.

The curves shown in Figure 1 are estimated from empirical investigations carried out in the city of Vienna. The second curve (left hand side) was estimated in pedestrian zones in Vienna, where no car traffic is allowed. As can be seen, if the structure is pedestrian-friendly the acceptance to walk longer distances can be increased by +70% from some 220m to 380m (to hold the acceptance level of 50% pedestrians constant).

So people are accepting longer distances if the built structure encourages people to walk. Additionally more destination points can be reached in pedestrian zones within the same distance because less space is used for car lanes and on street car parking lots.

Another, interesting investigation was carried out by Institute of Transportation - Center for Transport Planning and Traffic Engineering, University of Technology Vienna (Emberger & Knoflacher, 1995). Here the relation between car mode choice and distance to/from the parking place for the city of Vienna was under investigation (see Figure 2).

On the x-axis of Figure 2 there one can see the distance of the parking place either from the origin of a trip or from the destination of a trip. Every dot represents the average distance of parking place distance for an individual district in Vienna (for info in Vienna there exist 23 districts). As it can be seen the average distance of parking spaces are less than 250 meter, the majority of Viennese inhabitants park their car less than 100 meters away either from their home or from their destination. For information, the average distance to the next public transport stop in Vienna is about 250 to 300 meters, which

Figure 2. Private car mode choice in relation to parking distance, Vienna: Data source household survey Vienna 1986 and 1992
Source: Emberger & Knoflacher, 1995.

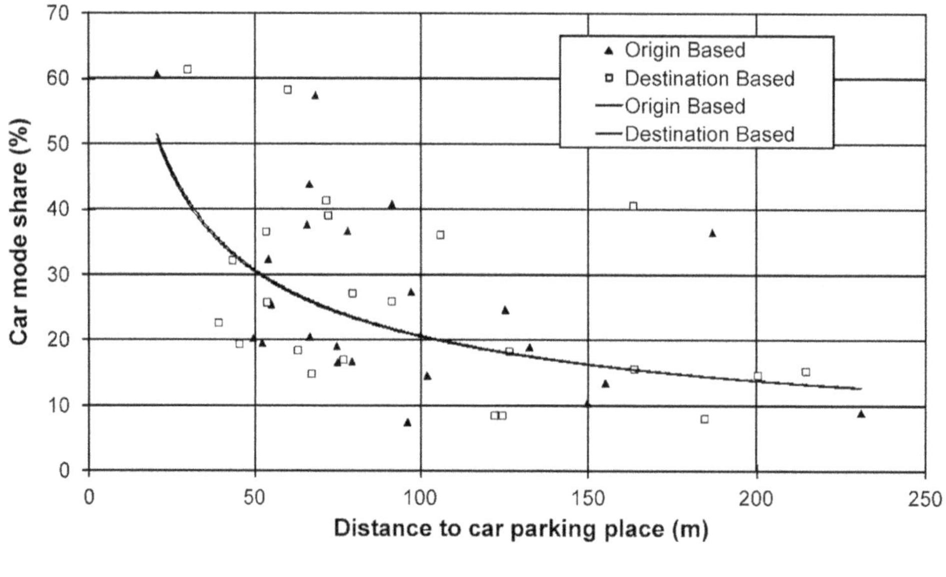

is internationally seen a very good value. On the y-axis one can see the modal split value for car use in the corresponding district for all trips in the data sample.

As it can be seen in Figure 2 the percentage of car usage in Vienna decreases with the distance of the car parking place either from the origin or from the destination. When a car parking place is located close to the origin/destination the car mode share is well above 30% - in the case the car parking place is further away – say 200m less than 20% are using the car anymore. This indicates strong empirical evidence that the distance to the parking space has a strong influence on the mode choice (especially in an urban context like Vienna).

The above carried out investigation is unique because in the data set used (household travel survey Vienna 1986 and 1992) there was the question included "How far away in meter is the parking place at your home location?" and "How far away is the parking place from your destination?" This kind of questions are not any longer asked in household surveys – and therefore no actual data are available anymore.

Unfortunately the authors could not find any further research investigating the impact of distance to and from the parking place on mode choice and car travel demand. When searching for the influence of parking on mode choice there exist a great number of studies which investigate the impact of pricing policies on mode choice (Albert & Mahalel, 2006; Kelly & Clinch, 2009; Marsden, 2006; Milosavljević & Simićević, 2016; Simićević, Vukanović, & Milosavljević, 2013). Interesting is here that in a lot of cities around the world parking fees already in place and still the transport problems are not solved. This is due that only very high parking fees would deliver a significant reduction of car use, but the implementation of such high parking fees is very difficult as far as politics are concerned. Another issue which is directly related with high parking fees is the discussion of social inclusion, since high parking fees will have higher impacts on low income person groups compared to high income groups. Also here is an ongoing scientific discussion, which is not solved, yet (Lucas, 2012; Preston, 2009; Rojo, dell'Olio, Gonzalo-Orden, & Ibeas, 2015).

WHAT DETERMINES TRAVEL BEHAVIOR?

Travel behavior, i.e. the decisions whether, when and how to travel, is determined beside intrapersonal subjective attitudes by two factors mainly, which can be directly influenced by transport engineers / city planners, namely:

1. The costs for a journey, and
2. Travel time necessary for carrying out a journey.

Ad Cost for a Journey

The influence of the cost part for journeys was/is decreasing in the past and becomes more and more negligible, especially in richer societies. The reasons for that are that car users tend to "ignore" costs associated with car ownership, especially when these costs are just occurring once (for example the purchase price of the car) or not regularly such as car insurance (which has to be paid once or twice a year), or maintenance costs (which are just payable when the car is in the garage). Therefore these not regularly occurring costs are not taken fully into consideration when a person carry out his/her mode choice decision. On the other hand car drivers are more sensible to costs which they can link directly to a certain trip such as parking fees and road pricing (cordon or distance based).

A huge amount of studies was carried out in the past to estimate elasticity values for different pricing measures, which confirm that directly paid charges have a higher impact on mode choice than indirect ones. In (Litman, 2013) a comprehensive overview of existing studies regarding transport policy instruments related elasticities can be found.

In the case of public transport user costs seem to be more relevant as the perception of costs are directly linked to a specific trip. However in cities with modern public transport systems, such as Vienna, London, etc. these costs are not perceived any longer so intensive because for example a yearly ticket in Vienna costs about 356 Euros per annum which is neglect able compared to other mode choice influencing parameters, such as access, egress or waiting/changing times.

Ad Travel Time

Travel time and changes in travel time play an important role for the individual trip destination and mode choice (Ojeda-Cabral & Chorus, 2016; Román et al., 2014; Wheat & Batley, 2015).

German research from the 1990ies shows that the different parts of a journey are perceived differently (Walther, 1973, 1991; Walther, Oetting, & Vallée, 1997). Everyone will confirm from her/his own experience that access and egress times as well as waiting and changing times are perceived as being more unpleasant than in-vehicle times. In the intra-personal decision making process for mode choices these times are subjectively weighted higher than their measured objective, physical time.

Figures 3 and 4 show the results concerning this subjective evaluation of the different parts of a motorized trip:

For example a public transport trip consists of the access time from home to the public transport stop, the waiting time, the in-vehicle time, the changing and waiting time, the in vehicle-time again and finally the egress time to the destination.

A car trip consists of the walking time from home to the parking place, the in-vehicle time and the walking time from the parking place at the destination to the final destination.

Walther proved in his studies that there exist different weighting factors for the different parts of a journey either for a public transport trip or car trip. It can be directly seen in Figure 3 and Figure 4 that a one unit change in access/egress time has a much higher behavioral effect than a one unit change in in-vehicle time. Therefore the access, changing and egress times for a journey can be seen as the pivotal points to behavioral changes.

Within the same research also "value of time" factors to convert the different cost components of a journey to time units were determined. The subjectively perceived time plus the costs transferred to time units measure the disutility of realizing a journey at a given time with a given mode of transport. The probability of choosing one alternative is the ratio of the reciprocal disutility of the alternative to the sum over all alternatives.

A small example is used to illustrate the effects of the application of subjective weighting factors.

Figure 5 shows on the x-axis the time in minutes and the distance, assuming a walking speed of 4 km per hour. The in-vehicle time (black line) has, as mentioned before, a slope of 1. The perception of the other time parts of a trip (access/egress, changing and waiting time) follows exponential functions as derived in the Walther research. As it can be seen, in Viennese circumstances (see also Figure 2), the parking places are located between less than a minute walk and up to 250m which corresponds to an

Figure 3. Subjective evaluation of the different parts of a public transport trip (metro)
Walther et al., 1997.

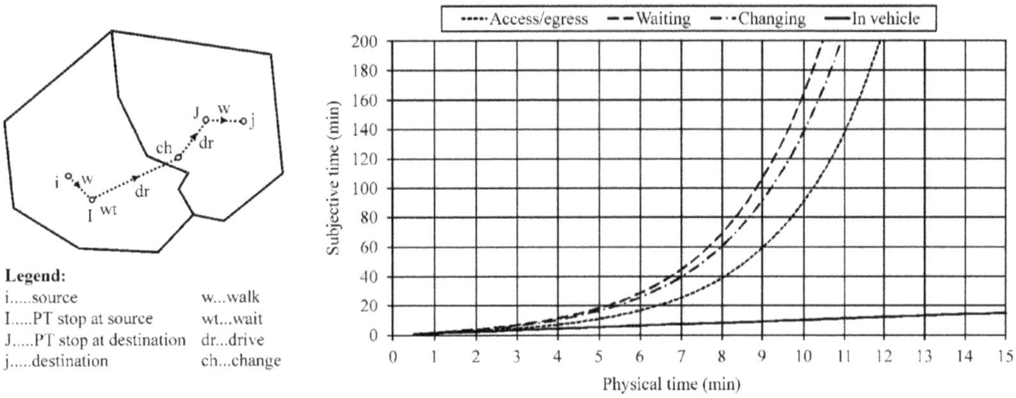

Figure 4. Subjective evaluation of the different parts of a car trip
Walther et al., 1997.

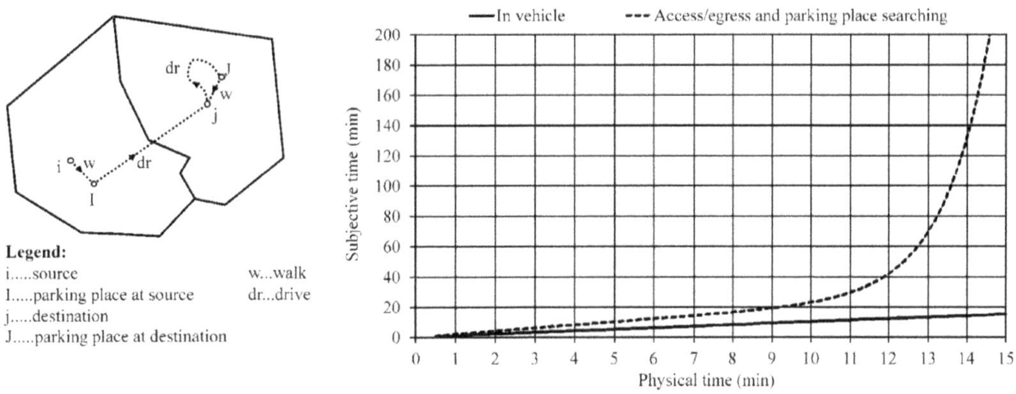

Figure 5. Comparison of objective travel time components versus subjective travel time components
Source: Own calculations.

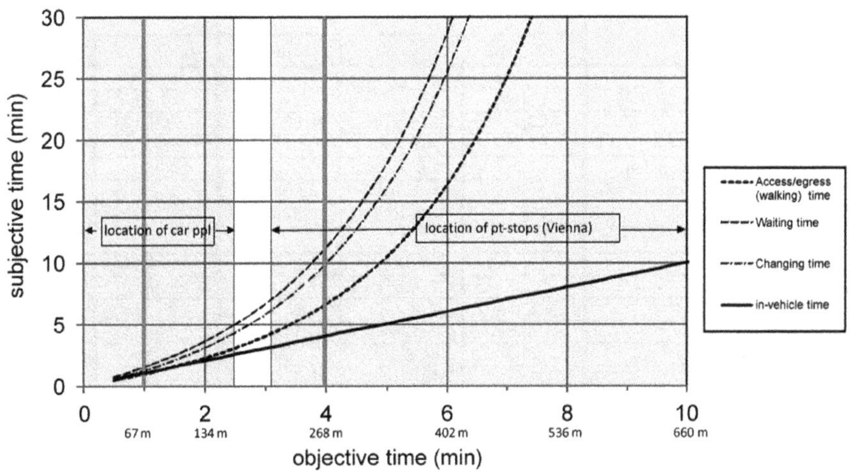

objective walking time of less than 4 minutes. This fact is expressed in the graphic with the slight red box and marked with “location of car parking place (ppl)”. In Viennese context public transport stations are located in average between 200m (~ 3 minutes objective walking time) and up to 700m in the case of subway stops, which correspondents to about 10 minutes walking time, in Figure 5 expressed with the left side box labelled “location of pt-stops (Vienna)”.

The red thick vertical lines express the values used for the following simple arithmetic demonstration example in Table 1.

In the example the following realistic assumptions are made (Viennese context): The trip length is 6 km, for both the pt-trip (public transport-trip) and the car trip example. The travel speed for car is assumed to be 20km/h for the pt-vehicle it is assumed to be 18 km/h. The assumed walking speed is 4 km/h, the parking place is therefore one minute away, to keep the example simple it is assumed that also at the destination the parking place distance is 1 minute or 67m. Additionally a parking place search time of 1 minute is assumed to reflect Viennese circumstances.

For the pt-trip an average distance from origin to pt-stop of 268m is assumed which corresponds to 4 minutes objective walking time. The same assumption was made for the distance from the pt-stop to the final destination. Further a waiting time of objective 3 minutes and a changing time of objective 3 minutes was assumed for the pt-trip.

In the first set of calculations headed “status quo” calculations were made based on existing Viennese context. Here the objective travel time for the car trip results to 21.01 minutes for the pt-trip the objective travel time is 34.04 minutes, which is 62% higher than the car trip time. When applying the subjective weighting factors the travel time for car goes slightly up to 21.82 minutes, this is due to fact that the parking place distance is so small (1 minute) that the exponential weighting factors have nearly no influence at all. For the case of the pt-trip the subjective travel time goes up from before 34.04 minutes to 48.05 minutes. When calculating the ration between the subjective car trip travel time and subjective pt-trip travel time there is an increase of plus 120% (21.82 versus 48.05 minutes). Here it can be seen that in-vehicle time is compared to access/egress, changing and waiting time not so important, and if someone (a transport planer) wants to improve the equality of opportunities between car and pt one have

Table 1. Demonstration example

General assumptions	status quo						eqidistance					
	base data		objective		subjective		base data		objective		subjective	
		Unit		Unit		Unit		Unit		Unit		Unit
trip distance car and pt	6	km					6	km				
car average speed	20	km/h	18.00	min	18.00	min	20	km/h	18.00	min	18.00	min
avergae car parking place distance orign	67	m	1.01	min	0.91	min	268	m	4.02	min	7.35	min
avergae car parking place distance destination	67	m	1.01	min	0.91	min	268	m	4.02	min	7.35	min
avergae car parking place search time destination	1	min	1.00	min	2.00	min	1	min	1.00	min	2.00	min
Sum Traveltime			**21.01**	**min**	**21.82**	**min**			**27.04**	**min**	**34.69**	**min**
pt average speed	18	km/h	20.00	min	20.00	min	18	km/h	20.00	min	20.00	min
average pt stop distance orign	268	m	4.02	min	7.35	min	268	m	4.02	min	7.35	min
waiting time (half of interval)	3	min	3.00	min	7.95	min	0	min	0.00	min	0.00	min
changing time	3	min	3.00	min	5.41	min	0	min	0.00	min	0.00	min
average pt stop distance destination	268	m	4.02	min	7.35	min	268	m	4.02	min	7.35	min
Sum Traveltime			**34.04**	**min**	**48.05**	**min**			**28.04**	**min**	**34.69**	**min**
relation car vs pt (car =1)			**1.62**		**2.20**				**1.04**		**1.00**	

to implement strategies to reduce the access/egress/waiting and changing times instead of improving the free float speed of the vehicles.

In the second set of calculations, headed "equidistance", it was assumed that the distance between the car parking place is the same as for the distance of the pt-stop, namely 4 minutes. In this case the objective travel time for the car trip goes up from 21.01 minutes to 27.04 minutes which is an increase of 29%. When applying the subjective weighting factors the subjective car trip time goes up to 34.69 minutes which results in a plus of 65% increase compared to the objective car trip time of 21.01 minutes.

The last calculation shows the impacts for the case that there is no waiting time at the origin pt-stop and no interchange between different pt-lines necessary. For e.g. in many cities there exist modern mobile phone based applications, such as "Quando"[2] in Vienna, which provide real time information about the pt-transport system for every pt-stop. This enables users to avoid waiting times at the origin pt-stop. Further as shown in Figure 5, the changing time is the subjectively seen most unattractive time when undertaking a pt-trip. So in this final example it is assumed that no change of pt-lines is necessary for the example trip. Comparing now the subjective weighted travel times for car (34.69 minutes) with the subjective weighted travel time for pt (34.69 minutes) it can be seen that both trips now consume the same travel time – equality of opportunities between these two means of transport is reached. Therefore it seems to be a good idea to design a pt-system in that way that it minimizes the number of interchanges.

Summing up – to increase the equality of opportunities between the car and the pt-transport systems it is recommend to adjust the access/egress times for the mode car and the mode pt-transport. This can be done by adaptation of the existing parking regulations and the construction of so called collective garages/parking lots which are located in the same distance as pt-transport stops. A further recommendation is to avoid the necessity of interchanges, and to provide a dense network of public transport stops instead of spending huge amounts of money to increase the in-vehicle speed of public transport systems.

DECISION SUPPORT TOOLS: FOUR STAGE TRANSPORT MODELS (FSM)

In nearly all urban agglomerations around the world FSMs are used to support decision makers regarding transport policy.

From a historical point of view FSMs were designed in the early 1950 and 1960 in the USA to support transport planners to develop the US high way network. These kind of models were at this time mono-modal, they only modelled the car mode.

The FSM is the primary tool for forecasting future demand and performance of a transportation system, typically defined at a regional or sub-regional scale (smaller scales often apply simplified models). The models system was developed for evaluating large scale infrastructure projects and not for more subtle and complex policies involving management and control of existing infrastructure or introduction of policies that directly influence travel behavior. The FSM require a huge amount of time for data collection, model estimation, calibration and subsequent forecasting exercises. During this time the activity and transportation systems change as do policies of interest, which then often requiring new data collection efforts and a new modeling effort. Little time is apparently available for systematic study of the validity of these models (McNally, 2008).

In the 1970 these models begun to incorporate also public mass transit. Their general intention was to forecast the demand for these two means of transport and the major objective of these kind of models was to identify infrastructure bottlenecks, which should be removed to keep the vehicle speed at a certain (high) level. In other words the main focus of transport planning at this time were to ensure the free flow speed.

An interesting summary of criticism on traditional FSM is provided in Kane and Behrens (2002), where a statement from Aktins from the year 1977 (Atkins, 1977) was provided: "we have a series of excessively complicated and expensive models using unsubstantiated and biased techniques to provide information of dubious accuracy for answering the wrong questions."

In response to these criticisms three new analytical methods emerged:

1. Land-use transport interaction models (which modelled the impact of transport changes on the land-use system over time, and vice-versa);
2. Disaggregate methods (which constructed travel-choice models for individuals rather than for households or zones), and
3. Micro-simulation methods, (an improvement on the aggregate assignment procedure, to take account of driver behavior at vehicle or 'platoon' level).

Despite these improvements during the 1970s and early 1980s, practitioners tended to remain with the familiar aggregate four-stage model. In 1986 Atkins compiled another compilation of criticism, where he addressed the redundancy, inefficacy and wastefulness of mainstream methods in transport planning again.

In the 1980s and 1990s further modelling developments emerged in response to the critics:

- Dynamic methods, and
- Activity-based methods.

Dynamic methods were developed in response to the 'static' nature of most preceding approaches. Travel analysis had been based on data from one cross section of time, but critics argued that since transport behavior across generations and sectors of the population vary over time, so transport responses will change over time also.

Activity-based methods emerged in the late 1970s and early 1980s in response to the limited behavioral theory underpinning the conventional aggregate FSM. In activity-based approaches the observation of the trip is replaced by a detailed consideration of the activity which leads to the trip. It was argued that it is not possible to understand (and so to change) travel behavior without a much deeper understanding of the human lives impacted by the transport system under study. Hence activity-based methods are crucially different from aggregate methods in that they focus on household and personal activity scheduling rather than trips as isolated events.

The final period of development during the 1990s saw fundamental changes in transport policy in the developed world, including the passing of the Clean Air Act Amendments (CAAA, 1990) and the Intermodal Surface Transportation Efficiency Act (ISTEA, 1991) in the US. These two pieces of legislation provided for (in the case of CAAA) actions (such as enhanced public transport, the provision of HOV lanes and congestion pricing) to counteract pollution problems, and, (in the case of ISTEA) unprecedented flexibility in the transfer of funds between highway and public transport investment. CAAA and ISTEA challenged the transport professional to be able to accurately forecast transport emissions, and in response to this the US government launched a substantial program of work in transport modelling: The Travel Model Improvement Programme (TMIP).

Despite these developments, and in the face of growing concern over the realism of the forecasts, mainstream practice in local government and in most consultancy practice has remained firmly entrenched in the use of aggregate four-stage models. Concern over this caused the at this time new Professor of Transport Policy at University College London to state in his inaugural lecture "our ability to treat the new policies analytically; to understand their effects; to assess their costs and benefits; is seriously hindered by our inheritance of a tool-kit that is bright, impressive, of unchallengeable intellectual achievement, and wrong" (Goodwin, 1997).

Such statements were part of growing dissenting voices against mainstream practice in the UK during the 1990s. The start of this dissent can be traced back to the 'New Realism' conference of 1991. The conference acknowledged the impossibility of increasing the capacity of the road network to address both the forecast traffic levels, and the environmental problems associated with such traffic. There was a realization (supported by the SACTRA report of 1996) that in congested areas increases in road capacity were likely to induce more traffic than would otherwise be the case. Thus the conclusion was that if supply could not be made to meet demand, then demand would have to be changed to meet supply (Goodwin, 1997). The aggregate methods still widely used in practice had not been designed to evaluate demand-management policies and fell far short of what was required for the development of 'strategic appraisal methods'.

Transport policy instruments which influence the demand for car use or influence the mode choice were at this time instruments which had a direct influence on in-vehicle speed. Later in the 1990ies the term Transport Demand Management (TDM) became very popular and pricing instruments such as road pricing, cordon charges, and parking levies were thought to reduce the demand.

These pricing measures could easily incorporated in the existing FSMs models since the only influenced the calculation of the general costs for a specific trip – so it was possible to model the impacts of pricing measures without any adaption of the existing FSM structure.

But the standard structure of traditional FSM is not designed for simulating the policy instrument parking place distance change. For e.g. in traditional FSM applications the distance from an origin to a parking place and the distance from a parking place to the final destination is an attribute of the underlying zone and is derived from an expert guess based on experience. The used distance is a constant value for each zone and mostly weighted with the constant factor "2", to reflect somehow that there is a difference between walking to/from a parking place and the in-vehicle time. This weighting factor is, taking the research results from Walther (see Figure 5) into consideration alright in existing circumstances where parking places are closer than 130m respective two minutes walk. But as soon someone wants to test the effects of equidistance with a FSM it underestimate the impact of distance on mode choice because of the exponential weighting factor functions and delivers therefore wrong results.

Summing up, although some of the above mentioned principles to weight the different part of a journeys are already partly utilized in similar ways in existing FSM-packages such as VISUM, EMME/2, etc. we here at TU Wien were not satisfied and decided to develop our own land use and transport model called MARS (= Metropolitan Activity Relocation Simulator), where the principles of exponential subjective weighting factors, beside others issues, were incorporated entirely.

WHAT IS MARS?

Background, Conceptual Framework, Theoretical Underpinnings, and Operational Aspects

MARS is an integrated, dynamic land use and transport model developed by the University of Technology Vienna (Emberger, 2012; Emberger & Ibesich, 2006; Jaensirisak, Emberger, Pfaffenbichler, & Ibesich, 2006; Mayerthaler, Haller, & Emberger, 2009; Pfaffenbichler, 2003; Pfaffenbichler, Emberger, & Shepherd, 2010). The basic underlying hypothesis of MARS is that settlements and activities within them are self-organizing systems. MARS is based on the principles of systems dynamics (Sterman, 2000) and synergies (Haken, 1983).

In the mid 1990ties sustainability became a major research concern within the EU and therefore within the EU-funded research, especially in the field of transport. Since transport is a complex field, heavily interwoven with day to day life of transport users, societies and economy, new tools to support decision making were looked for. Traditional modelling approaches used already in transport planning, so called 4-stage transport model (FSM) or even simpler models, such a traffic assignment models were not capable anymore to answer upcoming research question regarding future development of sustainable transport and sustainable societies. Therefore new approaches to develop decision support tools were initiated. At this time a research project, called OPTIMA (OPTIMA, 1996-1997) was launched and there the question which kind of transport policy instrument combination in space and time will deliver the most sustainable settlement structure in the long run was tried to answer. To be able to do so the modelling approaches in the field of transport research were combined with modelling approaches used in land use science. It was recognized that also the time dimension, in other words, the development path over time is important – so the modelling approach included a simulation period of 30 years to show how the final state in the year 30 of the underlying settlement structure was reached. The so called MARS (Metropolitan Activity Relocation Simulator) was born. Since then MARS was further developed and applied in a series of EU and national funded research projects (DISTILLATE, 2006-2008; FATIMA, 1997-1998; GHG Transpord, 2009-2011; PROSPECTS, 2000-2003; SPARKLE, 2004-2006; STEPS, 2004-2006; TRANSPLUS, 2003) and applied in numerous case study cities/regions worldwide.

Within MARS accessibility is used to estimate the attractiveness on one side for people moving from one place to another (= short term mobility - trips for commuting and other purposes) and on the other hand accessibility is used to estimate the future land use (= long term mobility – mobility of housing and workplace movements within the zones).

In more technical terms accessibility as it is used in MARS, is a weighted potential taking into account the transport supply between the origin and destination zones (general travel costs separated for different means of transport, such as walking, cycling, public transport and car traffic), the potential at origin (for e.g. the number of population) and the potential of the destination zone (e.g. the number of workplaces in a certain economic sector such as service or production).

The main innovation in MARS is that the different reaction speeds of the two systems (passenger transport system and land use system) are taken into consideration. These different speeds generate a dynamic process which is looking for a dynamic equilibrium, which is disturbed by external transport or land use policy measures, such as road capacity increases, public transport supply changes or land use instruments, such as e.g. a land value capture tax.

MARS calculates the accessibility for each zone for each means of transport for each simulation period and uses this information to redistribute the daily trips of population as well as the relocation of workplaces and housing within the case study area for a simulation period of 30 years.

MARS is implemented in a System Dynamics Software environment called VENSIM®. The MARS model consists of a transport and land use element and can be divided into a series of sub models as shown in Figure 6. The input data are stored in Excel® to enable an easy handling of the data. The data requirements of MARS are similar to data needs for traditional transport models supplemented with information to describe the land use of a zone.

MARS was/is already implemented for about 20 metropolitan areas around the world (Vienna- A, Madrid- E, Leeds – UK, Strasbourg- B, Ho Chi Minh City – VN, Chiang Mai- TH, Washington DC – USA, Porto Alegre – BR, etc). The setup process takes about 3-5 weeks depending on data availability and number of zones.

Besides other indicators and variables MARS calculates the number of trips on an origin destination basis, the travel times for each means of transport separated for access, in-vehicle time and egress time, the monetary cost for each trip, the fuel consumption, the noise and CO_2 emissions, the accident rate, etc. These values can be displayed for every simulation period, for each individual zone, for each zone pair (origin destination), for each means of transport, for peak and off peak period and can also be cumulated to a system wide sum.

Figure 6. Basic structure of the MARS sub-models

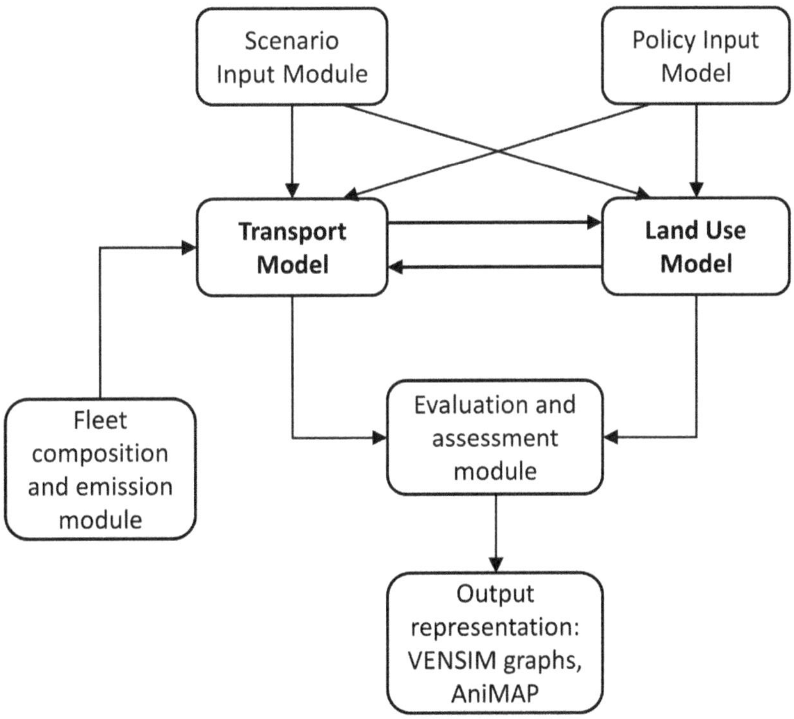

Relevance for Planning Practice

With the MARS policy input module policy profiles over time of certain policy instruments can be handled. The user can specify the start and end-point and the start and end level of any policy instrument implemented in MARS. For example, it is possible to test the impacts of increasing "Public transport fare levels" starting in simulation year 3, with a start value of +20%. Increasing the public transport fare linearly to 50% in simulation year 12 and keep it constant at +50% until the end year of the simulation. So it is possible to specify these kinds of policy profiles for many policy instruments simultaneously. This enables us to test and assess the impacts (synergy effects) of policy combinations over time. Depending on the underlying case studies more than 15 different policy instruments can be combined into one single simulation run.

MARS is a very flexible tool and can be used for different application types, such as:

1. Scenario testing,
2. Policy optimization,
3. Decision makers training.

When applied in the scenario testing mode MARS can be used to forecast the effects of certain policy instrument combinations on a set of output indicators such as energy consumption, mode share, CO_2 emissions, etc.

If MARS is used in the optimization modus a quantitative objective function and a set of changeable policy instruments has to be defined. MARS then will search for the optimal policy instrument combination which delivers the highest possible objective function value.

Finally, MARS can also be set up in the so called "flight simulator mode" where it can be used in workshops to enable attendees (stakeholders) to understand the long term impact of their decisions regarding policy implementations.

Strengths and Limitations

As mentioned before, MARS is implemented using System Dynamics. Within System Dynamics feedback loops which are widely present in the transport and land use system can be made explicit easily. The modelling language used respective the VENSIM programming environment is based on simple symbols and is therefore intuitively understandable also for programming laymen. MARS can be characterized to be a white box model, it can be seen which model entities are taken into consideration and how they are related, and also important, which entities and relations are excluded from the model. This way of presentation enables us/others to discuss implemented relations and change/improve MARS with relatively low effort.

MARS simulates the most likely development path of underlying land use transport systems of metropolitan areas and regions. Because of its low spatial resolution no local analyses are possible.

As it is true for all kind of models the quality of MARS simulations depends on the quality of the underlying input data set. Since for transport and land use modelling a broad variety of input data, such as population, land use data, model split data, travel time and travel cost data and their prediction of their growth rates for the next 30 years are necessary, these forecasts are essential for the forecast results of the MARS case studies. Beside these technical data issues also the lack of trained users of the MARS modelling suite is a potential limitation of the application of MARS. To overcome this limitation training courses for end users and maintenance contracts or MARS users are offered by the developers.

Detailed information regarding MARS and a free download of the software can be found under Verkehrsplanung and Verkehrstechnik (2016).

EQUIDISTANCE: MARS CASE STUDY VIENNA

As already mentioned before MARS was used in several research projects/studies (EIFER, 2010-2013; Emberger, 2012; Mayerthaler et al., 2009; Pfaffenbichler & Emberger, 2005; Pfaffenbichler, Emberger, & Shepherd, 2008; Pfaffenbichler et al., 2010) to test the sensitivity of different individual transport policies and combinations of these instruments on mode choice, GHG emissions, land use and other transport and land use related indicators.

Important is here that MARS was explicitly designed to simulated policy instruments and policy instrument combinations which are with traditional transport models difficult to investigate. In the context of this book chapter results are presented, where beside other policy instruments the impacts of an implementation of equidistance for the case city Vienna will be shown. (Pfaffenbichler & Emberger, 2005).

In this case study the following four scenarios, namely

1. Continue as before,
2. 5 Euro Cordon Charge City centre,
3. 1 Euro/Veh-km distance based charge, and
4. Equidistance were tested.

Scenario 1 presents an extrapolation of status quo, in scenario 2 the implementation of a 5 Euro cordon charge system analogous to London was tested, in scenario 3 the impacts of a very, very high distance based road charge were tested and in scenario 4 the impacts of relocating the car parking places to the same average distance than public transport stops were simulated (about 250m from origins and destinations). In all scenarios the subjective time weighting factors, as described in the former section were applied to estimate the general cost values for the individual mode choice decision.

To save space only the final outcome table of this research exercise is shown here in this publication (Table 2) (For the full text see Panel 3. Sustainable transport and land use, n.d.).

Table 2 summarizes these indicators, which are useful for a comparison of the four scenarios tested in the Vienna case study. It can be seen that the strategies "Continue as before" and "5 Euro cordon charge" are neither sufficient to achieve the greenhouse gas emission targets as defined in the Kyoto protocol (United Nations, 1998), which should be for Austria a reduction of -13% CO_2 emissions from the base year 1990 nor the official Viennese target for car use reduction, which is a reduction of the car share in Vienna until the year 2010 to 25% (Stadtplanung Wien, 1994a, 1994b).

The performance concerning CO_2-emissions of a 1 Euro per vehicle kilometer distance based charge and the new car parking organization is rather similar. The essential difference is that the distance based charge is very costly for the car users. They would have to pay about 28 billion Euro over the 30 years evaluation period. The new car parking organization achieves the same environmental benefits at the expense of time losses for car users worth 3.4 million euros in 30 years.

Table 2. Comparison of different indicators for the tested scenarios

Indicator	Continue as Before	5 Euro Cordon Charge	1 Euro/Veh-km Distance Based Charge	Equidistance
Modal Share Car				
2011	39.9%	36.1%	23.5%	20.2%
2021	41.8%	37.7%	24.6%	20.3%
Change CO_2-Emissions				
2001-2011	+8%	-4%	-50%	-41%
2001-2021	+15%	+2%	-47%	-40%
Car User Benefits (mio. Euro)[1)]				
Costs	+249	-3,511	-27,529	+43
Time	+607	+100	+327	-3,411
Land Consumed by Businesses Relative to Do-Nothing				
2011	+0.1%	-0.1%	-0.2%	-0.7%
2021	+0.6%	-0.3%	-0.8%	-2.7%
2031	+1.4%	-0.5%	-1.2%	-5.2%

1) Results of a cost benefit analyses. Values are discounted over the 30 year modelling period. The discount rate is assumed with 3.5%.

From this research exercise follows that the car parking organization is the only strategy which is strong enough to keep the share of car use in the long term years at about the same level as directly after implementation. In all other strategies the share of car use rises again in the later years. Additionally it seems that the equidistant car parking is the strategy which is most efficient concerning the consumption of land. Nevertheless it has to be mentioned that also with this strategy the amount of built up land increases.

The realization of the equidistance principle is, when introduced to politicians and decision makers always believed to be unrealistic and politically not viable. On the other hand there exist world wide a series of examples where the principle of equidistance is already in place, but under a different name, namely "car free" cities.

Further, to prove the realization of concept of equidistance, subsequent studies have be carried out at the Technische Universität Wien to show the potential transition path step by step within existing build structures (15th district (Gruber, 2009) and 9th district (Blasco, 2007) of the city of Vienna) and what has to be done to change from the existing parking place organization towards the concept of equidistance.

CONCLUSION

Transport related problems are man-made and transport planners and politicians are responsible for them. Undisputed there exists a demand for mobility in our societies, but how this demand could be fulfilled depends on the circumstances provided by transport policy. As long as the objectives of transport policy are to provide opportunities to travel fast and cheap (from an individual point of view) people will behave like these circumstances will allow.

At present, the existing transport system is not sustainable, the majority of burdens and costs and negative effects are externalized and not paid by the user/producer - instead it is paid by the system, the by the nature and by future generations.

One reason for that, beside a series of others, is the use of not adequate decision support tools, such as the application of the existing FSM transportation models.

To reach a sustainable transport system the overall energy consumption of the transport sector has to be reduced. One major step in this direction is the reduction of the overall transport system speed. A relatively easy way to achieve such a speed reduction in an urban context, is as shown, the reorganization of the existing parking place location. Parking places are from a systemic point of view the same as public transport stops, both are the interface between walking and motorized transport. Therefore they have to be treated in the same way! The harmonization of the distances to/from a car parking place and to/from a public transport stop is a cost effective policy instrument, which delivers, as shown in the Vienna case study, developments towards a more sustainable transport system, enhances the attractiveness of public transport and generates at the same time new opportunities in the road space for the (re-) development of local businesses and other social activities which hold societies and cultures together.

Unfortunately existing FSMs are not able to simulate these relations in system context and therefore transport engineers cannot see the potential behind the concept of this new parking organization.

Therefore tools such as MARS, which are designed to take all kind of system immanent feedbacks into consideration, seem to be much better adjusted to support decision makers, since they provide a fuller picture of the underlying transport and land use system.

REFERENCES

Albert, G., & Mahalel, D. (2006). Congestion tolls and parking fees: A comparison of the potential effect on travel behavior. *Transport Policy*, *13*(6), 496–502. doi:10.1016/j.tranpol.2006.05.007

Asian Development Bank. (2008). *The Guidelines and Toolkits for Urban Transport Development*. Retrieved from http://sti-india-uttoolkit.adb.org/mod4/se5/006_2.html

Atkins, S. (1977). Transportation planning: Is there a road ahead? *Traffic Engineering & Control*, *18*(2).

Attoney General's Chambers. (2010). *Singapiore Statutes Onlie - Legislation History*. Retrieved from http://statutes.agc.gov.sg/non_version/cgi-bin/cgi_getdata.pl?actno=2004-REVED-276&doctitle=ROAD%20TRAFFIC%20ACT%0a&date=latest&method=part&sl=1&segid=1104745820-005150

Beaumont, H. (2002). *SDS Maximum Parking Standards: Derivation of PTAL-based parking restraint*. SDS Technical Report Twenty.

BMVIT. (2002). *Generalverkehrsplan Österreich 2002 (GVP-Ö 2002) verkehrspolitische Grundsätze und Infrastrukturprogramm; Bericht Januar 2002*. Bmvit.

BMVIT. (2013). *Gesamtverkehrsplan für Österreich*. BMVIT.

City of Vienna. (2008). Vienna Garage Ordinance 2008. *Wiener Garagengesetz 2008 - WGarG 2008*. Retrieved from http://www.wien.gv.at/recht/landesrecht-wien/rechtsvorschriften/html/b1000000.htm

Commission of the European Communities. (2001). *White Paper - European Transport Policy for 2010: Time to decide*. Retrieved from http://ec.europa.eu/transport/white_paper/documents/doc/lb_texte_complet_de.pdf

Davidge, W. R. (1924). *The London Traffic Act, 1924*. Liverpool University Press.

Department for Transport (DfT). (2000). *Transport 2010 - The 10 year Plan*. Author.

Development & Building Control Division Roads Group. L. T. A., Singapore. (2005). *Handbook on VEHICLE PARKING PROVISION in development proposals*. Retrieved from http://www.distillate.ac.uk/about/about.php

EIFER. (2010-2013). *Software development and use Agreement Software regarding MARS - Application for Strasbourg and Mulhouse (France)*. EIFER.

Emberger, G., & Ibesich, N. (2006). MARS in Asia - How a model can help and influence decision makers. *Proceedings of CORP 2005, 11th International Conference on Urban Planning & Regional Development in the Information Society,* 193-200.

European Commission Directorate-General Energy and Transport. (2011). *White paper on Transport - Roadmap to a single European transport area — Towards a competitive and Resource efficient transport system*. doi:10.2832/30955

FATIMA. (1997-1998). *Financial Assistance for Transport Integration in Metropolitan Areas*. Retrieved from http://www.ivv.tuwien.ac.at/forschung/projekte/international-projects/fatima-1997.html

Garcia Blasco, M. (2007). *Eine Innovative Verkehrsorganisaion für den 9. Bezirk (Master)*. Wien: TU Wien.

GHG Transpord. (2009-2011). *EU-Project, 7th FP.* Retrieved from https://tiss.tuwien.ac.at/research/project.xhtml?projectId=8402

Goodwin, P. (1997). Solving congestion: When we must not build roads, increase spending, lose votes, damage the economy or harm the environment, and will never find an equilibrium. Inaugural lecture for the Professorship of Transport Policy, University College London.

Greater London Authority. (2008). *The London Plan, Spatial Development Strategy for Greater London Consolidated with Alterations since 2004.* Retrieved from www.london.gov.uk/thelondonplan

Gruber, J. (2009). Verkehrskonzept 15: zeitlich gestaffelte Umsetzung des Äquidistanzmodells am Beispiel Wien 15. *Verkehrskonzept, 15.*

Haken, H. (1983). *Advanced Synergetics - Instability Hierarchies of Self-Organizing Systems and Devices* (Vol. 20). Springer-Verlag.

Hong Kong Transport Department. (2002). *The Second Parking Demand Study*. Hong Kong: Author.

Jaensirisak, S., Emberger, G., Pfaffenbichler, P. C., & Ibesich, N. (2006). *Application of the Metropolitan Activity Relocation Simulator (MARS). Model in Decision Making Process for Sustainable Development.* Paper presented at the Technology and Innovation for Sustainable Development Conference - TISD, Khon Kean, Thailand.

Jittrapirom, P., & Emberger, G. (2011). Off-street residential parking organisation - a review of current practices. *Proceedings of the Eastern Asia Society for Transportation Studies*, *8*, 16.

Kane, L., & Behrens, R. (2002). *Transport Planning Models - An Historical and Critical Review.* Paper presented at the 21st Annual South African Transport Conference "Towards Building Capacity and Accelerating Delivery", South Africa.

Kelly, J. A., & Clinch, J. P. (2009). Temporal variance of revealed preference on-street parking price elasticity. *Transport Policy*, *16*(4), 193–199. doi:10.1016/j.tranpol.2009.06.001

Knoflacher, H. (1980). Zusammenhang zwischen Parkraumausstattung bzw. -organisation und Einwohner- sowie Arbeitsplatzentwicklung bzw. -verschiebung. *Straßenverkehrstechnik,* (1), 1-7.

Knoflacher, H. (1981). *Human Energy Expenditure in Different Modes: Implications for Town Planning.* Paper presented at the International Symposium on Surface Transportation System Performance.

Knoflacher, H. (1997). *Landschaft ohne Autobahnen. Für eine zukunftsorientierte Verkehrsplanung.* Wien: Böhlau Verlag.

Knoflacher, H. (2001). *Stehzeuge. Der Stau ist kein Verkehrsproblem.* Wien: Böhlau Verlag.

Kovacic, W. (1999). *Der österreichische Bundesverkehrswegeplan - Abgeschlossene und laufende Arbeitspakete.* Academic Press.

Litman, T. A. (2013). *Transport Elasticities: Impacts on Travel Behaviour.* Retrieved from Bonn, Germany: http://www.sutp.org/files/contents/documents/resources/B_Technical-Documents/GIZ_SUTP_TD11_Transport-Elasticities_EN.pdf

Lucas, K. (2012). Transport and social exclusion: Where are we now? *Transport Policy*, *20*, 105–113. doi:10.1016/j.tranpol.2012.01.013

Magistrat der Stadt Wien. (2003). Masterplan Verkehr Wien 2003. Wien: Im Auftrag des Magistrats der stadt Wien Magistratsabteilung 18 - Stadtentwicklung und Stadtplanung, Referat Verkehrsplanung und Regionalentwicklung.

Marsden, G. (2006). The evidence base for parking policies—a review. *Transport Policy*, *13*(6), 447–457. doi:10.1016/j.tranpol.2006.05.009

Mayerthaler, A., Haller, R., & Emberger, G. (2009). *Modelling land-use and transport at a national scale - the MARS Austria model.* Paper presented at the 49th European Congress of the Regional Science Assosiation International - Territorial Cohesion of Europe and Integrative Planning, Lodz, Poland.

McNally, M. G. (2008). The Four Step Model. In D. A. H. K. J. Button (Ed.), Handbook of Transport Modelling (2nd ed.). Amsterdam: Elsevier.

Milosavljević, N., & Simićević, J. (2016). User response to parking policy change: A comparison of stated and revealed preference data. *Transport Policy*, *46*, 40–45. doi:10.1016/j.tranpol.2015.11.007

Morikawa, S., Asao, K., Igo, T., & Kato, H. (2010). *Institutional system and current problems of car parking in Tokyo*. Japan: Academic Press.

Ojeda-Cabral, M., & Chorus, C. G. (2016). Value of travel time changes: Theory and simulation to understand the connection between Random Valuation and Random Utility methods. *Transport Policy*, *48*, 139–145. doi:10.1016/j.tranpol.2016.03.006

OPTIMA. (1996-1997). *Optimisation of Policies for Transport Integration in Metropolitan Areas*. Retrieved from http://www.ivv.tuwien.ac.at/forschung/projekte/international-projects/optima-1995.html

Panel 3. Sustainable transport and land use. (n.d.). Retrieved from http://www.eceee.org/library/conference_proceedings/eceee_Summer_Studies/2005c/Panel_3

Peperna, O. (1982). *Die Einzugsbereiche von Haltestellen öffentlicher Nahverkehrsmittel im Straßenbahn- und Busverkehr (Master thesis)*. Technische Universität Wien.

Pfaffenbichler, P. (2003). The strategic, dynamic and integrated urban land use and transport model MARS (Metropolitan Activity Relocation Simulator) - Development, testing and application (Doctoral thesis). Vienna: Vienna University of Technology. Retrieved from http://www.ivv.tuwien.ac.at/publications/online/MARS_smallest_size.pdf

Pfaffenbichler, P. C., & Emberger, G. (2005). A reform of car parking organisation as an instrument for creating sustainable urban regions. *ECEEE 2005 Summer Study*.

Pfaffenbichler, P. C., Emberger, G., & Shepherd, S. P. (2008). The Integrated Dynamic Land Use and Transport Model MARS. *Networks and Spatial Economics*, 3.

Pfaffenbichler, P. C., Emberger, G., & Shepherd, S. P. (2010). A system dynamics approach to land use transport interaction modelling: The strategic model MARS and its application. *System Dynamics Review*, *26*(3), 262–282. doi:10.1002/sdr.451

Planning Department of Hong Kong. (2006, June). *Parking Standard*. Retrieved from http://www.pland.gov.hk/pland_en/tech_doc/hkpsg/full/ch8/ch8_tbl_11.htm

Preston, J. (2009). Epilogue: Transport policy and social exclusion—Some reflections. *Transport Policy*, *16*(3), 140–142. doi:10.1016/j.tranpol.2009.04.003

PROSPECTS. (2000-2003). *PROSPECTS - Procedures for Recommending Optimal Sustainable Planning of European City Transport Systems*. Retrieved from http://www.ivv.tuwien.ac.at/forschung/projekte/international-projects/prospects-2000.html

Rojo, M., dellOlio, L., Gonzalo-Orden, H., & Ibeas, Á. (2015). Inclusion of quality criteria in public bus service contracts in metropolitan areas. *Transport Policy*, *42*, 52–63. doi:10.1016/j.tranpol.2015.05.002

Román, C., Martín, J. C., Espino, R., Cherchi, E., Ortúzar, J. D., Rizzi, L. I., & Amador, F. J. et al. (2014). Valuation of travel time savings for intercity travel: The Madrid-Barcelona corridor. *Transport Policy*, *36*, 105–117. doi:10.1016/j.tranpol.2014.07.007

Simićević, J., Vukanović, S., & Milosavljević, N. (2013). The effect of parking charges and time limit to car usage and parking behaviour. *Transport Policy*, *30*, 125–131. doi:10.1016/j.tranpol.2013.09.007

SPARKLE. (2004-2006). *Sustainability Planning for Asian Cities making use of Research, Know-How and Lessons from Europe*. Retrieved from http://www.ivv.tuwien.ac.at/forschung/projekte/international-projects/sparkle-2004.html

Stadt Wien - Stadtentwicklung. (2002). *Masterplan Verkehr Wien Positionspapier* (Vol. Nr. 43). Wien: Stadtentwicklung Wien, Magistratsabteilung 18.

Stadtplanung Wien. (1994b). *Verkehrskonzept Wien - Generelles Maßnahmenprogramm*. Author.

Stadtplanung Wien. (1994a). *step 1994 - Stadtentwicklungsplan für Wien*. Author.

Steininger, G. (2008). *Ziele und Vorgaben der österreichischen Verkehrspolitik in den Koalitionsabkommen der Bundesregierung - ein Zeitvergleich 1945-2007. Schriftenreihe des Instituts für Transportwirtschaft und Logistik - Verkehr, 02/2008. Institut für Transportwirtschaft und Logistik*. Vienna: WU Vienna University of Economics and Business.

STEPS. (2004-2006). *Scenarios for the Transport system and the Energy supply and their Potential effectS*. Retrieved from EU-Project, 6th FP: http://www.steps-eu.com/

Sterman, J. D. (2000). *Business Dynamics - Systems Thinking and Modeling for a Complex World*. McGraw-Hill Higher Education.

TRANSPLUS. (2003). *D4 Barriers, Solutions and Transferability*. Retrieved from http://www.ivv.tuwien.ac.at/forschung/projekte/international-projects/transplus-2000.html

United Nations. (1998). *Kyoto Protocol -Report of the conference of the partners on its third session, held at Kyoto from 1 to 11 December 1997, Addendum, Part two: Action taken by the conference of the parties at its third session, Framework Convention on Climate Change, FCCC/CP/1997/Add.1*. UN.

United Nations. (2015). *Framework Convention on Climate Change - Paris*. UN.

Verkehrsplanung & Verkehrstechnik. (2016). *MARS (Metropolitan Activity Relocation Simulator)*. Retrieved from http://www.ivv.tuwien.ac.at/forschung/mars-metropolitan-activity-relocation-simulator/overview/

Walther, K. (1973). *Nachfrageorientierte Bewertung der Streckenführung im ÖPNV*. Frankfurt/Oder: Fakultät für Bauwesen, TH-Aachen.

Walther, K. (1991). Maßnahmenreagibler Modal-Split für den städtischen Personenverkehr - Theoretische Grundlagen und praktische Anwendung (Vol. 45). Aachen: Academic Press.

Walther, K., Oetting, A., & Vallée, D. (1997). *Simultane Modellstruktur für die Personenverkehrsplanung auf der Basis eines neuen Verkehrswiderstands* (Vol. 52). Aachen: Academic Press.

Westminster City Council. (2010). *Unitary Development Plan*. Retrieved from http://www.westminster.gov.uk/services/environment/planning/unitarydevelopmentplan/

Wheat, P., & Batley, R. (2015). Quantifying and decomposing the uncertainty in appraisal value of travel time savings. *Transport Policy*, *44*, 134–142. doi:10.1016/j.tranpol.2015.06.010

KEY TERMS AND DEFINITIONS

FSM: Four Stage Model.
GHG: Greenhouse gases.
LUTI: Land Use and Transport Interaction Model.
MARS: Metropolitan Activity Relocation Simulator.
PPL: Car parking place.
PT: Public transport.

ENDNOTES

1 A detailed comparison of existing parking regulations around the world can be found in Jittrapirom and Emberger (2011).

2 Please see for further details http://www.wienerlinien.at/eportal3/ep/channelView.do?pageTypeId=66526&channelId=-46623 and http://www.wienerlinien.at/eportal3/ep/tab.do?tabId=0

Chapter 7
Effects on Car Ownership Rates Resulting from Increased Parking Lots in Residential Areas:
The Case of Gated Communities

Leyla Alkan
Gazi University, Turkey

ABSTRACT

New development areas impose higher parking supply than needed, and these excessive parking requirements impose the rise of car usage in cities. In this study, it is aimed to analyze the relationship between development patterns of residential areas and supply of parking spaces by focusing gated communities in Ankara. The results of the study revealed that the number of garage parking spaces is positively related to the developments of gated communities, and the ratio of having private open parking space is lesser for the non-estate housing. The results of the analysis also revealed that the ratio of having a garage is higher for new buildings, and house prices are affected significantly by the presence of both garage and open parking spaces. In the study, it is interesting to note the rise of number of road motor vehicles in Turkey after the developments of gated communities started to emerge in the 1990s. Also, the results of the study revealed that there is a positive relationship between the emergence of gated communities and the rise in the number of motor vehicles in Ankara.

INTRODUCTION

With the increasing urban population in Turkey, environmental problems such as traffic congestion, noise and air pollution have witnessed a significant rise, especially in the country's central residential areas. In Ankara, the capital of Turkey, even the minimum parking requirements cannot be satisfied in these old residential areas with the increasing rate of car ownership, with off-street parking all but absent in most

DOI: 10.4018/978-1-5225-2116-7.ch007

central residential areas, and on-street parking restricted due to limited space and narrow streets. Most residents struggle to find a single space for on-street parking, and so traffic accidents occur frequently. The limited parking brings with it also a security problem, as the risk of theft is greater when cars are parked on the streets. This has led to a greater demand for parking in the vicinity of the home, which has become a significant problem for many of the residents of Ankara's central areas, making them less attractive for occupation. As a result, households have started to look for new options and new locations in which to reside at the periphery of the city.

Gated communities constitute a significant proportion of such newly developed residential areas, where the purpose is to protect residents from such problems as violence, noise, etc. by having strictly controlled entrances for the passage of pedestrians and automobiles. These communities feature many services, such as sports facilities, green areas, off-street parking, etc., and their design is car-oriented, which privatize public spaces by prohibiting the access of public transport. Most of these residential estates provide a level of parking areas equal to around two spaces per dwelling in order to satisfy the higher parking demands of their residents, creating a greater than needed supply of private parking.

In this study, it is assumed that there is an association between development patterns of residential areas and the travel behaviors of households; that the higher supply of parking spaces generates a higher rate of private car ownership, and more vehicle trips than roads can carry; and consequently, that more cars create more traffic congestion. The aim of the study is to analyze whether residential parking supply has a causal influence on car ownership and usage, with particular focus on gated communities in Ankara. It is argued that gated communities are car-oriented developments with limited public transport connectivity, and encourage car usage by supplying more parking spaces than are required.

In order to test these assumptions, this study uses two data sets. The first of these is based on an analysis of data related to housing in both local residential areas and gated communities on sale in different neighborhoods of Ankara, selected randomly from one of Turkey's largest real estate websites, which provides such information about dwelling units as price, parking availability, production date, etc. The second data set is based on information garnered from the website of the Turkish Statistical Institute, which provides information on the number of road motor vehicles by years. Cross-table analyses are used to define the relationship between the excess parking supply and the sprawl of residential areas, and to investigate whether the emergence of gated communities has influenced the rise of car ownership by providing more parking spaces. Additionally, correlation analyses are used to examine the effects of the presence of parking areas on house prices.

The article begins with a review of previous literature related to residential parking demand, while the following section provides a summary of gated communities, followed by a definition of the data and general descriptive statistics. In the fourth section, the results of the analyses are presented, and the final section concludes the study.

RESIDENTIAL PARKING DEMAND

Modern cities under the influence of car oriented policies suffer from uncontrolled urban growth, and consequently settlements return to mono-functional urban forms (Knoflacher & Ocalir, 2011). Urban sprawl is considered by most environmental scientists and urban planners to be a serious environmental problem (Davis et al., 2010), and parking policy is an important link between transport and land-use policies (Marsden, 2006) and urban growth. There is growing awareness that urban form has significant effects on travel behaviors and car usage.

According to Willson (1995), high minimum parking requirements affect mode choice, site density, land value and urban design, which may lead to changes in the urban form and the transportation system. He argues that when parking supply exceeds demand, it tends to be treated like a costless good, and because parking is free to the motorist, solo driving is encouraged, increasing both traffic congestion and air pollution. He indicates further that in a lower-density environment, the automobile provides advantages in door-to-door travel times, although higher levels of automobile commuting can produce more traffic congestion per unit of development. In this regard, according to him, allowing parking requirements to control the amount and location of growth actually makes some of the problems of growth even worse, since it leads to an urban form in which the automobile is the only convenient mode of travel. Shoup (1997) argues that planners require developers to provide more parking, however minimum parking requirements make the original problem even worse. According to him, minimum parking requirements have transformed many residential streets into garagescapes, and the minimum parking requirements have forced development to subsidize cars. As a result, planners must then limit the density of development (and of people) to limit the density of cars. According to Shoup (1999), minimum parking requirements increase development costs, and shape land use, transportation and the urban form, while also increasing the supply and reducing the price of parking, but not its cost. Urban planners rarely consider the cost of parking requirements, just as developers rarely have the luxury of not considering this cost (Shoup, 1999). Knoflacher (2006) argues that the provision of parking at home turns people into car drivers, meaning that parking facilities are needed also at the workplace and where people go to shop and for recreation. According to him, providing parking at home and at all of these destinations requires large amounts of space, and if the walking distance to the public transport stop is further than to where the car is parked, the average person will use their car. Knoflacher indicated also that transport literature accepts the existing parking regulations that stipulate the need for parking places close to all activities and facilities, which goes against the idea of a sustainable city. Weinberger (2012) identified a clear relationship between guaranteed parking at home and a greater propensity to use the automobile for commuting. According to Guo (2013), parking supply can influence significantly household car ownership decisions. Guo argues that although all off-street parking types are significant, driveway spaces seem to be more important than garage spaces, and if parking policies can reduce the residential parking supply, a maximum off-street parking standard could significantly reduce car dependency and mitigate the associated externalities, such as congestion and greenhouse gas emissions.

The planning experience of Ankara contributes to these arguments. Ankara, the capital city of Turkey, was a high-density development and a rather compact form in the 1970s. Most businesses, services, and amenities were located in the city center, and urban growth took place at the fringe, resulting in continuous spatial growth. Although densities were still relatively high, the urban area stated to become difficult to manage and to serve effectively with public transport. The compactness of the urban form resulted in an intensification of such problems as traffic congestion and air pollution, and so planning studies from the 1970s onwards proposed two linear corridors of development for the future urban form of Ankara along the western and south-western corridors. There were notable differences in the nature of the residential developments along the two corridors. While the western corridor housed many large-scale mass-housing projects funded by the government, the south-western corridor was left to market forces, and saw much market-led residential development (Babalik-Sutcliffe, 2013).

Most transport studies of Ankara focus on the link between the public and private transportation system and land use policies, but overlook the link between parking provision and the rise in car ownership. Evidences strongly suggest that land-use characteristics have a significant impact on travel behaviors, which should be examined with focus on different residential neighborhood types. This study hypothesizes that gated communities, by providing a high level of parking supply, affect car ownership rates and usage in Ankara.

GATED COMMUNITIES

With the technical innovation in the transport sphere, increasing numbers of people have gained access to near effortless movement, which appears to have enhanced mobility and saved time. Technological innovation and economic effects have driven the development of a vast network of railways, roads, motorways and airports to move people, goods and information faster (Knoflacher, 2006). As a consequence of these developments in transportation system, new housing projects have dispersed out from the center of the city to the periphery to satisfy the demands of people who want to escape Turkey's crowded and unsafe city centers. Some of these developments take the form of exclusive housing projects, such as gated communities, which are car-oriented developments that promise a comfortable lifestyle with several amenities close at hand.

The number of gated communities for high level income groups has increased dramatically in Ankara in recent years. While there is a broad diversity in these housing developments in Ankara in terms of their type, size and the services and facilities they provide, all are expensive, and are equipped with high-end luxury technologies (Alkan, 2015). Gated communities can be defined as physical areas that are fenced or walled off from their surroundings, with the entrances to these areas controlled by means of gates (Landman, 2000a). These are closed urban residential areas where public space has been legally privatized by restricting access. They include private property, individual houses and common but shared private areas, for example playgrounds and sports facilities. They have security elements such as walls, fences, gates, barriers, alarms, guards, cameras, etc., and by and large, the infrastructure and services are of a high quality. They are designed with the intention of providing security to their residents and to prohibit access to non-residents. Their residents come mostly from the upper- and middle-income groups (Roitman, 2005).

Different gated communities have different physical impacts upon the urban space, including the closure of streets and the fragmentation of space (Roitman, 2005). Gated communities have private roads that are closed to general traffic with gates across the primary access. By surrounding the area with fences, walls or other natural barriers, public access is limited. The gating of emergency exit or access routes has turned many private roads into veritable cul-de-sacs, forbidding through traffic. Signs typically post lower than normal speeds, and alert motorists that they are entering private property. Parking is carefully controlled, and the streets in gated enclaves are narrow, quiet, well maintained and safe (Grant, 2005).

Gated communities are enclosed neighborhoods that provide a physical separation between a specific area and its environs as zones or pockets of restricted access within the urban fabric. This forces motorists and pedestrians to take alternative routes, which can often be longer. Gated communities, in this regard, have an impact not only on daily public activity patterns, but also on the urban form and it's functioning. Road closures may cause some other problems, such as the functioning of the emergency services, and many of the road and neighborhood closures may not allow sufficient turning space for large cleansing and maintenance vehicles (Landman, 2000b).

This phenomenon of gated communities, as a new stage in the urbanization experience, has become the most popular and attractive form for new residential areas for among the higher-income group, and this is true also for Turkey (Kurtulus, 2011). There are two types of gated developments in Ankara:

- Multi-unit housing developments surrounded by high fences or walls. The majority of these new investments, built on large suburban land plots as a group of detached or semi-detached houses, share some common private facilities such as playgrounds, green areas, and sports facilities (swimming pools, tennis courts, etc.). These developments are relatively isolated from the rest of the city, in that the facilities are provided only for the residents, and tend to be located far from the city center.
- High-rise residential towers. These tend to be located in the more central neighborhoods, where the building plots tend to be smaller. Most of these towers are mixed-use, containing both residential units and offices or commercial spaces, offering residents the opportunity to work within walking distance of their homes. For this reason they tend to be constructed in close proximity to the business districts.

This study contains information about both types of developments in Ankara.

DATA AND GENERAL DESCRIPTIVE STATISTICS

The data used in the analysis was collected from one of the largest real estate websites (Housing asked prices in Ankara, 2016), providing information about dwelling units on sale between January 2016 and February 2016 in different neighborhoods in Ankara. The sample comprised 552 dwelling units, spread across eight districts. For each dwelling unit, data was garnered on the asked prices and the different attributes of each dwelling unit. Table 1 presents a description of the variables in the data.

The first explanatory variable is the DISTRICT in which the housing unit is located. The analysis was made for eight central districts in Ankara, namely Altındağ, Çankaya, Etimesgut, Gölbaşı, Keçiören, Mamak, Sincan and Yenimahalle. The second explanatory variable denotes the neighborhood, and the data contains cases from 206 different neighborhoods. DISTANCE refers to the approximate distance (Kilometers) of the housing unit from the city center (Güvenpark/Kızılay). ROOM indicates the number of rooms per housing unit, while SIZE is the floor area of the dwelling. Another explanatory variable is AGE, which denotes the age of the building. The data also contains information on the number of bathrooms in the dwelling unit, which are expected to be increased in newer housing stock. The next variable is NO_OF_STOREYS which indicates total number of the storeys in the housing unit, and PRICE is the estimated price of the dwelling unit.

There are two types of dwelling unit in the data that are classified under the variable of TYPE. The first one is detached or semi-detached housing, and the second one is apartment unit. Different neighborhoods feature different housing patterns in Ankara. In central locations there are few developments of detached or semi-detached housing, as land prices are higher in these locations. Instead, the common housing pattern in central areas is apartment blocks, located on single plots. Clusters of detached or semi-detached housing are more common in the suburban neighborhoods, and there are also clusters of

Table 1. Variable description

Name	Description
DISTRICT	District in which the housing unit is located
NEIGHBORHOOD	Neighborhood in which the housing unit is located
DISTANCE	Approximate distance of the housing unit from the city center (km) (Güvenpark/Kızılay)
ROOM	Total number of rooms in the dwelling unit
SIZE	Floor area of the dwelling unit (m^2)
AGE	Age of the building in which the unit is located
BATHROOM	Number of bathrooms in the dwelling unit
NO_OF_STOREYS	Total number of storeys in the dwelling unit
PRICE	Estimated Price of the dwelling unit (,000 TL/m2)
TYPE	Type of dwelling unit (1= Detached or Semi-detached, 2=Apartment Unit)
SITE	Type of residential site (0= Non-Estate Housing, 1=Estate Housing, 2= Gated Community)
GARAGE	Dummy equal to 1 if housing unit has a garage
OPEN_PARKING	Dummy equal to 1 if housing unit has private open parking spaces

mixed types of housing units, including detached-semi-detached houses and apartment blocks located together on a single large plot.

As a result of increasing crime rates in the city center of Ankara there has been a rise in demand for housing units with controlled entrances. This has led a proportion of the population to move to newly developing housing areas in the suburbs, where this demand is being satisfied, although there is also an increasing trend in the central neighborhoods for new apartment blocks that are constructed with security fences and controlled entrances. This study contains information about whether the housing unit has controlled entrances, with housing units coded as Estate Housing if the housing unit features controlled entrances, and non-estate housing if there is not controlled entrance. In addition to these two categories, there is a third case in the data that is classified under the heading of gated communities. In fact, a gated community can be defined as a form of housing estate, although with more strictly controlled entrances, with higher walls or fences, and featuring security guards and security cameras throughout the community. In addition to controlled entrances, these communities tend to offer additional amenities such as sport facilities, swimming pools, playgrounds, gymnasiums, etc., creating a more secured and isolated environment than estate housing. SITE in Table 1 refers the categorization of these three groups in the data.

The purpose of this study is to examine whether developments of gated communities have a causal influence on parking supply by focusing on different residential developments in Ankara, and so the data also contains information on the private parking supplied with different housing types. GARAGE is a dummy variable that refers to whether the housing unit has a private garage, while OPEN_PARKING denotes whether the housing unit has private open parking lots.

Table 2 provides information on the districts of Ankara in which the data was collected, and Figure 1 shows the spatial distribution of these districts in Ankara. Çankaya is the central metropolitan district in which Kızılay, the central business district, is located. Altındağ is the old city center, and has several archeological, historical and cultural places. Mamak is another district that is close to the CBD. Yenima-

Table 2. Districts

	Frequency	Percent	Valid Percent	Cumulative Percent
Altındağ	32	5.8	5.8	5.8
Çankaya	235	42.6	42.6	48.4
Etimesgut	74	13.4	13.4	61.8
Gölbaşı	30	5.4	5.4	67.2
Keçiören	53	9.6	9.6	76.8
Mamak	34	6.2	6.2	83.0
Sincan	32	5.8	5.8	88.8
Yenimahalle	62	11.2	11.2	100.0
Total	552	100.0	100.0	

Figure 1. Central districts of Ankara

halle is located in the north of the city, and has Ankara's largest industrial area on its edge. Keçiören is a crowded district in the northern part of the city. Etimesgut is host to large housing projects, and Sincan and Gölbaşı are suburban districts that are far from the city center.

Table 3 presents the approximate distance between the neighborhood in which each housing unit is located and the city center. The distance is the shortest distance between the headman's offices in the neighborhood and Güvenpark in Kızılay. The table shows that the group means for DISTANCE are slightly different for all groups, with the distance is being highest for Estate Housing and lowest for Non-Estate Housing. In Ankara, detached or semi-detached housing projects are usually more popular on the outskirts of the city in the form of estate housing, where the population density is lower; apartment housing and multi-story towers, on the other hand, take advantage of smaller plots closer to the city center, as discussed previously.

Table 3. Approximate distance between the housing unit and the city center (km)

Distance (km.)	N	Minimum	Maximum	Mean	Std. Deviation
Non-Estate Housing	310	0.93	30.47	8.55	6.67
Estate Housing	121	2.42	42.46	13.50	6.75
Gated Community	121	2.16	22.56	11.51	5.96

Güvenpark/Kızılay.

In recent years the characteristics of all types of housing unit have started to change in Ankara. Of particular interest is the fact that the increasing number of high rise towers in the city center has brought with it a dramatic rise in the number of studio-type apartments (with only one bedroom), although the average number of rooms in properties has been climbing in recent years. The number of bathrooms is also on the rise in new developments. Prior to the 2000s it was the norm to have only one bathroom, shared by all the family, however most houses now have at least two (Alkan, 2014), with at least a separate bathroom for the bedrooms. Table 4 shows that there is no significant difference in the number of rooms of the different housing types, though gated communities tend to have more bathrooms.

The average floor area and the total number of storeys has also tended to rise over the years (Alkan, 2014). Table 5 shows that the average floor area is greater for the dwelling units in gated communities, while the averages are similar for Non-Estate and Estate Housing. Table 6 presents more remarkable differences in terms of the total number of storeys in the buildings. As would be expected, gated communities have the highest average in terms of the total number of storeys, while the figure is lowest for non-estate housing.

Table 7 provides information on the AGE of the buildings across the different housing sites, and indicates that the development of gated communities is a more recent phenomenon than estate housing.

Table 4. Total number of rooms and bathrooms in the dwelling unit

Room and Bathroom		N	Minimum	Maximum	Mean	Std. Deviation
Non-Estate Housing	Room	310	1	9	4.40	1.42
	Bathroom	310	1	4	1.54	0.71
Estate Housing	Room	121	2	8	4.38	1.39
	Bathroom	121	1	4	1.80	0.71
Gated Community	Room	121	1	10	4.48	1.90
	Bathroom	121	1	7	2.15	1.15

Table 5. Floor area of the dwelling (m^2)

Size	N	Minimum	Maximum	Mean	Std. Deviation
Non-Estate Housing	310	45	400	154.42	72.42
Estate Housing	121	40	400	162.89	77.12
Gated Community	121	30	750	206.67	142.32

Table 6. Total number of storeys in the housing unit

No_of_Storeys	N	Minimum	Maximum	Mean	Std. Deviation
Non-Estate Housing	302	1	15	3.95	2.01
Estate Housing	120	2	28	8.61	4.87
Gated Community	117	2	46	15.49	9.77

Table 7. Age of the building in which the unit is located

Age	N	Minimum	Maximum	Mean	Std. Deviation
Non-Estate Housing	310	0	48	10.81	10.65
Estate Housing	121	0	38	7.65	8.04
Gated Community	121	0	22	2.88	4.68

The data suggests that gated communities have existed for 22 years in Ankara, although the average age of the buildings is around three years.

Table 8 presents the distribution of types of buildings. In the data, detached or semi-detached houses are uncommon in non-estate housing, and estate housing and gated communities have similar ratios in terms of the type of buildings they contain.

Table 9 provides information on the estimated price of the dwelling units, garnered from a real estate website (Housing asked prices in Ankara, 2016) in Turkey, providing information about dwelling units on sale between January 2016 and February 2016. As expected, prices are highest in gated communities, and lowest in non-estate housing. As discussed previously, households in Ankara started to look for housing units with controlled entrances due to the rising crime rates in the city, which brought about an increase in the price of dwelling units with additional security features. In addition to security, gated communities also provide facilities for their residents, resulting in additional cost, and consequently, higher prices in the housing market, as the table illustrates.

Table 8. Type of dwelling unit

Type		Frequency	Percent
Non-Estate Housing	Detached or Semi-detached	14	4.5
	Apartment Unit	296	95.5
Estate Housing	Detached or Semi-detached	20	16.5
	Apartment Unit	101	83.5
Gated Community	Detached or Semi-detached	18	14.9
	Apartment Unit	103	85.1

Table 9. Estimated price of the dwelling unit

Price (,000 TL)	N	Minimum	Maximum	Mean	Std. Deviation
Non-Estate Housing	310	37	2450	279.18	217.21
Estate Housing	121	50	980	356.05	202.47
Gated Community	121	88	2500	774.08	575.30
Price (TL/m²)	**N**	**Minimum**	**Maximum**	**Mean**	**Std. Deviation**
Non-Estate Housing	310	607	16333	1838.87	1171.02
Estate Housing	121	455	5227	2248.20	959.19
Gated Community	121	1388	10000	3882.30	1646.95

RESULTS AND DISCUSSION

In this study it is assumed that a greater supply of parking spaces generates a higher rate of private car ownership, with the aim of the study being to analyze whether residential parking supply has a causal influence on car ownership and usage, with particular focus on gated communities in Ankara. In order to define whether there is an association between development patterns of residential areas and the provision of parking areas, a Chi-Square test was conducted as a first step.

Relationship between Types of Building Site and Parking Spaces

First of all, an examination is made of the relationship between the types of building site and the presence of a garage. For this purpose, the null hypothesis is "there is no association between different types of building sites and having a garage". To test this hypothesis of no association (or independence), a Chi-Square test was carried out, and is presented in Tables 10 and 11.

In Table 11, the Pearson Chi-Square value is 112.966 with a df (degree of freedom) = 2 and a p-value is 0.000 (<0.05). Hence the null hypothesis of no association is rejected against the alternative hypothesis that there is at least one of the groups that is different. However, the results of the chi-square test in Table 11 do not provide information about which group/s differ/s from the others.

In order to find which group/s is/are different to the other groups, chi-square tests of 2x2 tables are computed for each pair of groups (such as, Non-estate Housing x Estate Housing, Non-Estate Housing x

Table 10. Crosstable analysis of SITE and GARAGE

			GARAGE		Total
			No	Yes	
SITE	Not Estate Housing	Count	225	85	310
		%	72.6	27.4	100.0
	Estate Housing	Count	80	41	121
		%	66.1	33.9	100.0
	Gated Community	Count	21	100	121
		%	17.4	82.6	100.0

Table 11. Chi-Square Tests of SITE and GARAGE

	Value	df	Asymp. Sig. (2-sided)
Pearson Chi-Square	112.966[a]	2	0.000
Likelihood Ratio	116.218	2	0.000
Linear-by-Linear Association	96.184	1	0.000
N of Valid Cases	552		

a. 0 cells (0.0%) have an expected count of less than 5. The minimum expected count is 49.54.

Gated Community, Estate Housing x Gated Community). The results of the 2x2 chi-square tests showed that Gated Community is the group of difference, which is visible also in Figure 2. The ratio of having a garage is less than the ratio of not having a garage in both Non-Estate and Estate Housing; however, the ratio of having a garage is much higher in Gated Communities. That is, 82.6 percent of gated communities in the data have garage space.

In the next step, a chi-square test is conducted to analyze the null hypothesis "There is no association between different types of building site and having a private open parking space". The results of the test are illustrated in Tables 12 and 13.

In Table 13, the Pearson Chi-Square value is 24.084 with df = 2 and the p-value is 0.000 (<0.05). That is, the null hypothesis of no association is also rejected for OPEN_PARKING.

Figure 2. Relationship between SITE and GARAGE

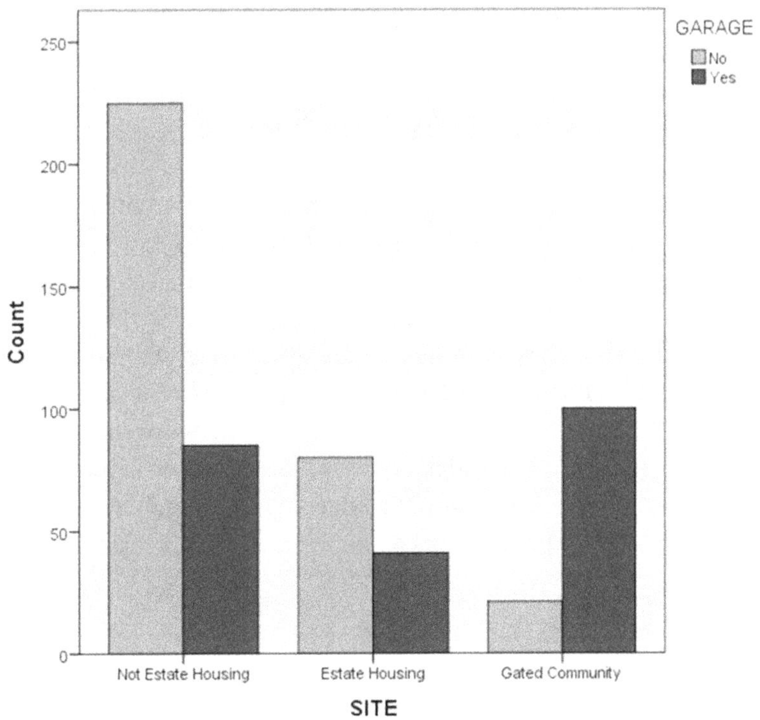

Table 12. Crosstable analysis of SITE and OPEN_PARKING

			OPEN_PARKING		Total
			No	Yes	
SITE	Non-Estate Housing	Count	72	238	310
		%	23.2	76.8	100.0
	Estate Housing	Count	12	109	121
		%	9.9	90.1	100.0
	Gated Community	Count	7	114	121
		%	5.8	94.2	100.0

Table 13. Chi-square tests of SITE and OPEN_PARKING

	Value	df	Asymp. Sig. (2-sided)
Pearson Chi-Square	24.084[a]	2	0.000
Likelihood Ratio	26.439	2	0.000
Linear-by-Linear Association	22.670	1	0.000
N of Valid Cases	552		

a. 0 cells (0.0%) have an expected count of less than 5. The minimum expected count is 19.95.

Figure 3 illustrates the relationship between the type of building site and open parking areas, and the results of chi-square test per groups showed that there is no significant difference between Estate Housing and Gated Community in terms of the presence of private open parking areas. However, the results revealed that Non-estate Housing is different to the other two groups. That is, as the ratio of having open parking spaces is much lower for Non-estate Housing than the other two groups.

Relationship between Types of Houses and Parking Spaces

The study also examined whether there is an association between different types of housing and having a garage.

The chi-square test in Table 15 illustrates that the null hypothesis, that there is no relationship between TYPE and GARAGE, is accepted. The data revealed no significant association between different types of housing and having a garage. As Table 14 and Figure 4 illustrate, the ratio of having garage spaces are similar for both Detached and Semi-detached housing and apartment units.

Tables 16, Table 17 and Figure 5 illustrate the relationship between the type of housing and open parking areas. As in the previous case, the null hypothesis that there is no association between TYPE and OPEN_PARKING is accepted. The results of the analysis reveal that there is no relationship between type of housing and having an open parking space.

Figure 3. Relationship between SITE and OPEN_PARKING

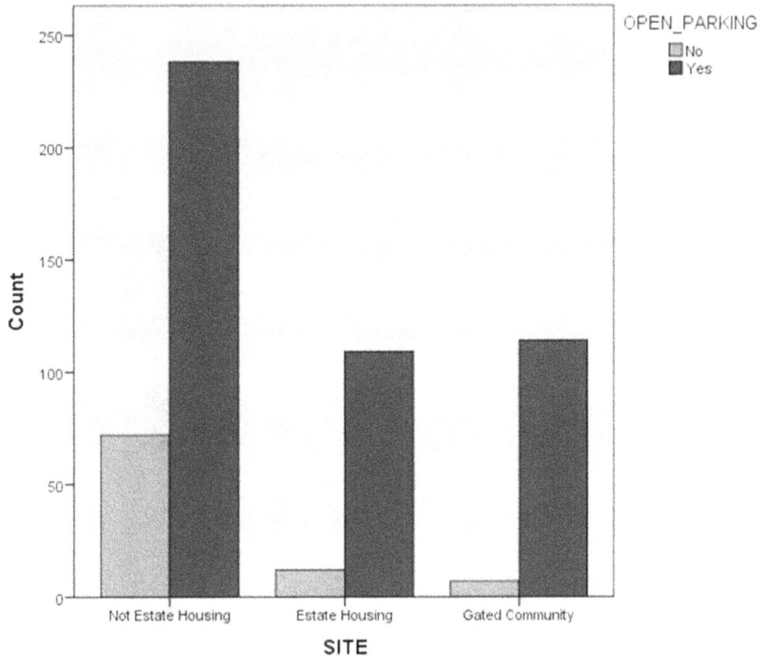

Table 14. Crosstable analysis of TYPE and GARAGE

			GARAGE		Total
			No	**Yes**	
TYPE	Detached or Semi-detached	Count	29	23	52
		%	55.8	44.2	100.0
	Apartment Unit	Count	297	203	500
		%	59.4	40.6	100.0

Table 15. Chi-square tests of TYPE and GARAGE

	Value	df	Asymp. Sig. (2-sided)	Exact Sig. (2-sided)	Exact Sig. (1-sided)
Pearson Chi-Square	0.257[a]	1	0.612		
Continuity Correction[b]	0.129	1	0.720		
Likelihood Ratio	0.255	1	0.614		
Fisher's Exact Test				0.658	0.358
Linear-by-Linear Association	0.256	1	0.613		
N of Valid Cases	552				

a. 0 cells (0.0%) have an expected count of less than 5. The minimum expected count is 21.29.

b. Computed only for a 2x2 table

Figure 4. Relationship between TYPE and GARAGE

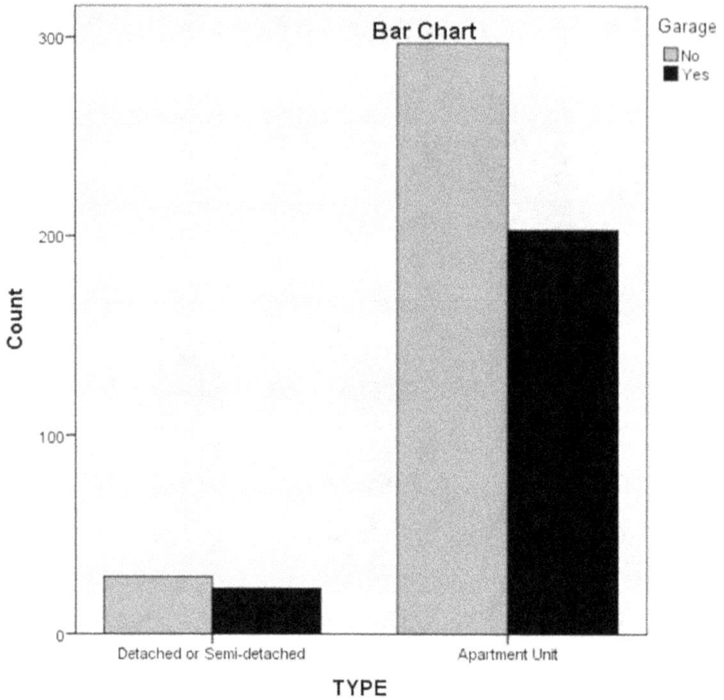

Table 16. Crosstable analysis of TYPE and OPEN_PARKING

			OPEN_PARKING		Total
			No	Yes	
TYPE	Detached or Semi-detached	Count	8	44	52
		%	15.4	84.6	100.0
	Apartment Unit	Count	83	417	500
		%	16.6	83.4	100.0

Table 17. Chi-square tests of TYPE and OPEN_PARKING

	Value	df	Asymp. Sig. (2-sided)	Exact Sig. (2-sided)	Exact Sig. (1-sided)
Pearson Chi-Square	0.051[a]	1	0.822		
Continuity Correction[b]	0.001	1	0.977		
Likelihood Ratio	0.051	1	0.821		
Fisher's Exact Test				1.000	0.503
Linear-by-Linear Association	0.050	1	0.822		
N of Valid Cases	552				

a. 0 cells (0.0%) have an expected count of less than 5. The minimum expected count is 8.57.

b. Computed only for a 2x2 table

Figure 5. Relationship between TYPE and OPEN_PARKING

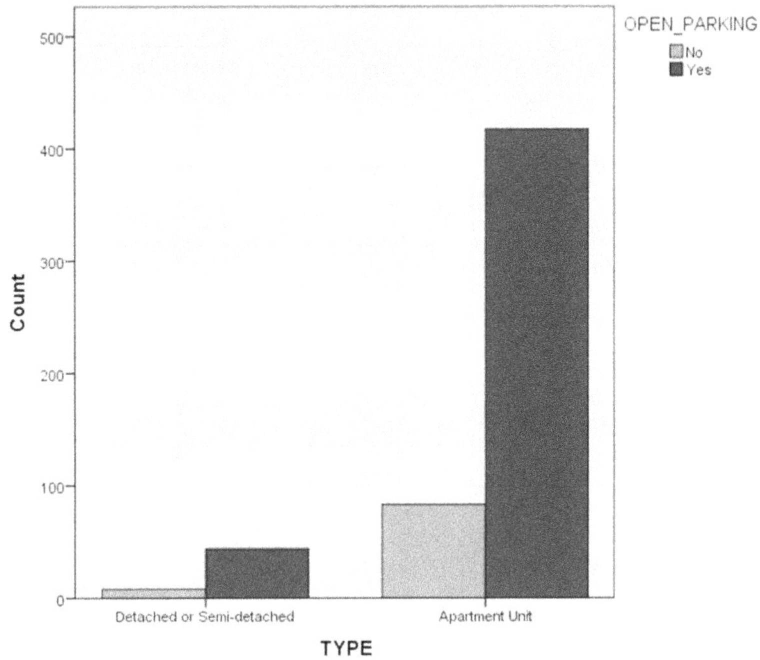

Relationship between Distance to the City Center and Parking Spaces

As discussed previously, the demand for parking at home remains as a significant problem for most of the residents of Ankara's central neighborhoods, and this parking restraint makes central residential areas less attractive. As a consequence, households have started to seek new options and new locations at the periphery of the city, although this will make them more car-dependent. In order to examine this assumption, the relationship between the distance to the city center and parking spaces are examined in the next analysis.

Tables 18 and Table 19 provide the results of the chi-square test on the relationship between distance to the city center and having a garage. In order to run the test, five categories of DISTANCE were assigned and recoded as DISTANCE_GRP in the data. The first group denotes a distance of less than 5 kilometers to the city center; the second group, a distance of between 5.01 and 10 kilometers; the third group, a distance of between 10.01 and 15 kilometers; the fourth group, a distance of between 15.01 and 20 kilometers, and the fifth group, a distance of 20 kilometers or more. The results of the chi-square test reveal a significant difference between groups in terms of the distance to the city center and having a garage.

In order to find out which group/s is/are different from the other groups, chi-square tests of the 2x2 table are interpreted for all pair of groups. In Figure 6 it is apparent that group 5 is statistically different to the other four groups, while there is no statistically significant difference between the other four groups of distance.

The relationship between the groups of distance and open parking areas was examined (Figure 7) by running a chi-square analysis in Table 20 and Table 21, revealing that the difference between groups is not statistically significant in terms of the presence of open parking spaces.

Table 18. Crosstable analysis of DISTANCE_GRP and GARAGE

				GARAGE		Total
				No	Yes	
DISTANCE_GRP	1	≤5 km.	Count	89	62	151
			%	58.9	41.1	100.0
	2	5.01-10 km.	Count	100	86	186
			%	53.8	46.2	100.0
	3	10.01-15 km.	Count	49	28	77
			%	63.6	36.4	100.0
	4	15.01-20 km.	Count	38	42	80
			%	47.5	52.5	100.0
	5	≥20.01 km.	Count	50	8	58
			%	86.2	13.8	100.0

Table 19. Chi-square tests of DISTANCE_GRP and GARAGE

	Value	df	Asymp. Sig. (2-sided)
Pearson Chi-Square	24.925[a]	4	0.000
Likelihood Ratio	27.560	4	0.000
Linear-by-Linear Association	4.588	1	0.032
N of Valid Cases	552		

a. 0 cells (0.0%) have an expected count of less than 5. The minimum expected count is 23.75.

Relationship between Age of the Building and Parking Spaces

Another important variable is the age of the buildings, as it is assumed that minimum parking requirements arose from the vision with ample free parking, and parking requirements brought this vision into reality, with every new building taking parking requirements into account (Shoup, 1997). Table 22 and Table 23 examine the relationship between the age of the building and the presence of a garage in order to test whether new buildings have increased the supply of private parking areas.

In order to run a chi-square test, AGE was recoded into seven age groups, as shown in Table 22. The results of the chi-square test indicated that there is statistically significant difference between the different age groups and their access to a parking garage. Again, the chi-square tests of 2x2 tables are interpreted for all pairs of groups in order to find which group/s is/are different from the others (See Figure 8).

The results of 2x2 chi-square test imply that there is no group difference between groups 3, 4, 5, 6 and 7, and that groups 1 and 2 have different group scores in terms of having a garage. As also shown in Table 22, the ratio of properties with access to a garage is higher for new buildings, which confirms the assertion that new buildings increase the supply of private parking areas in cities.

The above assertion in Figure 9 is not valid when the relationship between the years and the open parking areas are examined. The chi-square test results presented in Table 24 and Table 25 show that there is no statistically significant difference between groups of years in terms of having open parking areas.

Figure 6. Relationship between DISTANCE_GRP and GARAGE

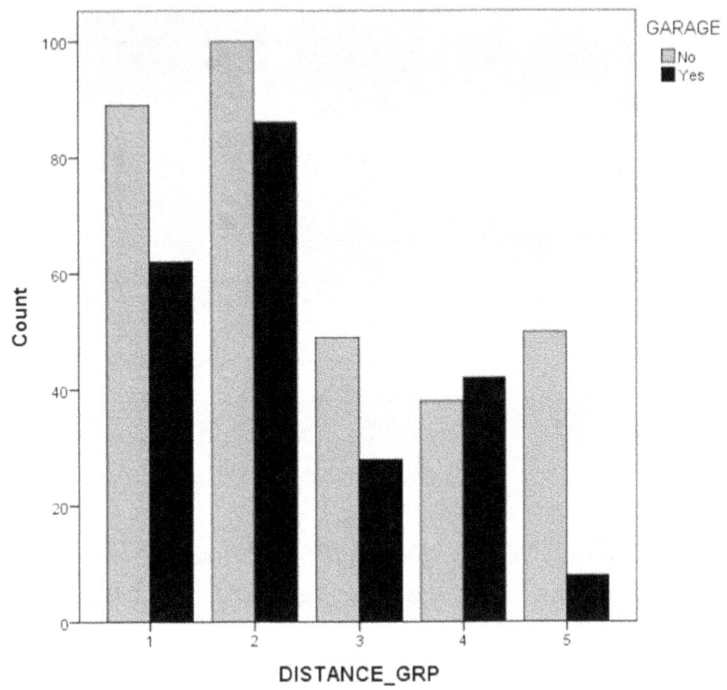

Figure 7. Relationship between DISTANCE_GRP and OPEN_PARKING

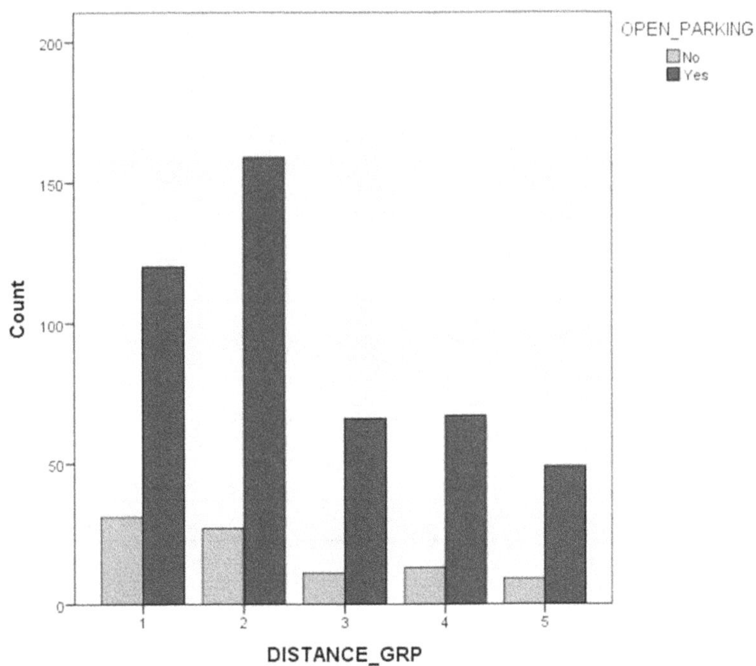

Table 20. Crosstable analysis of DISTANCE_GRP and OPEN_PARKING

				OPEN_PARKING		Total
				No	Yes	
DISTANCE_GRP	1	≤5 km.	Count	31	120	151
			%	20.5	79.5	100.0
	2	5.01–10 km.	Count	27	159	186
			%	14.5	85.5	100.0
	3	10.01–15 km.	Count	11	66	77
			%	14.3	85.7	100.0
	4	15.01–20 km.	Count	13	67	80
			%	16.2	83.8	100.0
	5	≥20.01 km.	Count	9	49	58
			%	15.5	84.5	100.0

Table 21. Chi-square tests of DISTANCE_GRP and OPEN_PARKING

	Value	df	Asymp. Sig. (2-sided)
Pearson Chi-Square	2.631[a]	4	0.621
Likelihood Ratio	2.556	4	0.635
Linear-by-Linear Association	0.744	1	0.388
N of Valid Cases	552		

a. 0 cells (0.0%) have an expected count of less than 5. The minimum expected count is 9.56.

Table 22. Crosstable analysis of AGE_GRP and GARAGE

				GARAGE		Total
				No	Yes	
AGE_GRP	1	≤5 years	Count	122	162	284
			%	43.0	57.0	100.0
	2	6–10 years	Count	53	42	95
			%	55.8	44.2	100.0
	3	11–15 years	Count	47	10	57
			%	82.5	17.5	100.0
	4	16–20 years	Count	54	7	61
			%	88.5	11.5	100.0
	5	21–25 years	Count	20	0	20
			%	100.0	0.0	100.0
	6	26–30 years	Count	16	4	20
			%	80.0	20.0	100.0
	7	≥31 years	Count	14	1	15
			%	93.3	6.7	100.0

Table 23. Chi-square tests of AGE_GRP and GARAGE

	Value	df	Asymp. Sig. (2-sided)
Pearson Chi-Square	90.458[a]	6	0.000
Likelihood Ratio	104.762	6	0.000
Linear-by-Linear Association	75.371	1	0.000
N of Valid Cases	552		

a. 0 cells (0.0%) have an expected count of less than 5. The minimum expected count is 6.14.

Figure 8. Relationship between AGE_GRP and GARAGE

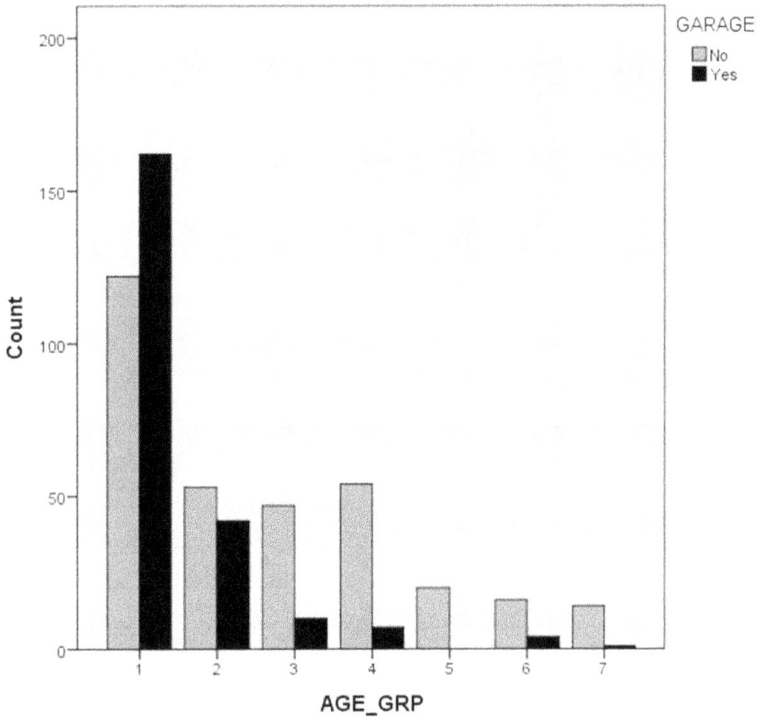

Relationship between House Price and Parking Spaces

Little research has been done to date into the effect of parking space at home on house prices, and literature is yet to show a clear relationship between off-road parking and house prices, although there are number of studies indicating a positive and significant effect of the presence of a garage on house prices. For instance, McMillen and Redfearn (2010) claim that house prices increase with the presence of a garage, which is a finding that is mirrored in the empirical analysis of Haurin and Brasington (1996). In this study, the relationship between house prices and parking spaces is examined in Table 26, which shows that the mean house price values are higher for buildings with either garages or private open parking spaces.

Figure 9. Relationship between AGE_GRP and OPEN_PARKING

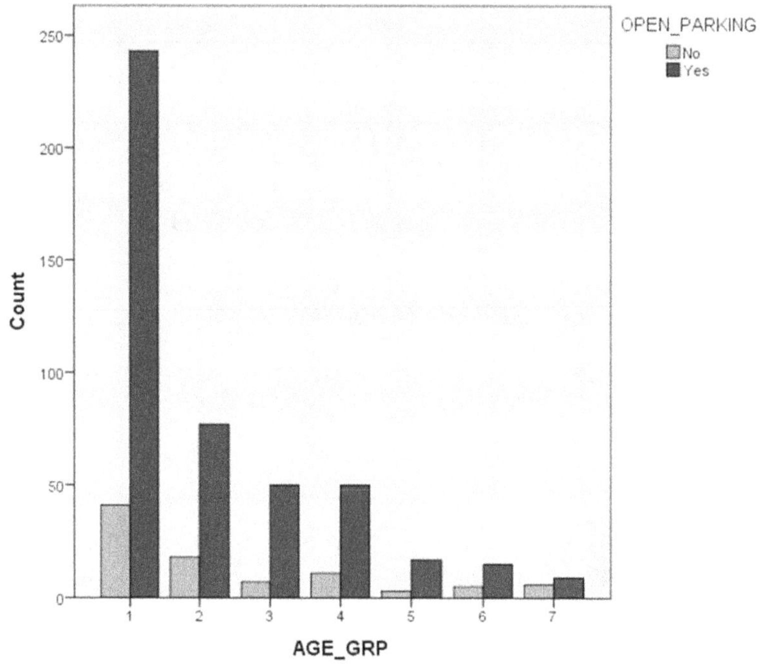

Table 24. Crosstable analysis of AGE_GRP and OPEN_PARKING

				OPEN_PARKING		Total
				No	Yes	
AGE_GRP	1	≤5 years	Count	41	243	284
			%	14.4	85.6	100.0
	2	6–10 years	Count	18	77	95
			%	18.9	81.1	100.0
	3	11–15 years	Count	7	50	57
			%	12.3	87.7	100.0
	4	16–20 years	Count	11	50	61
			%	18.0	82.0	100.0
	5	21–25 years	Count	3	17	20
			%	15.0	85.0	100.0
	6	26–30 years	Count	5	15	20
			%	25.0	75.0	100.0
	7	≥31 years	Count	6	9	15
			%	40.0	60.0	100.0

Table 25. Chi-square tests of AGE_GRP and OPEN_PARKING

	Value	df	Asymp. Sig. (2-sided)
Pearson Chi-Square	9,232[a]	6	,161
Likelihood Ratio	7,848	6	,249
Linear-by-Linear Association	4,201	1	,040
N of Valid Cases	552		

a. 3 cells (21.4%) have an expected count of less than 5. The minimum expected count is 2.47.

Table 26. Effects of presence of GARAGE and OPEN_PARKING on house price

PRICE (,000 TL)		N	Minimum	Maximum	Mean	Std. Deviation
GARAGE	No	326	37	2500	280,32	247,967
	Yes	226	79	2500	583,66	464,897
OPEN_PARKING	No	91	37	1000	304,45	212,371
	Yes	461	50	2500	424,26	405,785

In order to test the statistical significance of the relationship, a correlation test is conducted for both variables in Table 27 and in Table 28. The results of both tests show that house prices are affected significantly by the presence of both a private garage and open parking.

Relationship between Car Ownership and Gated Communities

Transportation engineers and planners have assumed that there is a link between land-use development patterns and the travel behavior of individuals (Bhat & Guo, 2007). In this study, it is assumed that the development of gated communities has affect car ownership propensity as developers increase parking provisions in new housing developments, as shown also in Figure 7.

As long as existing planning and parking guidelines prescribe parking close to human activities, private car use will increase (Knoflacher, 2006). As the number of parking spaces increases, the chance of owning cars also increases. Table 29 shows the number of road motor vehicles in Turkey by years, and

Table 27. Correlation between PRICE and GARAGE

		PRICE (,000 TL)	GARAGE
PRICE (,000 TL)	Pearson Correlation	1	0.390*
	Sig. (2-tailed)		0.000
	N	552	552
GARAGE	Pearson Correlation	0.390*	1
	Sig. (2-tailed)	0.000	
	N	552	552

*Correlation significant at the 0.01 level (2-tailed).

Table 28. Correlation between PRICE and OPEN_PARKING

		PRICE (,000 TL)	OPEN_PARKING
PRICE (,000 TL)	Pearson Correlation	1	0.116*
	Sig. (2-tailed)		0.006
	N	552	552
OPEN_PARKING	Pearson Correlation	0.116*	1
	Sig. (2-tailed)	0.006	
	N	552	552

*Correlation significant at the 0.01 level (2-tailed).

Table 29. Number of road motor vehicles in Turkey

Years	Number of Cars	Increase on Previous Year
1966	91 469	-
1967	112 367	20 898
1968	125 375	13 008
1969	137 345	11 970
1970	137 771	426
1971	153 676	15 905
1972	187 272	33 596
1973	240 360	53 088
1974	313 160	72 800
1975	403 546	90 386
1976	488 894	85 348
1977	560 424	71 530
1978	624 438	64 014
1979	688 687	64 249
1980	742 252	53 565
1981	776 432	34 180
1982	811 465	35 033
1983	856 350	44 885
1984	919 577	63 227
1985	983 444	63 867
1986	1 087 234	103 790
1987	1 193 021	105 787
1988	1 310 257	117 236
1989	1 434 830	124 573
1990	1 649 879	215 049
1991	1 864 344	214 465
1992	2 181 388	317 044

continued on next page

Table 29. Continued

Years	Number of Cars	Increase on Previous Year
1993	2 619 852	438 464
1994	2 861 640	241 788
1995	3 058 511	196 871
1996	3 274 156	215 645
1997	3 570 105	295 949
1998	3 838 288	268 183
1999	4 072 326	234 038
2000	4 422 180	349 854
2001	4 534 803	112 623
2002	4 600 140	65 337
2003	4 700 343	100 203
2004	5 400 440	700 097
2005	5 772 745	372 305
2006	6 140 992	368 247
2007	6 472 156	331 164
2008	6 796 629	324 473
2009	7 093 964	297 335
2010	7 544 871	450 907
2011	8 113 111	568 240
2012	8 648 875	535 764
2013	9 283 923	635 048
2014	9 857 915	573 992
2015	10 589 337	731 422

Source: TurkStat, 2016.

Table 30 shows the number of road motor vehicles in Ankara after the year 1994 as there is no specific data available before 1994 for Ankara case. Table 29 shows a significant increase after the second half of the 1980s that accelerated after the 1990s. As discussed in previous sections, gated communities started to emerge in around the 1990s, as Table 7 shows. It is interesting to note the rise in the number of road motor vehicles after these years also in Table 29. In 1966 there were fewer than 100,000 licensed vehicles on the road in Turkey, but by 1986 this had increased to over 1,000,000, and reached 10,589,337 in 2015.

In order to examine whether the emergence of gated communities has influenced the rise of car ownership by providing more parking spaces, a correlation test is used. The correlation test shows the relationship between the increase in the number of road motor vehicles in Ankara (Table 30) and the years when the gated communities was constructed. The result of the analysis reveals that the correlation between these two variables is insignificant at a 95 percent Confidence Interval level. That is, the emergence of gated communities has influenced the rise of car ownership by providing more parking spaces in Ankara.

Table 30. Number of road motor vehicles in Ankara

Years	Number of Cars	Increase on Previous Year
1994	404 521	
1995	424 129	19 608
1996	450 960	26 831
1997	495 497	44 537
1998	537 971	42 474
1999	580 176	42 205
2000	647 043	66 867
2001	665 234	18 191
2002	674 336	9 102
2003	694 758	20 422
2004	696 157	1 399
2005	738 960	42 803
2006	783 198	44 238
2007	820 355	37 157
2008	854 691	34 336
2009	887 703	33 012
2010	924 360	36 657
2011	986 688	62 328
2012	1 035 719	49 031
2013	1 093 843	58 124
2014	1 147 113	53 270
2015	1 222 519	75 406

Source: TurkStat, 2016.

CONCLUSION

This study has aimed to analyze the relationship between the development patterns of residential areas and car ownership, assuming that a higher supply of parking spaces in new residential developments will generate a higher rate of private car ownership. To this end, the study analyzes whether gated communities encourage car usage by supplying more parking spaces for the case of Ankara.

In order to test these assumptions, this study uses two data sets: 1) Housing data for both local residential areas and gated communities, selected randomly from one of Turkey's largest real estate websites; and 2) Data is from the website of the Turkish Statistical Institute, which provides information on the number of road motor vehicles.

Cross-table and correlation analyses are used to identify the relationship between excess parking supply and sprawling residential areas, and to investigate whether the emergence of gated communities has influenced the rise of car ownership by providing more parking spaces. Correlation analyses are then used to examine the effects of the presence of parking areas on house prices.

The results of the study reveal that the number of garage spaces is positively and significantly related to the development of gated communities. That is, the ratio of access to a garage is higher for gated communities than both estate housing and non-estate housing. The results of the study illustrate also that the presence of open parking areas varies according to the types of housing site. The ratio of having private an open parking space is lower for non-estate housing in the data, although there is no statistically significant difference between estate housing and gated communities in terms of their access to private open parking spaces.

The study assumes also that car parking provision may be linked to the age of the building, and the results revealed that the ratio of access to a garage is higher for new buildings, which confirms the assertion that new buildings increase the supply of private parking areas in cities. That said, this assertion proves to be invalid when the relationship between the years and the open parking areas is examined, in that there is no statistically significant difference between groups of years in terms of access to open parking areas. Older houses that are more centrally located are less likely to have garage parking, and in these central areas in Ankara, older houses usually feature open parking spaces in front of the building, but with limited capacity. In these areas, residents usually park on the road, and often prefer to use public transport than an automobile for journeys to the city center, as they do not have guaranteed parking close to their homes.

The results presented in this paper show that house prices are affected significantly by the presence of both garage and open parking, with increased parking supply having the potential to increase the price of housing in Ankara.

As discussed in the previous sections, gated communities started to emerge in around the 1990s in Ankara, and it is interesting to note the rise in the number of road motor vehicles on the road after these years in Turkey, especially after the second half of the 1980s. Also, the result of the analysis revealed that there is a positive relationship between the emergence of gated communities and the rise in the number of motor vehicles in Ankara.

For Ankara city as a whole, urban sprawl is increasing, and for every new housing unit constructed, developers satisfy also minimum parking requirements. More parking areas result in more cars, and more cars increase motorization, and as a result, the number of possible trips per person per day increases.

REFERENCES

Alkan, L. (2014). 1980 sonrası konut politikalarının mekansal yansımaları: Ankara örneği. *Journal of Urban Studies*, *12*, 103–131.

Alkan, L. (2015). Gated communities and social sustainability: A case study of Ankara.*2nd International Multidisciplinary Scientific Conference on Social Sciences & Arts Sgem 2015*, 423-430. DOI: doi:10.5593/SGEMSOCIAL2015/B41/S15.052

Babalik-Sutcliffe, E. (2013). Urban form and sustainable transport: Lessons from the Ankara case. *International Journal of Sustainable Transportation*, *7*(5), 416–430. doi:10.1080/15568318.2012.676152

Bhat, C. R., & Guo, J. Y. (2007). A comprehensive analysis of built environment characteristics on household residential choice and auto ownership levels. *Transportation Research Part B: Methodological*, *41*(5), 506–526. doi:10.1016/j.trb.2005.12.005

Davis, A. Y., Pijanowski, B. C., Robinson, K., & Engel, B. (2010). The environmental and economic costs of sprawling parking lots in the United States. *Land Use Policy*, *27*(2), 255–261. doi:10.1016/j.landusepol.2009.03.002

Grant, J. (2005). Planning Responses to Gated Communities in Canada. *Housing Studies*, *20*(2), 273–285. doi:10.1080/026730303042000331772

Guo, Z. (2013). Does residential parking supply affect household car ownership? The case of New York City. *Journal of Transport Geography*, *26*, 18–28. doi:10.1016/j.jtrangeo.2012.08.006

Haurin, D. R., & Brasington, D. (1996). School Quality and Real House Prices: Inter- and Intrametropolitan Effects. *Journal of Housing Economics*, *5*(4), 351–368. doi:10.1006/jhec.1996.0018

Housing Asked Prices in Ankara. (2016). Available at: http://www.hurriyetemlak.com/

Knoflacher, H. (2006). A new way to organize parking: The key to a successful sustainable transport system for the future. *Environment and Urbanization*, *18*(2), 387–400. doi:10.1177/0956247806069621

Knoflacher, H., & Ocalir, E. V. (2011). Eco-methodology for urban and transport planning for the future eco-technology. In O. Yalciner Ercoskun (Ed.), *Green and Ecological Technologies for Urban Planning: Creating Smart Cities* (pp. 154–173). IGI Global. doi:10.4018/978-1-61350-453-6.ch009

Kurtulus, H. (2011). Gated communities as a representation of new upper and middle classes in İstanbul. *İ.Ü. Siyasal Bilgiler Fakultesi Dergisi*, *44*, 49–65.

Landman, K. (2000a). Gated communities and urban sustainability: taking a closer look at the future.*2nd Southern African Conference on Sustainable Development in the Built Environment.*

Landman, K. (2000b). *An overview of enclosed neighbourhoods in South Africa*. Pretoria: CSIR Publication.

Marsden, G. (2006). The evidence base for parking policies—a review. *Transport Policy*, *13*(6), 447–457. doi:10.1016/j.tranpol.2006.05.009

McMillen, D. P., & Redfearn, C. L. (2010). Estimation and hypothesis testing for nonparametric hedonic house price functions. *Journal of Regional Science*, *50*(3), 712–733. doi:10.1111/j.1467-9787.2010.00664.x

Roitman, S. (2005). Who Segregates Whom? The Analysis of a Gated Community in Mendoza, Argentina. *Housing Studies*, *20*(2), 303–321. doi:10.1080/026730303042000331790

Shoup, D. C. (1997). The High Cost of Free Parking. *Journal of Planning Education and Research*, *17*(1), 3–20. doi:10.1177/0739456X9701700102

Shoup, D. C. (1999). The trouble with minimum parking requirements. *Transportation Research Part A, Policy and Practice*, *33*(7-8), 549–574. doi:10.1016/S0965-8564(99)00007-5

TurkStat. (2016). *Road Motor Vehicle Statistics*. Author.

Weinberger, R. (2012). Death by a thousand curb-cuts: Evidence on the effect of minimum parking requirements on the choice to drive. *Transport Policy*, *20*, 93–102. doi:10.1016/j.tranpol.2011.08.002

Willson, R. W. (1995). Suburban Parking Requirements: A Tacit Policy for Automobile Use and Sprawl. *Journal of the American Planning Association*, *61*(1), 29–42. doi:10.1080/01944369508975617

Chapter 8
From Regional Austrian Parking Ordinances to Sound Guidelines

Tadej Brezina
Vienna University of Technology, Austria

Josef Michael Schopf
Vienna University of Technology, Austria

ABSTRACT

Provision of parking space is a constant challenge in urban transport planning because streetscape is a limited resource and therefore highly contested between different uses. In Austria, building regulation competence is subsidiary appointed from the national level to nine provincial parliaments. This leads to nine different parking ordinances. Whereas historically all ordinances have been focusing only on car parking by means of rigid regulations, only recently selected provincial parliaments added detailed bicycle ordinances their building codes. After a detailed overview of the Austrian situation and selected international innovations, the authors identify long-run goals that parking organization in cities should aim at and suggest an improvement of parking ordinances so that city planners and city administrations will have at hand appropriate design tools for mobility regime improvement.

STATIONARY TRAFFIC: A FUNDAMENTAL PROBLEM

Public spaces in residential areas were used for living space with varied functions over millennia. During the industrialisation of the 19th century, the public space in Europe changed from a living to a transportation space. Until the start of the 1930s, flowing transport enjoyed preferential treatment. By contrast, stopping and parking was only permitted at specially designated areas. In the 1930s, streetscape also became a transport space on an entirely legal basis, while precedence was given to flowing traffic. Holding and parking was now permitted wherever it was not expressly banned. As a result, streets changed fundamentally in their appearance (Figure 1) (Schopf, 2016).

The now prevalent imbalance between transport system, residential structures and environmental targets was subsequently co-determined strongly by the organisation of stationary traffic. The prescribing of mandatory parking spaces in the practice that has been in place for decades now is a fundamen-

DOI: 10.4018/978-1-5225-2116-7.ch008

Figure 1. Streetscape changed from living space to storage space for vehicles
© Schopf.

tal problem of transport development (Schopf & Brezina, 2015). Parking spaces are the sources and destinations of motorised individual transport; they are assigned directly to the structures in terms of space, hugely influence the choice of transport mode to the detriment of the ecomobility (transport by foot/bicycle, public transport, car sharing) and thereby contradict all official environmentally relevant targets (Schopf et al., 2015).

By contrast, the growing traffic and structural problems were mostly addressed with measures in flowing traffic, in particular by installing additional infrastructure. In particular with the rolling out of IT-based methods, traffic planning and safety focus on the flowing traffic and its efficient processing. Great effort is put into fighting symptoms. This is understandable, as problems in flowing traffic are also most easily identifiable by laymen. The results of this approach become apparent in the urban sprawl, the dissolution of urban structures, the outflow of purchasing power from the historic city areas, traffic problems as well as the varied consequences in the social, environmental and health areas (ÖVG, 2005). In this process, old recommendations (Buchanan, 1964; OECD, 1981) point to the key parameter of urban traffic and urban development: garage and parking space management.

According to Riedel (2014), parking space policy therefore needs to be a key component of every effective traffic policy in (major) cities. The quality of an urban transport system depends crucially on the successes or failures in the organisation of parking spaces.

The Austrian minister of the environment also considers the organisation of stationary traffic to be a major influence in choosing the mode of transport and therefore in altering the environmental effects of transport (see Schopf & Brezina, 2015). Nevertheless, stationary traffic is often missed as a cause of the problems, as it does not cause any direct emissions. The indirect effects are in turn all the greater, because the accessibility of the relevant means of transport and the availability of parking spaces at the relevant destination are often crucial for using a certain mode of transport. As regards the pressing target, to reduce CO_2 emissions in traffic, the management of parking spaces and suitable regulative tools are therefore particularly significant.

In this chapter, the framework conditions are therefore briefly analysed and the legal regulations regarding the number and arrangement of parking spaces and garages for Austria, as examples for many regulations of this kind in Europe, are then collected and assessed. The discrepancies between the targets and principles of regional planning laws and environmental legislation as well as the associated provisions of building codes are analysed. Subsequently, results of research in this area, which are processed in technical sheets and directives, are presented. Progressive new regulations in the international context are also illustrated. On these bases, recommendations for future-looking regulations regarding parked vehicles are compiled.

BACKGROUND: LEGAL REGULATIONS IN AUSTRIA

The legal fields of regional planning and construction are the responsibility of the individual federal states in Austria. For that reason, corresponding regional planning and structural laws also exist for each federal state. In some federal states, the decision-making competencies for some sub-aspects of parking space legislation were handed over to municipalities.

It must be observed that today's parking space rules are not based on scientific considerations, but were randomly set by the German Reich Garage Rules (Reichsarbeitsministerium, 1939) and became the practice used in the assignment of spaces and development. The target of the Reich Garage Rules was not to solve transport problems, but to promote motorisation:

- **Preamble:** *The promotion of motorisation is the ... proven aim.*
- **Section 2:** *Anyone erecting residential buildings, workplaces or similar buildings [...] must create for the existing vehicles of the residents, and those to be expected, [...], a garage place of a suitable size, location and property [...] on the development premises or nearby.*

However, the existing regulations of the Austrian building codes, if they are used within the meaning of the regional planning targets, already include effective options for efficient parking space management. The regional planning laws and building codes govern, on the one hand, the competitive situation between the transport modes and, on the other hand, the design and structures with a long-term effect such as buildings, their arrangement and the public space (Schopf & Brezina, 2015).

Regional Planning Laws

If the targets and principles of the nine Austrian regional planning laws are analysed, and if they are related to the subject of parking and parking spaces, a breakdown into primary and secondary targets can be identified (see Table 1).

The primary targets discuss the main feature of parking spaces, the use of space. They demand the saving use of the resource of land that is only available to a limited degree.

The secondary targets are no less important than the primary ones, however, they refer to the sizes that are derived from the space usage and therefore have a direct relationship to surface parking. They relate to the urban sprawl, protection of the townscape and landscape and the prevention of pollution on people. It is understandable that there are many more targets and principles set out in the regional planning laws, which relate to parking space in the wide sense. For the purposes of clarity and comparability,

Table 1. Targets in the Austrian regional planning laws

Primary targets	Saving use of land.
Secondary targets	• Avoidance of urban sprawl. • Maintaining or improving the townscape. • Protection of the population from the negative effects of sound emissions and the like.

Source: Schopf & Brezina, 2015.

the four target areas in Table 1 were chosen. Remarkable here is the fact that the primary target of the saving use of land, with the exception of Vienna, is considered in the remaining eight regional planning laws. The saving, efficient handling of ground and space for a function within a residential structure is a key principle in a network of complex functions and interests that need to co-exist in a city on very little space and share it (Schopf & Brezina, 2015)

Private vehicles require space between 9.45 sqm and 11.50 sqm per vehicle for parking on average (FGSV, 1991). However, considerable additional space needs to be added for the parking manoeuvres. Depending on where the parking spaces are (roads or separate parking space) this additional need for space comes at the expense of the community road surface or must be added to the net surface need.

Transporting goods and persons is the primary function of vehicles. This means that the parking of vehicles is a secondary product of the primary transport function, with the exception of the loading and unloading functions required for the transport process. As the vehicles do not exercise a function during parking, or only the strongly secondary function of storage and availability, this conflicts with the other functions in the surface, which are mostly of a primary nature (Schopf & Brezina, 2015):

- Transport with other modes of transport (in particular weaker road users such as pedestrians and cyclists).
- Trade and gastronomy (both from the provider and consumer side).
- Leisure (in particular small-scale recovery functions).
- Learning and experience (especially for children and adolescents).

However, the majority of all parking spaces are on public ground, e.g. in the various Vienna districts they range from 42 to 66% (Dorner, Herry, & Schuster, 1997), and are distributed varyingly in the urban fabric. A relatively small parking space share is still ensured with concentrated parking in existing parking garages or parking houses.

The regional planning principle of saving use of land and the logic of putting primary over secondary functions, as they have a higher-valued function, results in the mandatory conclusion that the storing of vehicles for the purpose of "keeping them available for usage" does not need to take place in the public space. Concentrated parking in the form of space-saving multistory or underground garages corresponds to the primary target of regional planning laws (Schopf & Brezina, 2015).

Building Codes and Parking Space Ordinances

As already stated, the German Reich Garage Rules are the example of all relevant parking space regulations also in Austria. Table 2 specifies an extract of the parking space requirements in accordance with the corresponding state construction regulations. In general, the details of the prescriptions differ in the various federal states.

Table 2. Extract overview of the parking space requirements according to the state building codes

	Traffic Sources				
	Single-Family Home	**Block of Flats**	**Office and Administration Building**	**Shop, Business Premise**	**Workshop and Industrial Plant**
Burgenland	1 per flat	1 per flat	MC	MC	MC
Carinthia	MC	MC	MC	MC	MC
Lower Austria	1 per flat	1 per flat	1 per 40 sqm of FS	1 per 30-50 sqm of SS[b)]	1 per 5 emp.
Upper Austria	1 per flat	1 per flat	1 per 30 sqm of surface	1 per 30 sqm of surface	1 per 60-100 sqm of FS or 1 per 5 emp.
Salzburg	1.2 per flat	1.2 per flat	1 per 30 sqm of FS	1 per 30-50 sqm of FS	1 per 60 sqm of FS
Styria	1 per flat	1 per flat	1 per 5 emp.	1 per 50 sqm of FS	1 per 5 emp.
Tyrol	MC	MC	MC	MC	MC
Vorarlberg	1 per flat	0.8 to 1.3 per flat[a)]	"According to the planned need..."	1 per 15-60 sqm of SS	1 per 80 sqm of CR
Vienna	1 per 100 sqm of FS	1 per 100 sqm of FS	1 per 80 sqm of CR	1 per 80 sqm of CR	1 per 80 sqm of CR

Adapted from Karajan, 2001; Bundeskanzleramt, 2015.
CR = communal room, emp = employee, MC = municipal competence, FS = surface, SS = sales surface, FS = floor space
a) The same paragraph specifies that for blocks of flats 0.6 parking spaces and 0.7 garage spaces must be planned per flat.
b) 1 per 30 sqm of sales surface for shopping and specialist retail centres.

In addition to the specific number of parking spaces to be built, the building regulations – exceptions are Burgenland and Carinthia – also state the maximum distances at which parking spaces required according to the building regulations may be built if it may prove to be too expensive on the construction premises, too complicated structurally or not possible at all. Although the existing laws as a rule are interpreted such that the settlements must be equipped with passenger car parking spaces regardless of the actual traffic circumstances (the eco mode supply is not taken into account) some federal states do specify some freedom. This becomes manifest, for instance, in the possible removal of parking spaces not located adjoining the construction premises or in imprecise wordings that open up various interpretations, such as "unprofitability" when building the parking spaces. However, the usage of this discretion depends on the interpretation practice of the authorities and committees at the municipality level. If the parking space requirement cannot be fulfilled for structural or monument preservation reasons, the laws (with the exception of Burgenland and Upper Austria) specify an equalisation levy. This levy must be invested by the public authorities in line with the laws in garage construction and partly also in public transport (e.g. in Vienna) (Knoflacher, Schopf, & Brezina, 2007).

In the federal states of Carinthia and Tyrol, the building codes provide municipalities with the competence for parking space prescription for all transport sources, in Burgenland for subareas. The municipalities in Tyrol indeed show different regulations regarding the necessary parking spaces in the residential area. The prescriptions range from one parking space to three parking spaces per flat. The prescriptions are therefore in part significantly higher than is the case in the federal state regulations. The Tyrol state government took this situation into account by amending the parking space ordinance (Tyrol, 2015). It sets upper limits for the municipality regulations that the municipalities may lower further, while applicants for planning permission are not bound by the upper limits. The parking space obligations depend on the municipality category, the location and the usage surface. The upper limits range from

- One parking space per flat in the municipality category I (cities) with a flat in the main residential area with up to 55 sqm in residential surface, up to
- 3.5 parking spaces per flat in the municipalities category III (small municipalities), for a flat on the outskirts with more than 110 sqm in residential surface.

Parking place requirements are therefore still at a high level and constitute only a minor relief compared to the previous condition.

The (high) parking space obligations are to be considered from two aspects as a rule. On the one hand, the street space should serve as a place for meeting and spending time for people and not as a storage space for vehicles. For that reason, storing of vehicles that is broadly outside the public street space is to be welcomed (FSV, 2008). On the other hand, the parking space organisation should not promote the usage of passenger cars, because ensured and - at least in the perception - "free" parking spaces at the origin and destination of a trip promote the use of the passenger car. In general, both the parking space volume and motorisation rate (Figure 2) as well as the passenger car availability correlate with passenger car usage (Figure 3) (Schopf, 2016).

Under the building codes, the legislator forces applicants for planning permission to provide parking spaces and thereby promotes or hugely prefers motorised private transport and creates an additional incentive for its use. When organising and financing stops of public transport, however, the applicants for planning permission are not involved. Minimum standards for public transport links are also missing. Also not legally included in the planning and construction of traffic generators was the planning

Figure 2. Correlation between motorisation rate and share of flats with a parking possibility for Austria 1985–2001

Pfaffenbichler & Schopf, 2011; Adapted from Janik, 2002.

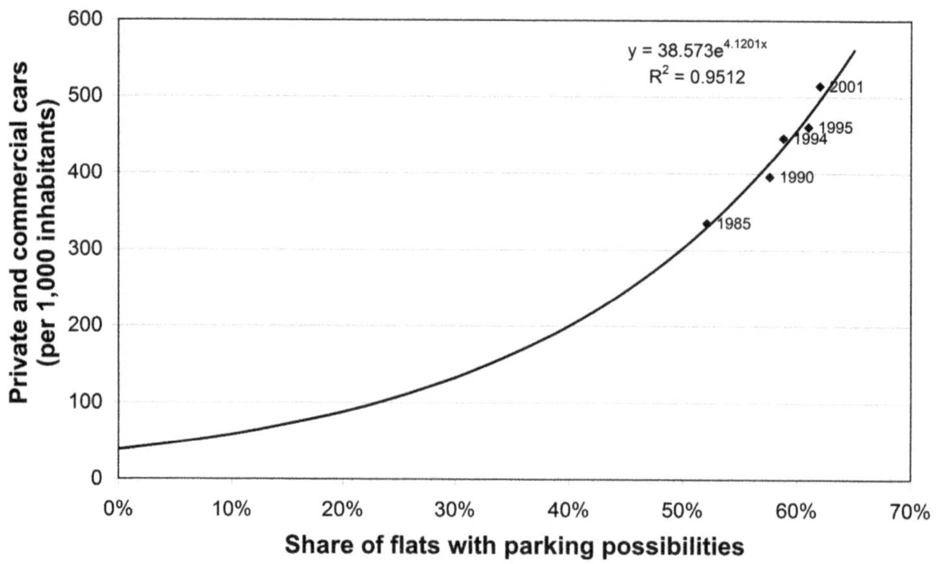

Figure 3. Choice of means of transport depending on passenger car availability in household
Adapted from Haider, 2010.

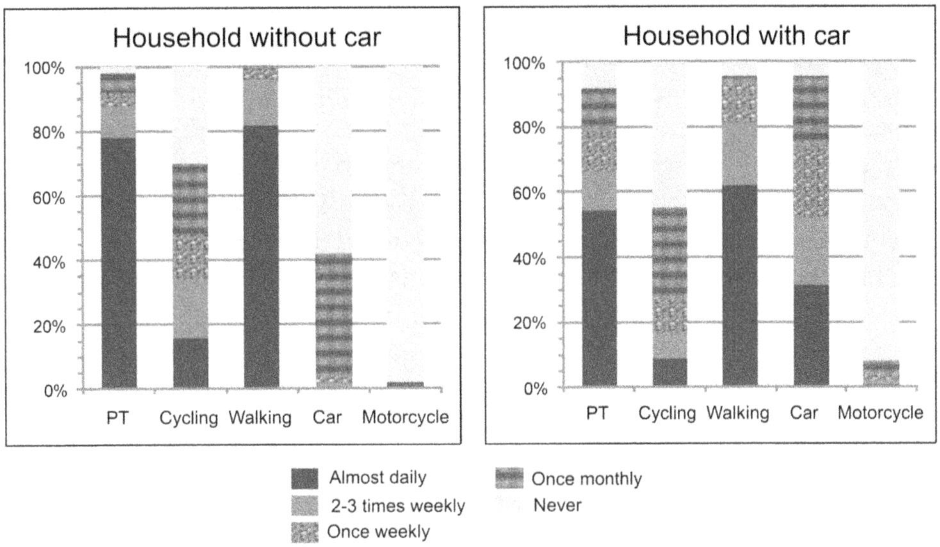

for cycling. At least there are some parking space requirements for bicycles in some federal states. In addition, mobility management is currently not rewarded and taken into account in the parking space obligation. This contradicts the climate strategy (BMLFUW, 2006) of strengthening environmentally sound modes of transport. The problem regarding the responsibility of applicants for planning permission in the residential area becomes clear from Figure 4. Before extending a house (on the right in the picture), the now sealed surface was an exemplary green space, as in the neighbouring yards. After the conversion, mandatory parking spaces had to be built, which remained unused for years because it was

Figure 4. Pursue a balanced parking space policy – sealed surface instead of green space
© Schopf.

possible to park for free on the street. In front of the relevant house (Figure 5) – and not only there – the street space in turn is degraded as a vehicle storage space. Two problems become visible here: on the one hand, it is not the passenger car owner who is obligated to ensure a parking space outside the public street space, but the applicant for planning permission. On the other hand, he creates the parking spaces on site, which is neither conducive to the equality of opportunities between the modes of transport nor to the green space (Schopf, 2016).

Regarding the not legally prescribed assignment of stops to traffic generators, it applies on a trivial basis that without an assignable stop there is no public transport. As an additional factor in the competitive relationship between car and public transport, it must be noted that humans do not react in accordance with measured time, but on the basis of perceived time. As such, the acceptance of walking distances to the stop decreases rapidly with length (Figure 6). For fair intermodal competition, the mere existence of stops is therefore not sufficient. Additionally, to ensure equality of opportunities, a comparable access route length to public transport as well as the private car is necessary (Knoflacher, 1987). Preference for the private car is driven to its extreme when a new kind of living is promoted (Car Loft, 2016):

Innovative car lofts enable garden and garage on all floors in the centre of town.

The consequences of the reigning parking space organisation has been materialising for many structures. The assignment of structures and parking spaces resulted in a location independence of the structures and, as a result, in the urban sprawl of locations and in the ruin of the location-based economy (Stift, 2002). Further consequences of the parking place regulations were already flagged up as negative by the Austrian Energy Strategy (BMLFUW, 2010):

Figure 5. Street space as vehicle parking space
© Schopf.

Figure 6. Acceptance of walking distance lengths – route to mode of transport and modal split car/public transport for Vorarlberger Medienhaus
Knoflacher & Schopf, 2001; Adapted from Walther, 1973.

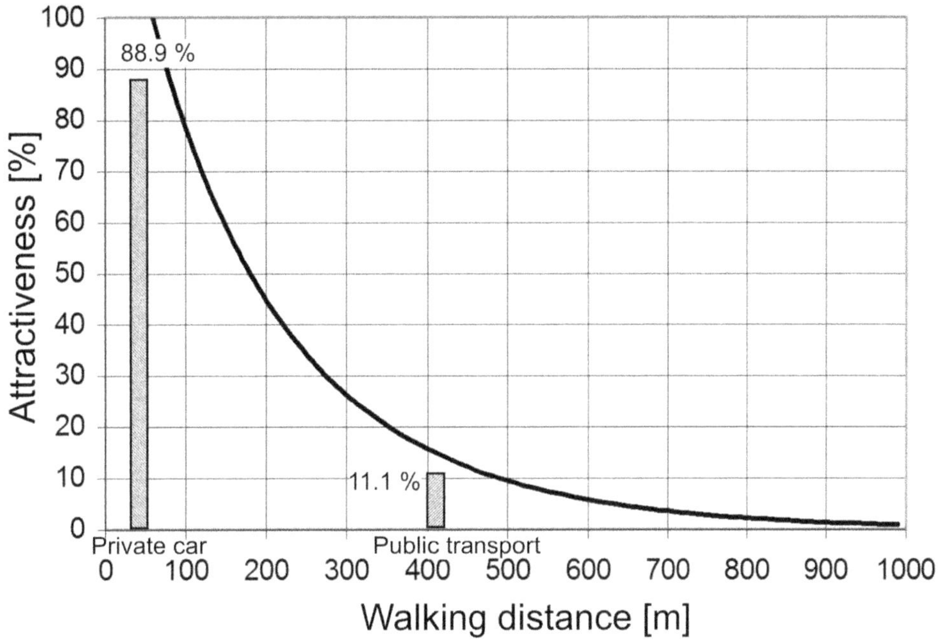

Due to the legal parking space obligation, parking spaces are built independently of the need. The costs for the construction of parking spaces in residential construction have an impact on living costs and are therefore borne by all flat owners, also by those who do not use their parking space.

As such, the users of eco-friendly transport modes subsidise on the one hand car traffic and, on the other hand, the construction costs of residential buildings are considerably increased. Table 3 shows the cost comparison of passenger car and bicycle parking space (without property costs). For underground garages, their share in the overall costs for an average flat is about 11%, thereby increasing the costs when purchasing or renting a flat accordingly (Schopf, et al., 2015).

Parking space requirements not only relate to new buildings, they also make subsequent densification difficult, as the subsequent construction of parking spaces/garages is very costly. Particularly affected are smaller flats, as a parking space requirement per flat increases the construction costs in particular

Table 3. Cost comparison of passenger car and bicycle parking space without land plot costs – up to seven bicycles can be parked on a passenger car parking space

Type	EUR per Passenger Car Parking Space	EUR per Bicycle Parking Space
Level ground parking space	2.000 ÷ 3.500	200
Parking house/bicycle parking house	10.000 ÷ 15.000	625 ÷ 1.700
Underground parking	15.000 ÷ 25.000	2.000 ÷ 3.500

Dortmund, 2006.

here. The legal parking space obligation results in a cost increase in residential construction in any event. If the erection of traffic systems from funds of residential property promotion is already questionable, the financing of vacant parking spaces (derstandard, 2013) with funds that are actually earmarked for the creation of residential space under social and ecological aspects is inacceptable (Parlament, 2015).

NEW APPROACHES TO GOVERNING PARKING SPACES ORGANISATION

For the legal framework conditions of the parking space management, in various countries different strategies are pursued or different tools are used. Beginning in the 1970s, some European countries and cities switched to more differentiated parking space obligations (Sammer et al., 2005).

Austria

Background

Already in working paper of 2001 (FSV, 2001) foresighted principles were recommended for the planning of parking spaces or systems of stationary private traffic. In doing so, it was stated that the provision of parking spaces is by no means merely a question of the geometrical dimensions or the figures-based assignment to certain space uses, but that the planning must also take into account the effect mechanisms in the system. Already in 1981 relevant mechanisms were published internationally for the first time (Knoflacher, 1981) and have been continuously improved and confirmed in the meantime (e.g. Peperna, 1982).

Furthermore, possibilities of fiscal influence on mobility in the desired sense are recommended. It is noted as key to prevent land-uses that promotes passenger cars (e.g. consumer market on the outskirts), through regulation of the land planning/commerce rules if this changes the mobility in the desired way. The relevant working paper (FSV, 2001) is limited to "if then" causal links and sets out the technical conditions, measures and consequences, but also refers to framework conditions of the legal environment. It was principally recommended to carry out pilot applications of collective garages in accordance with the principle of equidistance and car-free/low-car living.

The working paper serves as the foundation for a directive and contributes to the subjects of:

- Stationary traffic: a fundamental problem.
- Parking space restrictions: legal and operational circumstances.
- Transport policy objectives and system effects.
- Market economy aspects.

As a further basis for governing stationary traffic, the bulletin "Equality of transport modes" (ÖVG, 2005) is available from the Austrian Association of Transport Sciences. The bulletin states as a target, in coordination with the requirements of the Austrian policy on transport and regional planning policy that an equal treatment of the various transport modes needs to be pursued in accordance with market economy principles.

In general, the sheet notes as a problem the fact that the traffic in the traditional municipality and transport policy was not organised as an overall system. Components interconnected in reality are optimized independently from one another without taking the effects on the overall system into account. Therefore, special reference is made to parking space management and to the fact that for the equal opportunity among motorised private traffic and public transport not only the number of parking places plays a role, but their position in relation to the comparable elements of public transport. The parking space therefore corresponds to the function of the public transport stop. However, this notion is not to be found yet in the building codes in any way. Due to the car's impact on the behaviour of people, only a physical structural change can bring about equal opportunities and a fair option of choice. To ensure this, the walking distance to the parked car at all destination and origin points needs to be at least equally long as the walking distance to public transport stop. (ÖVG, 2005)

Case Studies: Vienna, Dornbirn and Innsbruck

In some Austrian municipalities efforts to reform the parking space obligations can already be observed. In Vienna, for instance, the parking space obligation was changed from one parking space per 80 sqm of newly built residential units to 100 sqm. The current Vienna regulation thereby takes into account the falling degree of motorisation and the high degree of Vienna households without an own car. The Vienna regulation reduced the construction costs apportioned to the individual flats, which also becomes noticeable for the tenants in particular in connection with small flats (Chorherr, 2014). In addition, the dedication in projects allows an even lower parking space obligation if the applicant for planning permission is willing to set effective measures for alternative mobility. As such, the Vienna regulation is innovative and sustainable. In general, such an instrument should be included into all Austrian building codes. Furthermore, the conversion of the regulation that can currently be interpreted as a "can provision" into a "should provision" would need to be examined (Käfer et al., 2011).

In accordance with the Vorarlberg parking space regulation, for the City of Dornbirn upper boundary values for parking spaces were set in a core zone in addition to minimum values. In Tyrol, too, upper limits for the number of parking spaces were introduced. In this sense, for instance, the City of Innsbruck is willing to also accept projects with less than 1.1 parking spaces per residential unit, which are normal in an area of repeatedly discussed underground parking spaces. Even 0.5 parking spaces per residential unit are conceivable if a consistent concept and a corresponding public transport link exist (Huber & Yilmaz-Huber, 2013).

An initial reform in the area of parking spaces was already set in 2001, in that the sale of garage parking spaces separately from the flats became possible (Käfer et al., 2011).

Further Reform Needs

Additionally, energy policy priority:

- Reform and nationwide harmonisation of the parking space rules with the target of anchoring the principle of causation (proof of the parking space by the car owner).
- Differentiation through to the non-application of the parking space obligation in residential construction and for operations depending on the connection by public transport and existence of mobility management measures.
- Binding installation of bike parking facilities and infrastructures for e-mobility.

- Expansion of the parking space management in core cities and metropolitan areas with an alteration of parking fees towards the public transport tariff level.

Germany

Case Studies: Hamburg and Berlin

Even in car-loving Germany, there are many examples of reduced parking space obligations:

The Hamburg Senate decided to abolish the general parking space obligation for residential projects in Hamburg in 2013 (Hamburg, 2013). In the future, the creation of parking spaces for residential projects is to be transferred to the responsibility of the principal. As such, parking spaces which are actually needed by the residents are to be built only. The city expects a simplification and acceleration of residential construction in Hamburg from lifting the parking space obligation and assumes that principals will also build parking spaces as required without a legal obligation. Berlin abolished the passenger car parking space regulation for new buildings a couple of years ago and instead rolled out a bicycle parking space regulation (Berlin, 2012).

Concept by Topp

Separated markets for flats and parking spaces for Berlin were already proposed by Topp (1998). Car-free living and flexible forms of collective garages should form a practical symbiosis with collective garage on the outskirts and access restrictions. Tangible improvements of residential quality and urban planning ambience were expected. In addition, the target was to thereby also achieve a permanent cost reduction, especially for planning permission applicants. Naturally, living circumstances and opinions can change. People who can initially live very well without their own car may not want to settle for something different for the long term. In addition, there is a common concern over the marketability of the flat or house without a parking space in the event of any sale. A flexible parking space range should provide help in this case. On the one hand legally, as there is no individual ownership to the parking spaces and, on the other hand, in terms of concept due to usage flexibility and the conversion and demolition possibility of parking spaces. The decision for or against a parking space rent is, however, possible at any time, which loosens the compulsion for car ownership. Furthermore, part of the fixed costs of owning a car are switched to variable costs, making the costs visible month for month. Topp (1998) summarises:

- No coupling of flat and parking space, with a clear separation of costs.
- No ownership of parking spaces.
- Cost-covering rents for parking spaces.
- Removal from the "assigned" parking space.
- Parking space management in the streetscape.
- Breakdown of the parking space available on basic and additional offering.
- Adjustment of the supply to changing demand through addition or reduction of parking spaces.

Case Studies: Dortmund, Düsseldorf, and Freiburg

Depending on the site, Dortmund adopted different definitions of the number of passenger car parking spaces depending on the location (Bundesinstitut für Bau, Stadt- und Raumforschung, 2011). This also includes the restriction of passenger car parking spaces in order to prevent the deviation of retail businesses to peripheral and more surface-intensive locations and therefore at a greater distance to residential areas.

In Düsseldorf (2016), the city administration has removed one of the biggest obstacles to investments in residential construction. This should enable investors to do without the parking spaces previously prescribed in many cases for new builds and to thereby reduce costs. Overall, the parking space prescription depends on the development of public transport and the flat size (Table 4).

A sample project regarding parking space organisation is the Vauban district in the south of Freiburg in Breisgau. A major component of the project is a smart parking space concept that enables a reduction of the parking space number in accordance with the regional building codes (Fabian, 2009). The development plan specifies on what properties parking spaces may be erected. In the core of the residential area, no private parking spaces are permitted, and the residential streets do not have any public parking spaces. In mixed and commercial areas, private parking spaces are possible in underground garages. On the outskirts, two collective garages were erected. For visitors, public parking spaces (for about 10% of the flats) are planned. For all car-free households, parking space areas are kept available for a potentially later date. The obligation to prove parking spaces is suspended for as long as a car-free declaration is issued annually, the car-free settlement can also be seen in transport parameters. While the city of Freiburg has a motorisation rate of about 370 passenger cars per 1,000 inhabitants, it stands at 160 in the Vauban district (Fabian, 2009).

Table 4. Differentiated obligation to create parking spaces during residential construction

Public Transport Supply	Very Good	Good	Medium
Flat Size			
Smaller than 40 sqm	1 parking space per 3 residential units (33%)	1 parking space per 2 residential units (50%)	1 parking space per 1 residential unit (100%)
Up to 87 sqm	1 parking space per 2 residential units (50%)	3 parking spaces per 4 residential units (75%)	1 parking space per 1 residential unit (100%)
88 to 103 sqm	2 parking spaces per 3 residential units (66%)	1 parking space per 1 residential unit (100%)	1 parking space per 1 residential unit (100%)
Larger than 130 sqm	1 parking space per 1 residential unit (100%)	2 parking space per 1 residential unit	2 parking space per 1 residential unit
Student and Old-Age Pensioners Flats, Promoted Residential Construction			
Smaller than 40 sqm	1 parking space per 3 residential units (33%)	1 parking space per 3 residential units (33%)	1 parking space per 2 residential units (50%)
Up to 87 sqm	1 parking space per 2 residential units (50%)	1 parking space per 2 residential units (50%)	3 parking spaces per 4 residential units (75%)

Düsseldorf, 2016.

Switzerland

Case Studies: Basel and Berne

In Basel there is no obligation to construct passenger car parking spaces, while the number of passenger car parking spaces can be reduced in Zurich with a mobility concept (Schopf et al., 2015).

The definition of a reduced parking space obligation is intended in Berne depending on the relevant spatial situation and the development quality by public transport (Berne, 1997). In the old town, a reduction of up to 80% for residential buildings, for other uses up to 50% of the parking space number is possible.

The Swiss Environmental Law specifies as a measure of the clean air ordinance and noise protection ordinance a parking space management obligation within the meaning of the traffic generator levy (VCÖ, 2006).

Netherlands

Background and Rule in Action

In the Netherlands, the parking space obligation was adjusted to the circumstances already decades ago. The Dutch ABC regional planning model establishes a close link between the site and parking space obligations for operational uses (Netherlands Ministry for Residential Planning, 1991; Martens & Griethuysen, 1999):

- **A Sites:** As a rule, located close to the city centre; good availability of public transport, but difficult access by passenger car. Here, businesses with many employees and strong visitor or customer traffic should be based (e.g. authorities): parking space restriction to 10 (in densely populated agglomerations) passenger car parking spaces per 100 employees.
- **B Sites:** Easily reachable both by public transport and passenger car: defined as 20 (dense) or 40 (rest) passenger car parking spaces per 100 employees.
- **C Sites:** Currently well reachable by passenger car, location for businesses that need to be easily reachable by car in commercial traffic.

The target of the Dutch regional planning policy is to prevent the formation of C sites and to convert existing C and B sites. This should take place by connecting them to public transport and reducing the permanent parking spaces in these areas.

Japan

Background and Rule in Action

Parking space regulations play a major role in Japan. There are guidance rules in Japan, according to which the municipalities can define the parking space regulations. Although parking space replacement levies and restrictions are often discussed, they have not yet been applied. (Nakamura, Schäfer, & Trieb, 2005)

There are various laws and directives in Japan that specify the creation of parking options for stationary traffic or define possible tools and procedures for parking space management. Where required, a change to the legal bases is made in order to meet the current needs. This relates, for instance, to the dependence between the distance to the next closest railway station and the number of parking spaces to be built according to the parking space regulations as well as the transfer of the oversight of parking processes to private companies.

A key basis for handling parked cars is the Japanese Garage Law. It stipulates that when buying a passenger car a parking space must be proven. To reduce the parking space issue in the city areas already under huge strain from stationary traffic and to obtain an unhindered flow of traffic, this act was rolled out as early as 1962. This is a local act that every city may adjust to its own needs and circumstances.

The Garage Law obligates the buyer of a passenger vehicle to submit a plan that shows the parking space for the new vehicle to the police before buying a car. The parking space must be on the land plot or in a certain surrounding (in Nagoya 2 km) around the residential property. The parking space may not be in the public street area. After checking the documents, the police issues a ticket that needs to fastened to the car. In the event of doubt about the existence of the parking space, the police can commission a committee with the verification. Violation is punished with a driving ban of the vehicle affected and a fine. Furthermore, the act describes that the use of the public street area is prohibited as a permanent parking. Even if no ban sign prohibits parking, it is prohibited to park a vehicle during the day for more than twelve hours or for longer than eight hours over night. Violations are also punished with a fine.

Transferability to Europe

If a Garage Law would be rolled-out in Europe too, some benefits and drawbacks would arise (Nakamura, Schäfer, & Trieb, 2005):

- **Benefits:**
  - The municipalities could hand over responsibility for the provision of residential parking spaces to the residents.
  - The street space would be available to other demand groups (customers and visitors).
  - Corresponding free spaces could be used otherwise.
  - The share of permanent parkers in the street space would be reduced.
  - Residential parking and the associated corresponding administration work for the issuance of special parking permits for residents would no longer apply.
  - Private, unused parking space could be mobilised.
- **Drawbacks:**
  - Possibly too little parking space or parking space to be rented may be available, in particular in old development areas close to city centres.
  - Stronger checks would be necessary to ensure observance.
  - An increased administrative amount of work would be necessary to check the proof of the parking space or to issue tickets.

Whereas in Japan the responsibility for residential parking was handed over to the residents themselves under the Garage Law, this becomes possible in Austria in old development areas close to city centres mainly in the public street space. Residents often consider these parking spaces to be their property.

Existing private garages and parking spaces are partly not used. A typical Japanese car-parking-free streetscape is shown in Figure 7. The transfer of Japanese tools and measures to Austria appears to be practical as a rule, as need for action has existed here for decades (Nakamura, Schäfer, & Trieb, 2005).

California and Australia

Case Studies

According to the Parking Cash-Out in California, businesses that provide company parking spaces for free or not at cost must pay out this "parking space promotion" not used by those employees who come to work without a car. The reduction in car usage by this requirement amounted to about 12% (VCÖ, 2006).

For major parking spaces in Sydney, the rule is that shopping centres outside of town pay for their parking spaces. This ends the competitive advantage compared to the payable short-period parking zones in cities (VCÖ, 2006) and leads to improved usage of space by means of parking garages (see Figure 8).

AUSTRIAN DIRECTIVES FOR PARKING PLACES: AN EXISTING REFERENCE

A mandatory parking space prescription is counterproductive as a rule, if it is provided without differentiation of area types. It is not conducive to prescribe the same number of passenger car parking spaces for office workplaces in a central city area with good public transport links as for workplaces in a wholesale business on the city outskirts, for instance. This differentiation also applies for residential buildings, for car-free living for instance. An important step in direction of the system consideration regarding parked cars is made by the directives and regulations for road traffic – RVS (FSV, 2008).

Figure 7. Car-parking-free urban streetscape in Miyagi, Japan
© Takeru Shibayama.

Figure 8. Parking garage at a Sydney shopping centre
© Richard Choy.

The Directive's General Recommendations

A balanced parking space provision and management is considered to be an effective instrument in (FSV, 2008) as a rule in order to align traffic to local structures. Different local periphery conditions and objectives require that the directives can only be the basis for planning carried out with expertise. Standardisations are mostly difficult. In general, traffic and development concepts should form the basis of planning for parked cars. Although the entirety of Austrian building codes already specify a certain prescription flexibility at the moment, this is only difficult to implement for the building authorities in charge.

FSV (2008) also specifies that passenger car parking spaces should as a rule be outside the public street area, exceptions must be proven on the basis of usage. In this context, it is necessary to mention the fact that the growth in the width of the cars (Schuster, Sattler & Hoffmann, 2012) causes problems in the width dimensioning of the longitudinal parking stripes.

In addition to traffic spaces, the directive (FSV, 2008) additionally recommends designating concentrated parking locations for the improvement of existing urban situations. The distance from trip origin and trip destination to the parking space (collective garage) should not be shorter on average than to the stop of adequate public transport. Adequate public transport is described, for instance, by the Canton of Berne when rural areas have a public transport headway of at least 30 minutes (VCÖ, 2006).

In general, when calculating the parking space need, a parking space balance for city areas and the overall traffic system must be taken into account. Planning allocated to individual buildings is not permissible under this directive (FSV, 2008).

Number of Parking Places in the Directive

The directive (FSV, 2008) specifies that the maximum passenger car parking space number is determined by the use (Table 5) and other influencing factors such as location and connection circumstances (Table 6) and the existing or desired modal split of the area/structure considered. The indicative values apply until proof of a good connection is provided for environmentally friendly modes on the basis of traffic analyses, allowing to further reduce the parking space number (e.g. by business mobility management).

For industry, commercial and service businesses, retail businesses and shopping centres, the parking space maximum limit is calculated by multiplying the basic value (Table 5) with the share of passenger car vehicles according to land-use (Table 6). If possible, specific modal split values should be used, otherwise values can be used from the stated ranges. As a rule, the average values of the ranges from Table 6 are to be used. When using business mobility management the lower thresholds apply. In the event of bad supply with eco-friendly modes, the upper limits apply (Schopf, 2016).

As a rule, the number of parking spaces for residential uses (Table 5) depends on the motorisation grade. The directive (FSV, 2008) underlies a value of 530 passenger cars per 1,000 inhabitants. Deviating target values require interpolation or extrapolation. Within a city, the motorisation rate can show major fluctuations as well – city centre vs. outskirts. Parking should as a rule be organized in collective garages (see also ÖVG, 2005). When planning for car-free settlements and when good eco-friendly mode connections exist, the parking spaces number should be adjusted to these circumstances. A restriction of the parking spaces number according to corresponding traffic analyses and by way of an ordinance fixing requirements is permissible, and indeed to be worked towards (see also Sammer et al., 2005). If a waiver is given for private car usage, the parking space erection can be suspended for a certain period

Table 5. Extract from the table of parking space base values

Type of Usage	Parking Place Base Value (= 100% Private Car)	Minimum Number
	Residential Building	
Single-family home	1 per 60 sqm GFS	
Block of flats	1 per 60-80 sqm GFS	1 per flat
Dormitory (students/workers)	1 per 2 beds (where applicable, reserve for changed social structure)	
Old-age pensioners' home	1 per bed	5 per home
	Industrial and Commercial Businesses	
Staff	1 per 1.25 jobs	1 per business
Visitors	1 per 6 jobs	1 per business
Warehouse	1 per 80 sqm hall surface	
	Gastronomy Businesses	
Restaurant	1 per 3 seats	
Hotel	1 per 2 beds	
Motel	1 per 1.5 beds	
Youth hostel	1 per 5 beds	

FSV, 2008.
GFS = gross floor surface

Table 6. Framework values for the passenger car share in the modal split

Location	Land-Use	
	Customer-Intensive Services Operations and Retail Businesses	**Other Service Businesses and Retail Businesses as Well as Industrial and Commercial Businesses**
Core of large or medium-sized cities	20 to 40%	10 to 30%
Areas close to centre of major/medium-sized cities; Centre of small towns or suburban municipalities	20 to 60%	30 to 70%
Edge of large or medium-sized cities; Residential areas of small cities; suburban municipalities or villages; Industrial and commercial zones in general	50 to 80%	40 to 85%
Outside of the developed area (to be avoided), catchment areas of smaller to medium-sized places or highly motorised customers	70 to 90%	60 to 90%

FSV, 2008.

of time. When the requirement no longer applies, the provision of a parking space by the user takes place (see also Topp, 1998).

In order to take into account the rising mobility needs of mobility-restricted persons, the directive (FSV, 2008) stipulates disability-compliant parking spaces (Figure 9). As about 1% of all passenger cars registered in Austria are licensed to disabled persons, the parking spaces basic value for the disabled at 1% results from the values of Table 5. In particular for shopping centres, retail businesses, public build-

Figure 9. Arrangement of parking spaces for the disabled
FSV, 2010.

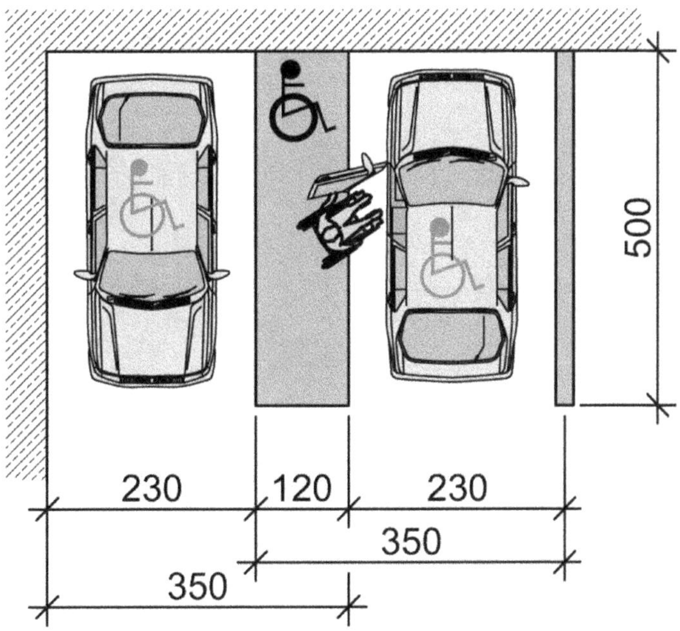

ings and hospitals, one disabled parking space is required. If some building codes require higher values, this is appreciated. Special circumstances, e.g. care homes in the catchment area, must be observed in any event.

Considering that the transport policy objectives assume an increase in the bicycle traffic share, (FSV, 2008) recommends the erection of a sufficient number of bicycle parking spaces as a promotion measure – both at the origins and the destinations of bicycle traffic (Figure 10). For that reason, an own framework value table for the number of the bicycle parking spaces per usage was included in the directive (FSV, 2008) (see Table 7). The framework value is based on a transport mode share of about 20% of bicycle traffic.

The difference in procedures between the states' parking space ordinances and the Austrian directives are illustrated in Figure 11.

SOLUTIONS AND RECOMMENDATIONS

Regional planning laws and building codes govern, among other things, the competition situation between modes of transport and therefore play a major role in transport policy. From past experience, the need for consistent (!) rules and foresighted building codes applies. The current foundations allow only for, at best, planning with many compromises and the known results (Schopf, 2016).

Between the current parking space requirements and the regional planning and climate targets, a general contradiction exists. The regional planning laws require, among other things, the protection of the land, the climate targets additionally ask for the promotion of mobility management and the reduc-

Figure 10. Bicycle parking spaces are in demand, here at Vienna State Opera
© Josef Michael Schopf.

Table 7. Extract from the table of the framework values for the minimum number of bicycle parking spaces

Type of Usage	Minimum Number of Parking Spaces
1. Residential	
1.1 Flat in general	1 per 50 sqm GFS
1.2 For visitors of private flat	1 per 300 sqm GFS
1.3 Home (dormitory)	1 per 2 beds
1.4 For visitors of residential unit	1 per 5 beds
1.5 Old-age pensioners home	1 per 5 beds
1.6 Hospital	1 per 4 beds
5.4 Visit of Hotels and Gastronomy	
5.4.1 Inns	1 per 8 sears
5.4.2 Hotels and B&Bs	1 per 20 beds
5.4.3 Youth hostels and youth guesthouses	1 per 10 beds
5.4.3 Camping and tenting sites	1 bicycle box per 600 sqm of lot surface
Further: Training/Work/Care/Leisure	

FSV, 2008.

Figure 11. Schematic comparison of ordinance vs. directive procedures

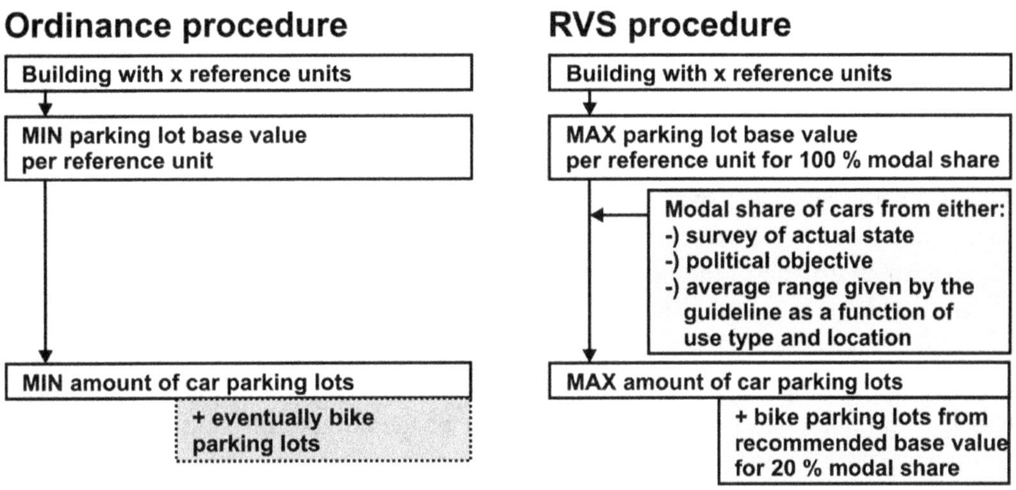

tion of CO_2 emissions. All of this is not ensured with the building codes. Adequate in terms of transport policy and market economy would be (Schopf, 2016):

- A zero prescription in the residential area with a parking spaces market.
- Permanent parking in collective garages in equal distance to public transport stops.
- A limited number of managed short-term parking spaces in public spaces.
- Parking space maximum thresholds for structures.
- A traffic generator levy.

A good share of the traffic problems observed results from the fact that the numerous aspects of the transport system have still not been examined comprehensively. For instance, the analysis of the building codes has shown that the current procedure does not suffice when handling the stationary traffic to fulfil future requirements in accordance with the "smart city" notion.

Parking space planning is not an individual matter. Parking space planning and organisation are a task of the community and the public. The provision and planning of all parking spaces must therefore be carried out by an organisation that ensures, as does public transport with the stops, that the parking places are planned, built and managed accordingly (ÖVG, 2005).

It is therefore proposed for future regulations that the parking spaces obligation be made generally dependent on the eco-friendly transport mode supply or motorisation rate. A complete abolition of the parking space obligation makes sense if an adequate management of public parking spaces is given and a market was created for collective parking spaces, i.e. a comprehensive parking space management is in action.

On the basis of a functioning parking space management, parking spaces can be arranged separately from the flats. As such, it is possible to respond more flexibly to situations and needs changing over life (see also Topp, 1998). The federal states could here, for instance, assume a practical guiding function in residential property promotion using promotion directives (see also Huber & Yilmaz-Huber, 2013). The measures in the area of stationary traffic can be summarised as follows (Schopf et al., 2015):

- The minimal requirement is the definition of parking space upper limits depending on the eco-friendly transport mode supply.
- A good connection by the eco-friendly transport modes should be part of residential construction promotion.
- An attractive minimum number of bicycle parking spaces is to be anchored in all building codes.
- Promotion of car sharing offers for every settlement unit.
- Collective garages/parking spaces with adequate density in relation to the stop density instead of house garages/parking spaces (equi-distance).
- Longer-term separation of the markets for flats and parking spaces in accordance with the principle of causation

A key step towards systematic consideration of parked cars is directive RVS 03.07.11 (FSV, 2008). It articulates a warning against simple standardisation of the number of parking spaces, both due to different framework conditions and due to its far-reaching consequences. The specific definition of the number of parking spaces therefore requires an expert, responsible and comprehensive planning approach. Deviation from the directives values are always possible in justified cases.

The directive should form the basis of the building codes in that it is employed directly by an ordinance or in that the existing building codes are modified on the basis of the directive, e.g. (Schopf & Brezina, 2015):

- Federal states change framework (+ harmonisation between the states).
- Municipalities use the leeway of the state ordinances.
- Municipalities change regulations in their own ambit (plan instead of stubborn calculation).

In light of the importance of foresighted solutions for parked vehicles, measures should be set rapidly in any event. Crucial for the understanding is the notion that this is the necessary organisation of a technical system and not a chicanery of people (ÖVG, 2005).

Some benefits of such an improved parking policy are likely to be:

- Areal parking management would close a substantial gap in municipal resource management.
- Additional funds for environmentally friendly modes of transport would be made accessible.
- Public transport subsidies by the public sector would be reduced due to increased ridership.
- Improvement of local shopping opportunities.
- Improved quality of life of urban streetscape due to less parked and less moving cars and a reassignment of surface use.
- Inner-city shopping and trade improve their position over peripheral shopping centres, with seemingly limitless parking space ressources.

E-MOBILITY: FUTURE RESEARCH DIRECTIONS

Electric vehicles will gain in importance in the medium future given the numerous electric-drive models announced and the numerous initiatives from the EU, the federal government and the energy supply companies (Schopf & Brezina, 2015). As a logical consequence, charging will be in stronger demand in the future. Charging stations in public street areas will only be possible to a limited degree. However, the fitting of garages with charging stations seems technically unproblematic. A quantitatively adequate retrofitting option, however, must be observed. In addition to the commercial questions in need of clarification, the safety restrictions existing in some building codes for garages need to be adjusted to the current state of the art (FSV, 2010). So far, only few Austrian states have adopted relevant regulations for electric vehicles. As such, electric vehicles and corresponding charging stations argue for being implemented in collective garages in principle. The optimal planning of this infrastructure will keep research busy in the near future.

CONCLUSION

The main aim of our chapter was to disclose the misconceptions in Austria's parking ordinances and the differences to the state-of-engineering guidelines. Besides that we did not only introduce the theoretical background for a massively improved parking policy but also gave selected inter-/national examples of alternative parking policies. In combination with the introduced guidelines cornerstones, the presented examples may serve as a stimulating narrative to be copied.

The essence of a parking policy improvement thus is leaving behind the rigid, demand stipulating regulations of the past and introducing tools for active and progressive transport regime design by means of parking space provision: transition from minimum to maximum numbers principle, enable adaptation to different and changing boundary conditions and political aims, introduce true-cost pricing, strive towards access equality among different modes.

The solution of traffic problems is mostly sought in flowing traffic. Though, a balanced parking space provision is one of the most effective instruments to coordinate traffic to local structures. The existing imbalance between the traffic system, settlement structure and environmental targets is strongly co-determined by the existing parking spaces. The lack of parking places in the city centres is pitted against the free peripheral parking spaces, which promote urban sprawl. Added to this is the undesirable effect of choice of location for trade and businesses, such as the development of peripheral shopping centres and the associated out flow of purchasing power from town centres. Unfortunately, most parking space prescriptions do not take into account the location, the connection and the principle of causation. If they were to do this, an incentive would arise to select well-connected sites. A commercially and ecologically sustainable settlement structure is dominated by limited passenger car parking spaces, which means that they can also be connected by public transport means by bike or on foot (Schopf & Brezina, 2015).

The handling of parked vehicles has a major impact on settlements. On the one hand, parked vehicles in the public streetscape use up valuable space, on the other hand, the choice of mode of transport and choice of location of the structures are hugely influenced by the provision of parking space. An innovative parking space management is therefore one of the most effective tools to coordinate traffic with settlement policy targets. (Schopf in Schopf & Brezina, 2015)

REFERENCES

Berlin. (2012). *Stellplatzregelung bei Neubauten.* Retrieved April 04, 2016, from http://soup.fin.io/post/276091733/Berlin-hat-die-Pkw-Stellplatzregelung-bei-Neubauten

Bern. (1997). *Baugesetz (BauG), Fassung vom 18.6.1997, Änderungen bis inklusive 25. März 2002.* Bern, Schweiz: Großer Rat Kanton Bern.

BMLFUW. (2006). *Anpassung der Klimastrategie Österreichs zur Erreichung des Kyoto-Ziels 2008 -2012 - Entwurf 17.06.2006.* Wien, Österreich: Bundesministerium für Land- und Forstwirtschaft, Umwelt und Wasserwirtschaft.

BMLFUW. (2010). Maßnahmenvorschläge. Wien, Österreich: Bundesministerium für Land- und Forstwirtschaft, Umwelt und Wasserwirtschaft; Bundesministerium für Wirtschaft, Familie und Jugend. Author.

Buchanan, C. (1964). *Verkehr in Städten.* Essen, Deutschland: Vulkan-Verlag.

Bundesinstitut für Bau-, Stadt- und Raumforschung. (2011). Ohne Auto einkaufen - Nahversorgung und Nahmobilität in der Praxis. Berlin, Deutschland: Bundesministerium für Verkehr, Bau und Stadtentwicklung, Werkstatt. *Praxis*, *76*, 29.

Bundeskanzleramt. (2015). *Landesrecht im Rechtsinformationssystem.* Wien, Österreich: Bundeskanzleramt.

Car Loft. (2016). *Was ist Carloft?* Retrieved April 1, 2016, from http://www.carloft.de/carloft-gmbh/

Chorherr, C. (2014). *Flexibilisierung der Stellplatzverpflichtung.* Retrieved April 04, 2016, from http://chorherr.twoday.net/stories/453140592/

derstandard. (2013). *Der "Qualitätswahnsinn" und seine neuen Feinde*. Retrieved April 04, 2016, from http://derstandard.at/1363710562791/Der-Qualitaetswahnsinn-und-seine-neuen-Feinde?seite=2#forumstart

Dorner, A., Herry, M., & Schuster, M. (1997). *Werkstattbericht Parkraumbewirtschaftung in Wien*. Wien, Österreich: Magistratsabteilung 18.

Dortmund. (2006). *Stellplatzkosten in Dortmund - Stand: 2006*. Dortmund, Deutschland: Stadtplanungs- und Bauordnungsamt Dortmund.

Düsseldorf. (2016). *ÖPNV-Lagegunst der Wohnstandorte nach Stadtteilen und ÖPNV-Einzugsbereiche ausgewählter Haltestellen in Düsseldorf*. Retrieved April 4, 2016, from https://www.duesseldorf.de/bauaufsichtsamt/baulexikon/stellplaetze.shtml

Fabian, T. (2009, November). *Quartier Vauban, Freiburg - Autoreduziertes Verkehrskonzept*. Paper presented at Innovatives Parkraummanagement – Chance für Wirtschaft & Umwelt (Lebensministerium), Bregenz, Österreich.

FGSV. (1991). *Empfehlungen für Anlagen des ruhenden Verkehrs (EAR 91)*. Köln, Deutschland: Forschungsgesellschaft für Straßen- und Verkehrswesen.

FSV. (2001). Grundlagen und Motive bezüglich der Organisation und Anzahl der Stellplätze für Fahrzeuge im Individualverkehr. Wien, Österreich: Österreichische Forschungsgesellschaft Straße-Schiene-Verkehr, Arbeitspapier Nr. 1.

FSV. (2008). *Organisation und Anzahl der Stellplätze für den Individualverkehr – RVS 03.07.11*. Wien, Österreich: Forschungsgesellschaft Straße-Schiene-Verkehr.

FSV. (2010). *Entwurfsgrundlagen für Garagen – RVS 03.07.32*. Wien, Österreich: Forschungsgesellschaft Straße-Schiene-Verkehr.

Haider, A. (2010). *Urbane europäische Ballungsräume – Potenzial für eine stadtverträgliche Mobilität (Unpublished master's theses)*. Wien, Österreich: Technische Universität Wien.

Hamburg. (2013). *Senat beschließt weitere Erleichterungen für den Wohnungsbau*. Retrieved April 04, 2016, from http://www.hamburg.de/pressearchiv-fhh/4128858/2013-10-29-bsu-wohnungsbau/

Huber, J., & Yilmaz-Huber, N. (2013). *Leistbarer Wohnraum in Tirol*. Innsbruck, Österreich: Arbeiterkammer Tirol.

Janik, W. (2002). Parkmöglichkeiten für Pkw - Mikrozensus Juni 2001. *Statistische Nachrichten*, *4*, 294–300.

Käfer. (2011). *Ordnungspolitik und energieeffiziente Raumstrukturen: Evaluierung von Instrumenten und Least-Cost*. Wien, Österreich: Klima- und Energiefonds, Blue Globe Report, 1/2011

Karajan, J. (2001). Wirkung und Festlegung von Stellplatzvorschreibungen für Wohnnutzungen. Innsbruck, Österreich: Institut für Straßenbau und Verkehrsplanung, Universität Innsbruck, Schriftenreihe, Heft 34.

Knoflacher, H. (1981). *Human Energy Expenditure in Different Modes: Implications for Town Planning*. Paper presented at the International Symposium on Surface Transportation System Performance, Washington, DC.

Knoflacher, H. (1987). Verkehrsplanung für den Menschen: Vol. 1. *Grundstrukturen*. Wien, Österreich: Orac Verlag.

Knoflacher, H., & Schopf, J. M. (2001). Modellvorhaben Sanfte Mobilitäts-Partnerschaft. *Internationales Verkehrswesen*, *53*(3), 86-89.

Knoflacher, H., Schopf, J. M., & Brezina, T. (2007). *Zusammenstellung und Analyse der Vorschriften und Richtlinien zu Anzahl und Organisation der Stellplätze im MIV für Österreich in Hinblick auf eine Verbesserung der Umweltverhältnisse*. Wien, Österreich: Bundesministerium für Land- und Forstwirtschaft, Umwelt und Wasserwirtschaft.

Martens, M. J., & Griethuysen, S. v. (1999). *The ABC Location Policy in the Netherlands - 'the Right Business at the Right Place'*. Retrieved April 04. 2016, from https://fenix.tecnico.ulisboa.pt/downloadFile/3779572236303/abc.pdf

Nakamura, H., Schäfer, P., & Trieb, A. (2005). Parkraumbewirtschaftung - Was macht Japan anders? *Internationales Verkehrswesen*, *57*(10), 434-437.

Niederländisches Ministerium für Wohnungswesen (1991). *The Right Business in the Right Place*. Den Haag, Niederlande: Niederländisches Ministerium für Wohnungswesen.

OECD. (1981). *Evaluation of urban parking systems*. Paris, France: OECD.

ÖVG. (2005). *Merkblatt Chancengleichheit der Verkehrsträger*. Wien, Österreich: Österreichische Verkehrswissenschaftliche Gesellschaft.

Parlament. (2015). *Preistreiber Stellplatzverordnung*. Retrieved April 04. 2016, from http://www.parlament.gv.at/PAKT/VHG/XXV/J/J_02362/fnameorig_362885.html

Peperna, O. (1982). *Die Einzugsbereiche von Haltestellen öffentlicher Nahverkehrsmittel im Straßenbahn- und Busverkehr (Unpublished master's theses)*. Wien, Österreich: Technische Universität Wien.

Pfaffenbichler, P., & Schopf, J. M. (2011). *Einfluss der Parkraumorganisation und der Anzahl der Stellplätze auf die Nutzung des motorisierten Individualverkehrs und die Erreichung verkehrs-, umwelt- und siedlungspolitischer Ziele (PAN)*. Wien, Österreich: Österreichische Energieagentur – Austrian Energy Agency.

Reichsarbeitsministerium. (1939). *Verordnung über Garagen und Einstellplätze (Reichsgaragenordnung)*. Berlin: Author.

Riedel, R. (2014). Grundsätze der Stadtplanung im Ruhenden Verkehr. *Zeitschrift für Verkehrsrecht*, *12a*, 453-461.

Sammer, G. (2005). *Instrumente zur Steuerung des Stellplatzangebotes für den Zielverkehr – IN-STELLA, Teil 1: Analyse nationaler und internationaler Umsetzungsbeispiele*. Wien, Österreich: Universität für Bodenkultur, Department für Raum, Landschaft und Infrastruktur.

Schopf, J. M. (2016). Der ruhende Verkehr in der RVS. In H. Knoflacher & H. Frey (Eds.), *Paradigmenwechsel im Verkehrswesen* (pp. 145–166). Wien: Österreichische Verkehrswissenschaftliche Gesellschaft.

Schopf, J. M., et al. (2015). *Wohnbau, Wohnumfeld und Mobilität*. Wien, Österreich: Verkehrsclub Österreich, Mobilität mit Zukunft 3/2015.

Schopf, J. M., & Brezina, T. (2015). *Umweltfreundliches Parkraummanagement – Leitfaden für Länder, Städte, Gemeinden, Betriebe und Bauträger*. Wien, Österreich: Bundesministerium für Land- und Forstwirtschaft, Umwelt und Wasserwirtschaft.

Schuster, A., Sattler, J., & Hoffmann, St. (2012). Benötigen wir ein neues Pkw-Bemessungsfahrzeug für den Entwurf von Anlagen des ruhenden Verkehrs? *Straßenverkehrstechnik, 56*(1), 5-10.

Stift, W. (2002). *Stop der Stadtvernichtung*. Wien, Österreich: Bundesgremium Textilhandel.

Tirol. (2015). *Stellplatzhöchstzahlenverordnung 2015*. Innsbruck, Österreich: Tiroler Landesregierung.

Topp, H. H. (1998). Getrennte Märkte für Wohnung und Stellplatz. *Internationales Verkehrswesen, 50*(7-8), 322-324.

VCÖ. (2006). *Gute Verkehrslösungen in Gemeinden und Städten*. Wien, Österreich: Verkehrsclub Österreich. VCÖ-Magazin.

Walther, K. (1973). Die Fußweglänge zur Haltestelle als Attraktivitätskriterium im öffentlichen Personennahverkehr. *Verkehr und Technik, 26*(11), 480-484.

Wikipedia. (2016). *Smart City*. Retrieved April 05, 2016, from https://de.wikipedia.org/wiki/Smart_City

Chapter 9
Sustainable Parking Planning for Smart Growth of Metropolitan Cities

Ozge Yalciner Ercoskun
Gazi University, Turkey

Ebru Vesile Ocalir Akunal
Gazi University, Turkey

ABSTRACT

This chapter identifies main problems of parking planning in demand-oriented conventional paradigm. Parking planning principles and cost-effective programs in supply-oriented sustainable paradigm are presented in the chapter listing many economic, social and environmental benefits. The chapter describes various parking strategies for developing a sustainable parking plan for smart growth of metropolitan cities, including cases of Istanbul and Vienna. A critical perspective is drawn for the parking planning policies for these city cases. Parking is recognized as an important factor influencing accessibility. Parking planning, parking policies and pricing play a critical role in local government decision making.

INTRODUCTION

The ongoing demand for the automobile in developed countries and the growth in automobile demand in developing countries has imposed many difficulties on sustainable transport. These have accelerated some studies on the influence of the built environment on climate change, air pollution, obesity, land use and parking policies worldwide. Specifically, sprawling urban development in the metropolitan cities has been criticized for its contribution to the rising demand for automobile use, land use, transport and parking policies. Researchers aim to prove that the built environment has effects on automobile ownership and offer guidance for policy design and parking planning implementation.

Parking planning is an evolving subject, a paradigm shift in how a problem is perceived and solutions put. The old paradigm is demand oriented which creates new parking lots when needed according to the automobile ownership boom. Parking facilities are subsidized by local governments, and that every

DOI: 10.4018/978-1-5225-2116-7.ch009

region should provide its own parking needs. The new paradigm is oriented to provide optimal parking supply. It strives to use parking facilities sustainably by sharing between different regions. Parking facility costs are charged directly to automobile owners, and financial rewards, subsidies or credits are provided to urban dwellers that reduce their parking and oriented to public transport and bicycle.

Smart Growth is based on New Urbanism and Transit Oriented Development which generates development policies for more sustainable transportation and land use by creating more compact and polycentric with multi-modal transportation systems. Smart growth is supported by sustainable parking planning. Parking planning reduces the amount of land required for parking facilities, reduces automobile use and increases public transportation. Parking plans for smart growth allow more sharing of parking facilities, shifts to alternative modes, and various types of parking pricing.

Parking problem is one of the largest negative results of rapid urbanization and conurbation of Istanbul. It decreases the quality of life and threats environment and natural resources in this metropolitan city. Istanbul Metropolitan Municipality prepared a master plan for solving this problem. Within this plan, problem analysis and site studies have been made, database was built, zoning has been created and parking supply and demand has been determined. After modeling with Geographic Information Systems, parking plan proposals are put by demand based estimation. These proposals will guide current parking regulations in this metropolitan city. As a second case, parking planning in Vienna has a different approach in new paradigm. It is supply oriented and district based which has strict regulations with parking stickers for visitors and residents of planned districts. Short term parking, seasonal parking and permanent parking permit have different requirements which push automobile users to pedestrian, bicycle or public transport. Electronic parking sticker projects started and applications are held by e-government procedures. The city aims to reduce automobile traffic and environmental pollution. An improvement was made on public transit, overall parking with economic accessibility for sustainable urban transportation. These projects are under strategic planning by transit oriented development for smart growth of Vienna.

This chapter investigates main problems with current parking planning in old conventional paradigm. Parking planning principles and cost-effective programs are put in new sustainable paradigm providing many economic, social and environmental benefits. This chapter describes various parking strategies, how to evaluate these strategies and develop a sustainable parking plan for smart growth of metropolitan cities, including cases of Istanbul and Vienna and resources for more information. A critical perspective is drawn for the parking planning policies for these city cases.

PARKING PROBLEMS AND LEGISLATION

Parking spaces attract cars; so they generate car traffic. Parking needs space, which is not available for other street uses. Nothing else has changed the traditional streetscape as dramatically as parked cars have done during the last few decades. (H.Topp, Uni. of Kaiserslautern, Germany)

The parking problems can be categorized into four main headings such as lack of strategic approach, parking legislation, off-street parking supply and on-street parking. The sub-problems under these headings are given in Table 1. These problems are caused according to direct and indirect impacts.

Table 1. Parking problems

Main Problems	Sub-Problems
Lack of strategic approach	-Rapid urbanization and increase in population -High rise buildings in city centers -Increase in household income and rapid automobilization -Automobile-oriented transport and public transport on roads -Fragmented institutional structure in municipality and having no strategic plan -Demand oriented conventional approach
Parking legislation	-Fee preference of constructors and municipalities -Public loan and use of parking fees -High urban rent and having no regional parking areas -İnadequate parking areas in master plans
Off-street parking supply	-Marginal parking management without license -Non-standard parking services -Wide use of open air parking areas -Temporary open parking areas -Small size of closed parking areas and inadequate public investments -İnvestment risks for private sector -İnadequate technological developments such as automatic parking
On-street parking	-The habit of irregular and free parking on streets -Resistance to parking fees -İnadequate controls -Lack of control mechanism of municipalities -İnefficient tow truck services -Occupying sidewalks, inaccessible for pedestrians -Long time parking near commercial corridors -Large amount of traffic circulating looking for a parking space contributing to congestion, fuel consumption and pollution -On-street parking causes safety and congestion problems

An increase in income correlates with higher car-ownership and more central residence. These rapid developments raise the parking demand. Parking was solved on streets but this was not a long-term solution and parking problem grew. Double and triple parking on roads is also done regularly, blocking the flow of traffic. This is a problem of attitudes and a failure to respect the fact that roads are a shared space where mutual consideration should be the rule. Driving on sidewalks is also a ubiquitous problem. Drivers parking on sidewalks in vehicles as large as trucks could be made to pay for all repairs caused by their damage. The placement of obstacles on streets by individuals and businesses blocking parking needs to be controlled. Parking legislation is demand-oriented and estimate-supply model is conventional which is inadequate for the parking problem solution. The constructors pay parking fees and do not build parking area but they transform this area to an apartment and it is possible according to this legislation. These applications provide barriers to parking area requirements and city centers have no vacant land for parking. Parking fees are paid to municipality, 25% of the fee is paid in advance and the rest of it is paid after the construction and it is applied to flat deed in Turkey. However, it is uncontrolled and flat owners do not pay attention. Sometimes the flat is rented and parking fees are not controlled in the municipalities. On-street parking is a problem at night time and causes social, economic and environmental problems. The adequate parking and creating regional parking areas should be prioritized for public interest.

SMART GROWTH, TRANSIT ORIENTED DEVELOPMENT (TOD) AND PARKING PLANNING

Smart Growth requires a balanced perspective to local development by encouraging economic improvement, a sustainable environment and resilient communities. The wise use of land around public transport stations is one approach to achieve smart growth objectives of metropolitan cities. Well planned station areas promote the sustainable city by (TOD Handbook, 2004):

- Emphasizing transportation modes and increasing transit ridership,
- Decreasing automobile dependency and greenhouse emissions,
- Creating a more compact urban form,
- Making better connections between offices and residential areas,
- Revitalizing shopping areas and older neighborhoods,
- Creating opportunities for affordable housing,
- Providing increased travel options for urban dwellers,
- Making walkable neighborhoods,
- Creating safer street environment,
- Increasing assessment values of vacant and underused land.

Transit-oriented development (TOD) proposes walkable communities for people and provides more transportation and housing alternatives. TOD is compact development around a transit station or within easy walking of a station and containing a mix of land uses such as housing, offices, shops and entertainment. New transit stations are planned according to regional transportation patterns and travel needs. By concentrating economic activity and housing around stations, TOD plays an integral role in shaping regional development, including both public and private investment. TOD can decrease personal transportation costs, encourage public transit ridership, reduce greenhouse gas emissions and air pollution, bring economic development, and make affordable high density housing by reducing developer expenditures on parking (EPA website).

Some strategies are designed to support the use of transit and to create a walkable transit friendly environment, reducing or eliminating the need for a private automobile. One of these is TOD friendly parking. Retail streets support a lively pedestrian realm. Shared parking between transit stations and other land uses has significant potential benefit. Transit can also be supported by the use of transit supportive zoning and overlay zones. TOD and Transit Overlay Zones allow for more density while reducing parking requirements. Walkable and bike-able environments are keywords to developing vibrant downtowns, city centers, and transit neighborhoods (MTC website, 2007).

Convenient parking and drop-off zones need to be planned for in all station area by TOD design. The motto is 'Parking - enough, but not too much!'. TOD reduces the amount of parking in the station area through increased transit riderships, decrease residential automobile ownership and shared off-peak parking at public park and ride areas. Minimum and maximum parking standards can be set near transit station. Parking lots can be located rear and sides of the buildings near the station (Figure 1). This keeps the station and building entrances oriented to the sidewalk and to pedestrians. Larger parking lots can be divided into smaller lots and separated by landscaped walkways. Structured parking consumes less land than surface parking and allows maximum development. If parking structures are located along

Figure 1. The digital signs showing the off-street parking capacity near the station in Cologne (Cologne website, 2017)

key walking routes, they can enhance the public environment with pedestrian-friendly facades (TOD Handbook, 2004).

PARKING MANAGEMENT

Parking is an essential component of the transportation system. It can encourage car ownership, driving in cities and urban sprawl as space is needed for parking lots. On the other hand, policies influencing parking supply and costs can reduce traffic and car use (Litman, 2008).

The management of parking space is a form of regulating the use of parking facilities along streets by way of a system of charges in order to restrict, or favor, certain user groups (Herry et al., 1999). Parking management can affect the relative price and convenience of driving, land use density, accessibility, and walkability. Policies implemented to improve sustainable transportation conditions should incorporate planning, regulation, information, technology, and economic measures (Dalkmann & Branningan, 2007).

Different parking types may have different conditions in terms of operation, regulations and functions. Parking types are grouped as (Rios et al., 2013, 27):

- **On-Street or Off-Street:** Determined by location on or off the public roadway (separate properties),
- **Paid On-Street:** Any on-street parking space requiring payment, either by regulation or informal means,
- **Free On-Street:** On-street parking free of charge,

- **Informal On-Street:** Part of paid on-street parking, referring to any parking that has an informal service and/or fee (paid to a person who has "taken" the space and watches the vehicles parked there),
- **On-Street, Informal Service, Regulated Rate**: This type of service was found in the field – even though the service is informal, a mechanism has been defined in some way to standardize the on-street parking rate,
- **Off-Street:** Parking that is located off the public roadway and must comply with certain regulations. May be operated by the public and/or private sector,
- **Use-Based Off-Street:** This type of parking is primarily linked to a specific land use, such as residential or commercial (e.g., a shopping center or office building).

Regulatory instruments place restrictions on car users. If managed well, they can make car use less attractive and result a mode shift to public transport. In order to increase the level of acceptance of parking management regulations, a full sustainable transport strategy should be implemented by the local authority, including the provision of viable and attractive alternatives (public transport, cycling, and walking), and awareness campaigns (Dalkmann & Branningan, 2007) .

Parking management strategies can be in different forms (OECD,2009);

- Limiting the number of spaces to establishing pricing,
- Tax policies such as charging for parking at public spaces,
- Levying taxes on privately offered spaces,
- Converting public spaces to another purpose and reducing the number of spaces within an area or entire municipality.

Some cities adopt efficient measures in parking management. Examples are as follows:

In Barcelona (Spain) a mobility pact, which is intended as a tool to enable the local administration, different associations and civic groups to build a mobility model for the Barcelona of today and tomorrow, is signed (COST Action Report 342, 2005). The basic principles of the Pact can be summarized as; sustainability, accessibility for all, a rational use of means of transport, improving safety and adequate planning processes. The Pact has some objectives on public transport, quality for pedestrians, promoting bicycle use, the system of distribution of goods in the city, traffic information, legal aspects and road safety in addition to the number of parking spaces and their quality.

In Zurich (Switzerland) multiple tools are implemented to manage parking process (Hofstetter, 2015). The "Urban Traffic 2025" program in Zurich (Switzerland) has a parking space policy aiming at advancing city-compatible modes of transport, improving the environmental situation and revaluing public space. Parking in the inner city is shifted from the street to car parks whilst keeping the total number of available parking lots constant. Between 1990 and 2013, the number of lots for street parking in the central areas went down from 4.605 to 3,667 whilst parking spaces in car parks rose from 3,017 to 4,134 (Vienna City Administration, 2014a).

In Portland, Oregon (USA) 1997, maximum and minimum requirements for managing new parking structures, according to zone codes have been implemented since 1997. Except for the densest commercial neighborhoods, including the city center, commercial districts, and central residential districts and any site fewer than 500 feet from a public transportation line that provides frequent service, the number of spaces were limited to promote efficient use of land, enhance urban form, encourage use of alternative modes of transportation, provide for better pedestrian movement, and protect air and water quality. Many builders and owners participated the program by providing shared spaces or bicycle parking. The use of public transportation jumped 20% to 25% in the early 1970s and to 48% in the mid-1990s (Dalkmann & Branningan, 2007).

In Utrecht (the Netherlands) parking system had a reform as a part of an air quality improvement and reduce car traffic program in the mid 1980's. Parking spaces were reallocated, parking charges were increased and new circulation patterns within the city center were created. These measures helped reduce travel to the city center by car by 15% (OECD, 2009).

Vienna (Austria) has also implemented a series of these parking management measures in order to obtain a sustainable transport system and is chosen as a case study to make a comparison with Istanbul (Turkey).

CASE STUDY AREA 1: ISTANBUL

The population of Istanbul was 10 million inhabitants in year 2000, only after 5 years, it increased to 15 million. The number of automobile was recorded as a million in 2000, increased up to 2.5 million in 2015. Istanbul is a city of 15 million people living at high densities in relatively small areas, with 2.5 million vehicles, 90% of which have to park on-street, and economic growth of 10% per year. Parking demand is intense as car ownership grows from its current level of around 150 per 1000 people. There is also very little off-street parking, and a tradition of 'informal' control of on-street parking in areas of high demand.

In 2005, the Greater Municipality of Istanbul built a private parking firm, ISPARK (Istanbul Parking Management Trade Inc.), to control on-street parking and to put a stricter parking policy in city center. ISPARK controls 51 off-street car parks with a total of 17,000 lots, and 10,000 on-street lots in 226 locations. This firm has many objectives such as:

- Provide a safe parking service,
- Create a 'parking habit' as part of urban culture,
- Improve awareness of sustainable parking,
- Create a parking inventory,
- Build new parking at key points,
- Implement user friendly parking payment systems,
- Increase off-street parking,
- Use parking to encourage public transport use.

ISPARK aims to shift parking from on-street to off-street in order to create more space for pedestrians, so pricing aims to make off-street parking more attractive than on-street. ISPARK developed automatic parking payment by mobile phone. It also builds new off-street parking although this is limited by high land prices, and building costs that vary between US$7,500 to US$12,000 per space. At Sisli, one of Istanbul's key business locations in the European part of the city, on-street parking costs 4 TL (€2.30) per hour to stimulate parking turnover and use by shoppers and business visitors. In the more residential suburb of Bostanci on the Asian side, parking all day next to the ferry terminal costs 3 TL, thus encouraging park and ride. ISPARK shows significant progress in administering Istanbul's parking more rationally and in creating acceptance of parking management among riders (Eltis website, 2014).

ISPARK and Grater Municipality of Istanbul-Department of Transportation started making Istanbul Parking Master Plan in 2013. The detailed supply analysis was drawn and an inventory was created in GIS environment by determining the on-street and off-street parking in road geometry. Off-street parking was classified in to two groups: 1) regional parking areas open to public, operated commercially 2) functional short-time parking areas open to users, not operated commercially. Functional parking areas were classified into sub-groups according to night time long residential parking, day time short retail parking etc. For İstanbul, some statistics are listed in Figure 2.

The supply was irregular and uncontrolled on-street parking; the empty lots were used as off-street parking which were unregistered. The inadequate parking capacity was analyzed in residential areas. Regional analyses were made in Istanbul Transportation Master Plan, the results of car ownership and parking in the traffic zones of this plan were used in Istanbul Parking Master Plan. Minimum and maximum values were found in automobile ownership and parking capacity, then main zones were categorized as problematic when two variables are high, potential problematic zones when one variable is high and unproblematic zones when two variables are low. Main zones were also categorized as residential, non-residential and mixed use according to land use density and twelve sub-zones were defined.

The number of problematic zones is 142, potential problematic zones are 523 and unproblematic zones are 118 according to traffic zones of Transportation Master Plan (Kulekci et al., 2015). New problematic areas can be added in the near future. After categorizing, the demand estimation model was developed in four steps. Parking demand was designed in two categories as static and dynamic. Static demand was

Figure 2. Parking categories and capacities in İstanbul, 2013 (Kulekci et al., 2015)

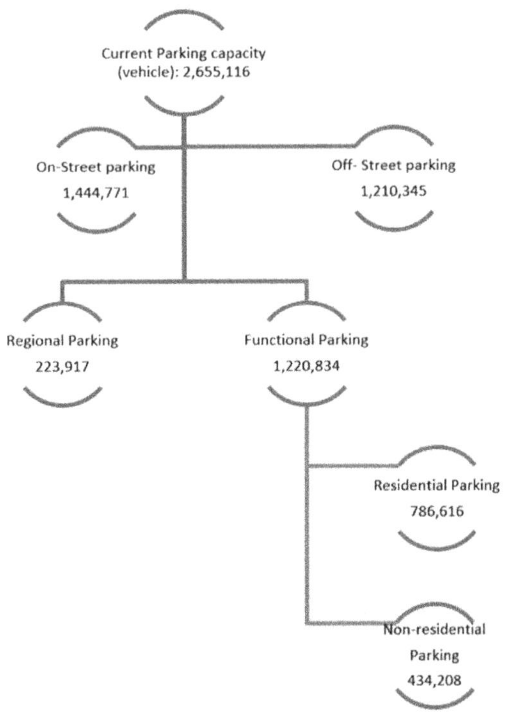

night time parking near residential areas, dynamic demand was related as short-term parking near office, school, retail etc. at day time.

The first step was to find static residential parking demand according to the car ownership data in Transportation Master Plan (Figure 3). The base assumption was one automobile per two dwellings and the highest limit was one automobile per dwelling. The second step was calculating static parking demand for offices at day time. The main parameter was office employees coming to their offices by their own automobiles. Third step was making static parking demand arrangement. The main parameters were parking demand of residential areas and offices and also day and night parking at mixed use zones as balance parking. The fourth step was calculating dynamic parking demand which are produced from the travel of office, retail, school, health etc. in short-time, this was added to static parking demand and total parking demand was estimated. The demand and supply layers are shown in Figure 4.

The parking supply was found according to the capacities of off-street functional parking, off-street general parking and on-street parking. The requirement was found as the difference between demand and supply (Figure 3). Every update in data source used in supply and demand was recognized by the demand estimation model and parking demand was also updated. The model is dynamic, when the input data is changed, the output products can be renewed. Parking requirement in Istanbul can be calculated at the block, neighborhood, district and city scales. Figure 5 shows the locational density map of parking requirement at city scale.

The flexible implementations according to the parking legislation could not provide adequate parking in the built environment, development plans propose very limited general and regional parking places in the plans and local governments could not build adequate parking areas. Large expropriation costs in built environment for supplying the off-street parking demand, limited land stock in public ownership, lack of vacant land in private ownership and implementation difficulties prevent creating new parking areas. One of the main objectives of the parking master plan was to transform the fragile parking demand

Figure 3. Parking estimation model scheme (Kulekci et al., 2015)

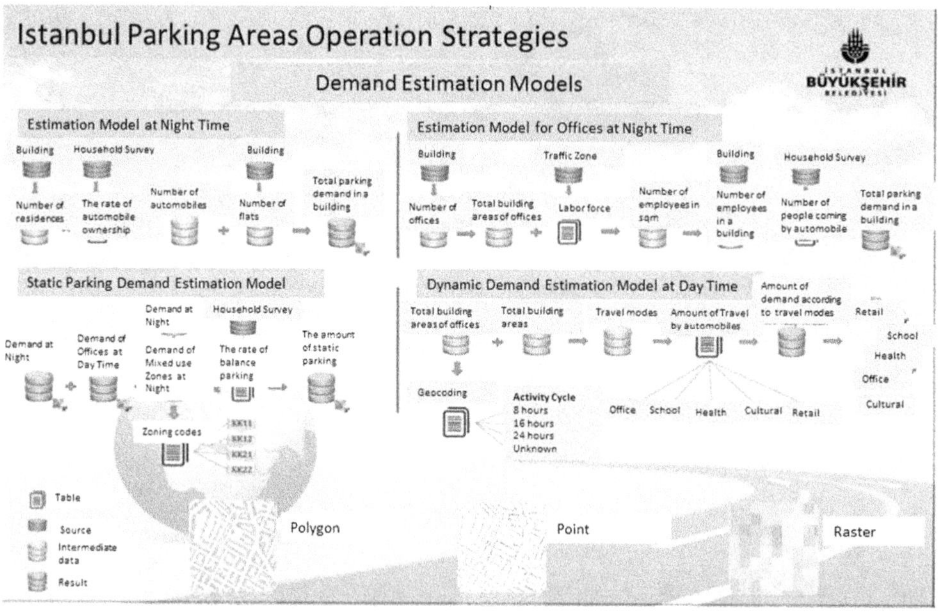

Figure 4. Demand and supply layer in parking requirement (Kulekci et al., 2015)

Demand layer	Data source	Data used
Built environment	Building	Name of building
		Object ID
		Residential-Office
		Number of floors
		Base area
Socio-economic structure	Istanbul Transportation Master Plan Household Survey	Automobile ownership
		Population
		Labor force
Travel Preferences	Istanbul Transportation Master Plan Household Survey	Travel rate by automobile
		Travel mode
		Amount of travel

Supply layer
Off-street functional parking
Off-street general parking
On-street parking

Figure 5. Locational density map of parking requirement at city scale (Kulekci et al., 2015)

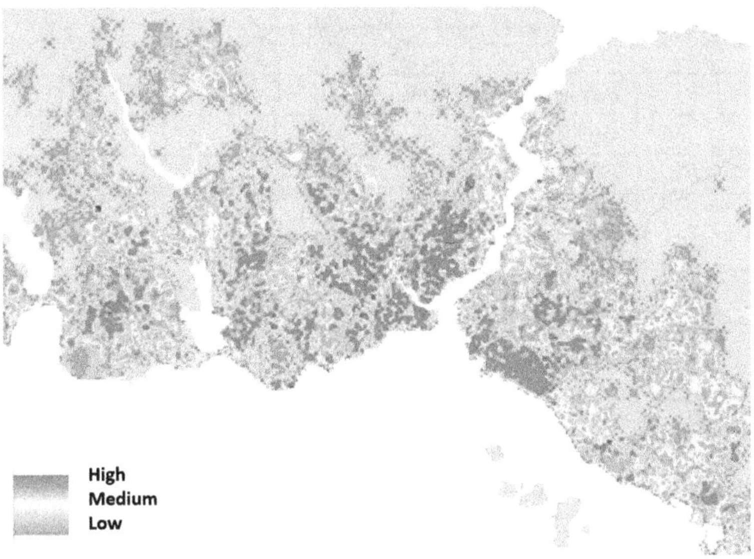

**For a more accurate representation of this figure, please see the electronic version.*

of Istanbul into resilient capacity. Greater Municipality of Istanbul launched 'Istanbul New Parking Area Planning with 100,000 Capacity' in 2014. New off-street parking areas were determined according to the detailed analyses listed in Table 2 (Kulekci et al., 2015).

Alternative parking areas were tested according to the appropriation criteria listed in Table 3. Parking demand level was determined after the results of network analysis in a walking area within 250 m. The building technique and capacity attachment was made according to the area geometry. If the selected area

Table 2. The site selection criteria of parking areas (Kulekci et al., 2015)

Zoning status	Parking
	Park/Green area
	Street/Square
Ownership	Greater Municipality of Istanbul/Abandoned areas
	Governmental/District Municipality
Current Status	Vacant Land
	Tree existence
Other public areas	Marketplace/Municipality service area
Expropriation Status	Vacant land/Low density area
Accessibility	Proximity to public transport stations
Limits	Registry status / Slope status
	Creek zones, flood risk area
	Urban transformation area
Size of the land	Min. 1000 sqm for conventional ones
	Min. 200 sqm for mechanical ones
	Practical geometry

Table 3. Appropriation tests for off-street parking (Kulekci et al., 2015)

Level of demand	Walking area within 250 m
Area geometry	Appropriate type selection
Capacity attachment	
Attribute attachment	Park and Ride
	General parking/Regional parking
	Functional parking
Parking ID	

was linked with light rail station, park and ride can be attached, if the area was in the retail or mixed-used zone, it was attached as general parking. If the selected areas were in the low density residential area, it was attached as regional parking and if the areas had only limited parking lots according to the buildings, then it was recorded as functional parking. The areas had parking ID after these tests (Kulekci et al., 2015).

In summary, the parking plan of Istanbul in 2015 is shown in Figure 6. Planned parking areas, park and ride parking areas, land use zones are drawn here. Figure 7 displays the building status of these parking areas such as completed, under construction, in the project and planned areas.

Figure 6. Istanbul Parking Master Plan 2015 (Ispark website)

*For a more accurate representation of this figure, please see the electronic version.

Figure 7. Status of parking areas in Istanbul (Kulekci et al., 2015)

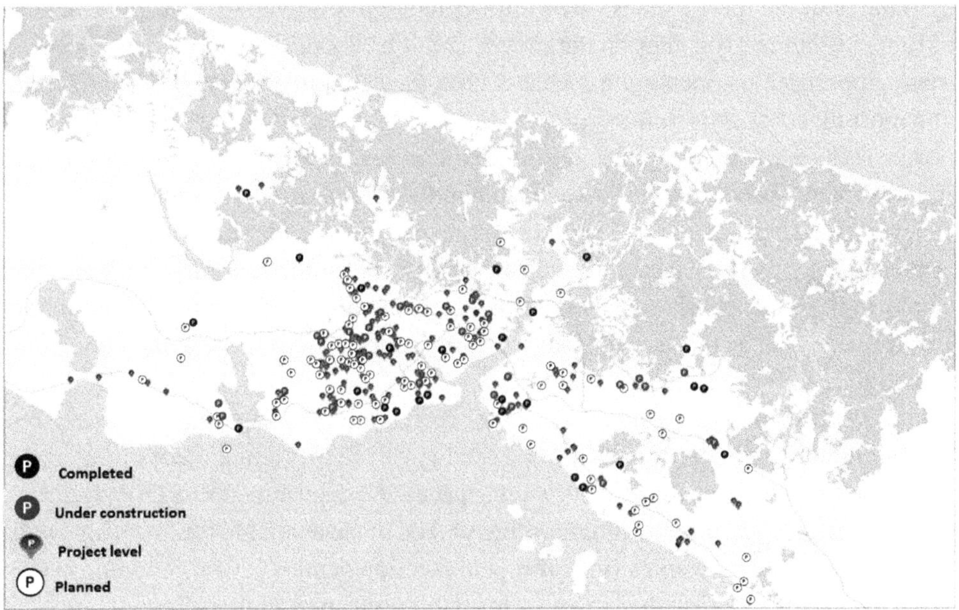

*For a more accurate representation of this figure, please see the electronic version.

CASE STUDY 2: VIENNA

In Vienna parking organization is shaped by some urban development and transport plans, policy papers and regulations:

The Urban Development Plan - 2025

In 2011, the Vienna City Council changed the City Administration with the preparation of a new Urban Development Plan, to be submitted for adoption by 2014. The task of urban development is to provide fundamental planning tools ensuring Vienna's innovative development in line with social, economic and ecological requirements. This task requires cross-border cooperation with neighboring provinces and countries, particularly in the framework of the European CENTROPE region. The Urban Development Plan -2025, which is a strategic plan, (Stadtentwicklungsplan - STEP) specifies whether an area is designated as green space or for traffic purposes, for commercial buildings or housing construction. In the future and in particular in densely inhabited (residential) quarters, priority will not only be given to traffic management or the availability of parking lots, but to atmospheric quality, so that public space can be recovered for other forms of use, such as communication, business, leisure, play and culture (Vienna City Administration, 2014b).

The Vienna Urban Mobility Plan sets out the ambitious goals of the City of Vienna for a viable transport system of the future, and it describes the steps to be taken in the next ten years so these goals can be reached. In the past, the mere expansion of infrastructure has proven to be insufficient. The motto of the Urban Mobility Plan is "Together on the move" and it banks on a variety of approaches which help seize the many opportunities opening up. This is how transport in Vienna will continue to work well and fulfil the most diverse needs in the future.

Public space reflects the ongoing changes in urban society, it is an environment for all. In the past, public space was mainly used as a transit space to handle the maximum of motorized traffic flows and for parking vehicles. Accordingly, the design of streets was for decades oriented towards cars. At present, more than 65% of street surfaces are used for the flow of motorized traffic and parking. However, as more people walk or use public transport or bicycles, the needs of non-motorized traffic and equal opportunities for it, as well as the aspect of high quality public spaces for rest and recreation have come to the focus of attention for citizens and planners.

Attractive and sufficiently large street spaces for recreation and lingering in densely built-up areas help maintain and improve the traditionally good quality of life in Vienna. Pedestrians and cyclists can be given more space by the reorganization of street spaces. As determined in STEP 2025, selected areas currently used as traffic lanes or for turning/parking will be made available for lingering, walking, public transport and cycling in suitable spots (including intersection centers). This can improve the quality of the street space for rest and defuse danger spots for public transport and cycling.

Some keynotes from the plan are (Vienna City Administration, 2014b):

- The recovery of public space is realized through a paradigm change in the planning and organization of traffic: parking space management already constituted an important step to reduce the number of cars parked on the streets and hence to make room for other, non-commercial forms of use.

- Atmospheric quality, aesthetic appeal and functionality of the recovered space are enhanced e.g. by high-quality design and art in public space, by creating encounter zones, by redesigning now obsolete parking lanes as temporary and permanent street gardens and sitting areas or cycle lanes, by reducing traffic surfaces or by setting up and upgrading multimodal traffic infrastructure facilities (bike parking spaces, spaces for all types of sharing systems, etc.). The activities proposed are developed and conducted in co-operation with the municipal districts.
- Unattractive, long distances to reach the nearest public transport stop, a convenient garage slot close by and a lack of alternative mobility services result in undesirable competitive advantages for cars in residential quarters, at company locations and in shopping malls. In the future, Vienna will therefore demand mobility concepts that highlight how car trips triggered by new projects can be made as short and eco-friendly as possible.
- The negative traffic effects of new, large-scale office sites and shopping malls will be significantly reduced by imposing the mandatory provision of efficient public transport links.
- The New Car Park Programme for Vienna promotes garage construction above all in locations where despite control measures of traffic policy, such as parking space management, the demand for parking slots is still massive. The subsidized construction of collective garages in residential quarters will focus above all on Gründerzeit quarters and densely built-up urban zones with a clear need for more parking lots. The construction of additional parking lots is to create a shift towards more permanent parking lots in garages and significantly more room to move in public space for people (cycling, walking, lingering) and public transport.

Parking Zones/ Short-Term Parking

The main parking control measure is residential parking zones (Figure 8). Vienna's program is based on areas. By preventing commuter parking in neighborhoods surrounding the center, Vienna has helped reduce the motor vehicle mode split by four percent while increasing public transit and biking mode

Figure 8. Restricted parking zones in Vienna
Source: Generated from (https://www.wien.gv.at) (June 2016)

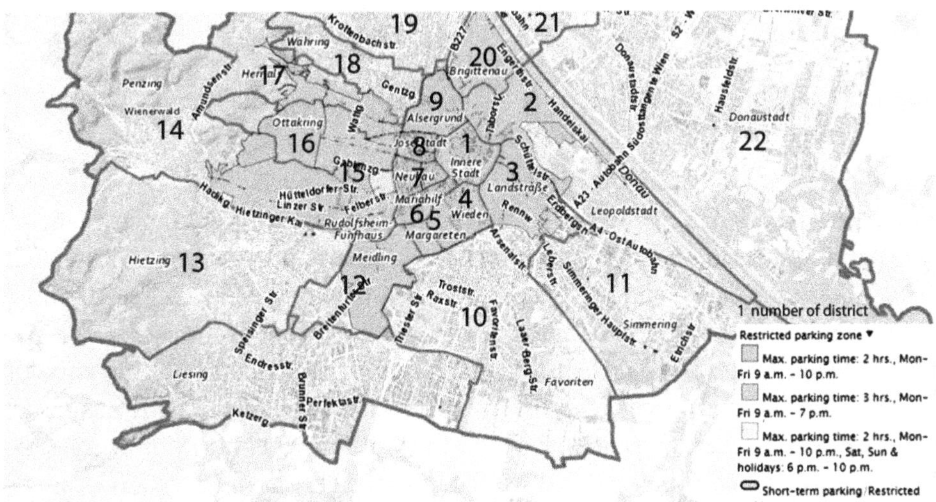

**For a more accurate representation of this figure, please see the electronic version.*

split by two percent each. In Vienna parking control is used as a tool to control traffic and transit use. Implementing new parking controls targeted towards reducing commuter traffic. Vienna replaced surface parking lost for traffic calming with underground parking. Drivers must pay to use these garages (Nash, 2004).

In Vienna, short-term parking may apply to individual streets or to entire areas. Parking space management has turned entire districts or large connected parts into short-term parking zones.

Area-wide short-term parking now covers districts 1 to 9, 15, 20 and parts of the 12th, 14th, 16th and 17th district. In the 15th district the area the area around Stadthalle is subject to special parking regulations. Against payment of a fee, longer-term parking is made available for local residents and time-limited parking for other user groups.

Herry et al. (1999) summarize their research findings on this system as;

- A definite reduction in the use of parking space along public streets;
- A significant reduction in the number of people wrongly parking their vehicles (according to the existing traffic code);
- A significant reduction in the volume of traffic created by drivers searching for a parking spot, resulting in lower traffic flows generally;
- A reduction in the volume of traffic generated by the search for parking space in the area under the control of parking management;
- No notable increase in demand for parking space in neighboring districts in Vienna and - overall it can be stated that the politically motivated system of parking control has been positively accepted by users as well as by commercial companies.

Every district has its own rules. For example;

- In districts 1 to 9 and 20 short-term parking applies on weekdays from Monday to Friday from 9 am to 10 pm.
- In districts 12, 14, 15, 16 and 17 short-term parking restrictions apply Monday to Friday on workdays from 9 am to 7 pm.
- In the zone around Stadthalle (part of the 15th district) short-term parking times are from 9 am to 10 pm on workdays and from 6 pm to 10 pm on weekends (Saturdays and Sundays) and public holidays.
- Some of the shopping streets in these districts, have a maximum parking duration of 1.5 hours for which there is no exception.

On-street parking spaces can be reserved to local residents in parts of Vienna's inner-city districts to improve the parking situation for residents in these neighbourhoods. Drivers with disabilities in their car can also park in the residents' parking spaces.

The pilot project in the districts of Mariahilfer, Neubau and Josefstadt has been successful among the local residents. The one-year pilot phase that has been running since June 2012 in parts of districts 6, 7 and 8 has shown that parking spaces reserved to local residents clearly remedy the parking shortage. In addition, the number of parking spaces reserved to local residents has been increased from ten to currently 20 percent. Between 2002 and 2005 Mariahilfer, the gender mainstreaming pilot district, was the first in Vienna to try equal opportunities in public space planning under the motto "Sharing the city fairly". The work processes linked with this project, which included evaluation tools for quality assurance and physical results, continue to be exemplary. Coordinated by the then "control centre for planning and building for every-day life and women", all seven City Administration units directly in charge of public space were involved in the pilot process; two external offices were brought in for support. In the course of this pilot process, roughly 1,000 metres of pavement were widened, three traffic lights were programmed for more lead time to pedestrians before motorized traffic turns left, around 40 crossing facilities and one lift were installed in public space, the pavement was made barrier-free in five spots, lighting was improved in 26 places, two pram ramps were installed on public stairs, two smaller squares were redesigned and additional seating in public spaces was set up in nine spots. The sum total of these small-scale measures led to an improvement of "network quality" of pedestrian routes in the district. Conditions for moving and resting in public space were markedly changed for the better.

In the future, the office of the district chair person or the district councils can propose areas where residents' parking spaces should be made available. They also have to submit a survey of parking spaces and parking capacity of the area concerned. The competent authority then examines whether the requirements for residents' parking spaces are met.

In line with the Vienna City Statutes, the districts have to cover the costs for new residents' parking spaces. Depending on the length of the residents' parking zone the costs per parking space were between EUR 200 and EUR 300 during the pilot phase.

A parking guidance system helps drivers find a garage quickly and without too much effort. Electronic technology is used to guide drivers to the nearest and most convenient garages and Park & Ride facilities. Dynamic displays provide information on the nearest garages and available spaces (Figure 9).

All revenues from parking management are earmarked, i.e. 100 percent of the fees and charges paid, are set aside for the Vienna traffic and transport system and thus benefit everyone in Vienna. In the

Figure 9. The online DSS (Decision Support System) to follow availability of parking areas
(https://www.wien.gv.at/english/transportation/parking/shortterm.htm).

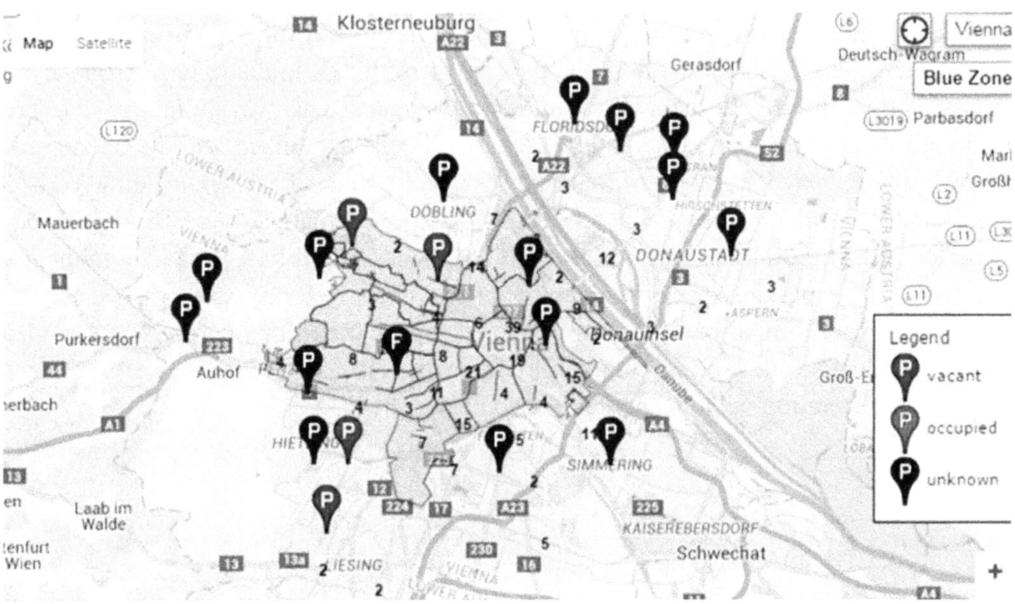

**For a more accurate representation of this figure, please see the electronic version.*

past years, the construction of all municipal neighborhood parking garages ("Wohnsammelgaragen") has been financed from parking management revenues. The funds have also been used to purchase new underground trains, and go into measures that enhance traffic safety in particularly accident-prone locations in Vienna.

The Viennese Urban Mobility Plan

The Viennese Urban Mobility Plan sets out the ambitious goals of the City of Vienna for a viable transport system of the future, and it describes the steps to be taken in the next ten years so these goals can be reached Vienna City Administration (2014a).

One of the most important approaches to a successful qualitative enhancement of street spaces is parking space management. On the one hand, parking in car parks reduces the demand for parking spaces on public streets. Gradually, permanent parking is to be shifted from the streets to car parks or permanently hired lots in public garages, in particular multi-car garages for residents in the neighborhood.

According to the 2014 Car Park Programme, standardized criteria are in place for funding car park construction in restricted areas 2014 Car Park Programme where these are needed most from the point of view of urban structures. In areas characterized by older buildings, high population density and few green areas multi-car garages for residents are to be fostered so there is more surface space for pedestrians, cyclists and public transport. As an update of the Transport Master Plan requirements, surface parking spaces are to be reduced at least at a ratio of 1:3. Moreover, in areas which are well connected to public transport, maximum numbers of parking spaces on private land are set.

Parking space management has also proven an effective instrument to shift transport to eco-mobility among the Viennese and commuters alike. This has cleared areas that are now available for other purposes. The evaluation of parking space management has shown that the capacity use for parking spaces in streets has declined by up to 30% when the system was introduced, especially due to a reduction of cars not licensed in Vienna. Apart from a marked decrease in traffic caused by drivers looking for a place to park, the number of parking offences has also declined. All things considered, the parking situation has generally eased in the districts where parking management and charges for on-street parking were introduced. In the future, this approach is to be adapted, developed further and combined with other existing instruments (e.g. resident parking, the park&ride strategy). In the long run, this also includes reviewing the option of a digital mapping of all parking spaces. This way, administrative procedures could be simplified and new services could be created. This is also linked with the increased use of a mobile application for paying short-term parking fees. The "Smart City Wien Framework Strategy" has set the objective "Reducing private motorized traffic from currently 28% to 15% by 2030" over the planning period of STEP 2025 and the Urban Mobility Plan. On this basis and in combination with a larger number of garage parking spaces, the number of parking slots for privately owned cars in existing streets will go down at a similar rate. To convert and repurpose the former parking spaces, ideas which are already in the pipeline are to be communicated to the districts and citizens; moreover, new ideas are being collected. Cooperation with business will also be sought.

CONCLUSION AND RECOMMENDATIONS

Sustainable parking planning and management play a crucial role in smart urban growth, both in enhancing accessibility and in competing urban congestion. Parking policy is an important tool in long stay and short stay parking in urban areas. Sustainable parking policy facilitates visitors, businessmen and customers on the streets to mitigate congestion and occupying wrong parking places.

As a comparison, case studies are summarized in Table 4:

Urban parking policy and legislations should be specific and consider land use planning and environmental issues. Sustainable parking standards can be given in the land use planning studies with a guide.

Table 4. A comparison of the case studies

Case Study 1	Case Study 2
Istanbul's parking planning policy is demand oriented old paradigm which is based on creating new parking areas when needed according to the automobile boom.	Vienna's parking planning policy is supply oriented new paradigm which is based on providing optimal parking supply and sustainably sharing with different modes.
In Istanbul, parking spaces are managed by a private company (ISPARK) owned by the Municipality.	In Vienna, parking spaces are managed by municipality itself.
Private company has the chance to maximize its profit with a demand oriented policy.	Municipality uses parking management as a tool to promote public transport, bicycle and pedestrians.
Different from European countries, in Turkey there is a nationwide law arranging parking conditions. This law is also valid in Istanbul. Parking legislation is demand oriented and demand estimation model is conventional.	In Vienna, parking conditions are arranged by local governments with regulations, just like other European cities. Optimal supply parking model is new and sustainable.
Parking facility costs are charged from automobile owners and constructors by a private company to maximize its benefit.	Parking facility costs are charged directly from automobile owners by municipality to reduce their parking and orient to public transport and cycling.
Istanbul's strategy is built on maximizing urban rent and giving more development to parcels facing on narrow roads and creating more parking problem in the dense urban zones. ISPARK wants to increase off-street parking rather than on-street parking.	Vienna's strategy is TOD friendly parking which allows shared parking between transit stations and bring walkable and bike-able environment.
Istanbul's Development Plan and Transportation Master Plan emphasize air quality however implementation is on increasing road amount by new constructions, tunnels and bridges	Vienna's 2025 Urban Development Plan gives priority to atmospheric quality in traffic management.
The amount of pedestrian roads is minimal according to total road space.	Street spaces transform into recreation, lingering and cycling.
Istanbul deals with parking capacities of off-street functional parking, off-street general parking and on-street parking. Demand at night and demand of offices at day time is considered.	Vienna has a parking zone system for short term and longer term parking. These zones have special parking regulations for residents and visitors.
Barrier-free pavements are built. No other projects are put for the disadvantaged groups.	Drivers with disabilities can park in the residents' parking spaces. Women play important role in the coordination of pilot pedestrian projects.
Historic peninsula project is ongoing and pedestrianized areas and cycling are encouraged. Smart parking system shows empty parking spaces and drivers are taken by transit services to nearest terminals.	Inner city center belongs to pedestrian, bicycle and public transport. It is a restricted parking zone and maximum parking time is 2 hours.
ISPARK developed nearest parking area, shortest path and automatic parking payment by mobile phone.	A parking guidance system helps drivers to the nearest and most convenient parking spaces and park&ride facilities.
Surface parking spaces are increased to prevent on-street parking and multi-level parking garages are encouraged vertically.	Surface parking spaces are reduced and replaced by underground parking or pedestrian area.

This should facilitate and adapt according to the regional differences. Sustainable policies minimize the carbon emissions and reduce automobile and integrated mode systems are encouraged. Traffic calming and TOD facilities will have positive impact on urban environment and air quality. Legal arrangements and urban legislation set parking offences with automobile drivers and define level of payment for parking with specific values. Local authorities have the power for enforcement but provide opportunity for private sector as well. A regional authority can coordinate parking standards and parking fees. Local authorities can generate parking policies according to these standards and fees. Enforcement of paid parking is part of traffic legislation and related to the police. Automobile fee areas differ in many cities. Priority zones for residents exist in European cities. A license which allows the resident to make parking in a given zone is one of the common practices in the localities. Individual parking meters as common devices control the payment on street parking. Cash payment was replaced chip based cards or credit cards as payment methods. Cities develop ultimate technologies such as payment by smart phones. Parking tariffs vary however on-street parking is more expensive than off-street parking. Local authorities can arrange these levels according to the income groups in the specific neighborhoods. Also park and ride facilities can assist parking problems of the central business district. These facilities should be accessed easily and provide ultimate technological solutions by other means of transport. Park and ride is better for commuter traffic but also available for the visitors of central business district.

Parking policies and rules should be explained to the public carefully. Parking policy can be improved by well-defined communication actions. Parking, mobility and local economy in the city center should be monitored regularly. Introducing new parking measures in a shopping street can change the customer habits and decrease the shop turnover. Commuter parking can be controlled easily. Resident parking can be directed well through the introduction of a residents permit parking scheme and visitors parking can be carried out paid parking scheme. Sustainable parking management should lead citizens to park at longer distances and use other transport modes.

REFERENCES

Cologne Website. (2012, December). Retrieved from http://www.koeln.de/koeln/ein_jahr_koeln_3__parkplatznot_in_der_zweiten_reihe_ist_doch_ immer_was_frei_670143.htm

COST Action Report No. 342. (2005). Parking policies and the effects on economy and mobility (Report on COST Action 342). European Cooperation in the Field of Scientific and Technical Research. The Netherlands: Technical Committee on Transport.

Dalkmann, H., & Brannigan, C. (2007). *Transport and climate change. In A Sourcebook for Policy-Makers in Developing Cities: Module (5th ed.).* Gesellschaft für Technische Zusammenarbeit–GTZ Eschborn.

Eltis Website. (2014, September 29). Retrieved from http://www.eltis.org/discover/case-studies/ispark-parking-turkey

EPA Website. (n.d.). Retrieved from https://www.epa.gov/smartgrowth/smart-growth-and-transportation

Herry, M., Rosinak, W., & Schuster, M. (1999). Parkraumbewirtschaftung in Wien: Vorgehensweise und Erfahrungen. *Internationales Verkehrswesen, 51*(6), 244–249.

Hofstetter, F. (2015). *A Public Parking Management System for Zurich* (Master Thesis). University of Zurich, Zurich, Switzerland.

Ispark Website. (n.d.). Retrieved from http://ispark.istanbul/projeler/otopark-master-plani/

Kulekci, O., Ayozen, Y. A., & Inac, H. (2015). *IstanbulOtopark Master Plani Calismalari*. ROTRASA Road and Traffic Safety Congress, Ankara, Turkey.

Litman, T. (2008). *Parking Management: Strategies, Evaluation and Planning*. Victoria Transport Policy Institute. Retrieved from www.vtpi.org/park_man.pdf

MTC Website. (2007, June). *Reforming Parking Policies to Support Smart Growth*. Retrieved from http://mtc.ca.gov/sites/default/files/Toolbox-Handbook.pdf,14/04/2016

Nash, A. (2004). Traffic calming in three European cities: Recent experience.*Proceedings of the 84th meeting of the Transportation Research Board*.

OECD. (2009). *Transport, Energy and CO2: Moving Towards Sustainability*. Paris: OECD International Energy Agency.

Rios, R. A., Vicentini, V. L., & Acevedo-Daunos, R. (2013). *Practical Guidebook: Parking and Travel Demand Management Policies in Latin America, Despacio and Institute for Transportation and Development Policy (ITDP)*. Interamerican Development Bank.

TOD Handbook. (2004, January). *The City of Calgary Land Use Policy*. Retrieved from http://www.reconnectingamerica.org/assets/Uploads/bestpractice031.pdf

Vienna City Administration. (2014a). Urban Mobility Plan Vienna, STEP 2025 Thematic Concept. Municipal Department 18 - Urban Development and Planning.

Vienna City Administration. (2014b). Vienna's Urban Development Plan-STEP 2025. Municipal Department 18 - Urban Development and Planning.

Chapter 10
Evaluating Alternatives of Transport Network Design of a Metropolitan City Using Hierarchical Cluster Analysis:
Case Study of Istanbul's 2023 Plan Networks

Darcin Akin
Gediz University, Turkey

ABSTRACT

The objective of this study is to examine how the share of public transit investments affects urban structure using spatial-temporal distribution of transport passenger flows over transport network alternatives. A methodology to model urban structure (identifying and classifying centers and subcenters) based on urban travel data (interzonal urban passenger flows via urban rail modes during morning peak-hour) was developed using hierarchical cluster analysis for the case study of Istanbul Metropolitan City in Turkey since the rail investment is the major determinant in the definition of the network alternatives studied. Effects of the alternatives of Istanbul's 2023 Transport Master Plan networks on urban structure were modeled and compared using hierarchical cluster analyses (HCA). Analysis of the travel patterns over the alternative transport networks did not yield significant differences under the given constraint that the number of total trips in the metropolitan city was constant for all scenarios.

INTRODUCTION

Types of urban transport investments have been long known by their impacts on the sustainability of transport and urban systems. With the increased share of public transport modes, the sustainability of a transport system increases since the cost (or travel time) of trip per passenger traveled decreases dramatically in the availability high-speed and high-capacity modes. Urban transport networks are generally

DOI: 10.4018/978-1-5225-2116-7.ch010

invested to various degrees of public transit and highway infrastructures. In that respect, 2023 Istanbul's master plan network alternatives had three alternatives developed in addition to the base scenario (existing and projects under construction with their projected termini by 2014 were counted). The 2023 network alternatives are defined based on their share of rail transit and highway projects, namely, i) rail-based, ii) highway-based, and iii) balanced or master plan scenario. The base network includes 29,861 km highway links (without the third Bosporus Bridge and its connections), and 231 km rail links. The rail-based network includes additional 357 km highway (without the third Bosporus Bridge and its connections) and 624 km rail links, the highway-based network includes additional 717 km highway (with the third Bosporus Bridge) and 343 km rail links, and the master plan network includes additional 717 km highway (with the third Bosporus Bridge) and 624 km rail links (ITMP, 2011).

The mode shares of the alternative future networks of Istanbul are tabulated in Table 1. It is seen that the highest share of public transit (48.77%) is achieved by the rail-based network. The second highest rate (48.04) belongs to the master plan network (ITMP, 2011). The master plan network includes the links of highway and rail systems included in the highway- and rail-based networks. The share of public transit in Istanbul is considered to be quite high and promising for sustainable transportation compared to developed countries where auto trips have the highest share of all trips. In addition, company and school shuttles can be considered as the modes of mass transportation. In that case, the share of mass transportation increases to 57.44% for the master plan network. For example, public transit constitutes 8.2%, 13.4%, 37.7%, 38.6%, 45.9%, 63%, 64.6% of motorized trips in Melbourne, Sidney, New York, London, Madrid, Seoul, Paris, respectively, in 2010-2011 (JOURNEYS, 2011).

Regarding the sustainability with respect to Istanbul's alternative transport networks given in Table 1, the highest average travel velocity (19.74 km/hr for auto trips and 19.63 km/hr for public transit trips) is yielded by the master plan network. The length is also the shortest with the same network (23.85 min. for auto travels and 24.75 min. for public transit trips). Thus, it is very rational to consider it as the most sustainable transport network among its alternatives including the mode-share data presented above.

The objective of this study is to examine how the design of transportation networks with respect to public rail transit investments affects urban structure using spatial-temporal distribution of transport passenger flows over the network alternatives. A methodology to model urban structure (identifying and classifying centers and subcenters) based on analyzing travel data (interzonal urban passenger flows via urban rail modes during morning peak-hour) was developed using hierarchical cluster analysis for

Table 1. Mode shares of motorized trips for Istanbul's alternative transport networks

Networks	Trips by Auto (%)	Trips by Company and School Shuttles (%)	Trips by Public Transit (%)	Total No. of Motorized Trips	Trips per Capita	Trip Length (Auto/Transit)-min.	Trip Velocity (Auto/Transit)-km/h
Existing (2010)	33.37	19.38	47.25	12,595,919	0.948	14.00/15.32	N/A
Base (2023)	42.92	9.37	47.71	25,907,698	1.505	25.00/26.52	N/A
Rail-based (2023)	42.02	9.21	48.77	25,907,698	1.512	24.55/24.76	19.15/19.03
Highway-based (2023)	42.78	9.44	47.78	25,907,698	1.507	23.89/25.02	19.40/19.12
Master Plan (2023)	42.55	9.40	48.04	25,907,698	1.509	23.85/24.75	19.74/19.63

Source: ITMP, 2011.
N/A: Not available.

the case study of Istanbul Metropolitan City in Turkey (Figure 1) since the rail investment is the major determinant in the definition of the network alternatives studied. Results of the analyses can be used to present the effects of different network alternatives on the sustainable urban transportation planning by confirming the expansion of urban city core as well as other urban and suburban developments and their emergence with their neighboring clusters.

Istanbul is the largest city in Turkey with 18.5% of the country's population and is the 21st largest city in the world with its 13,576,000 population in 2013. In 2014 the city's population exceed 14 million. The density of Istanbul is 2,523 people per square kilometer (6,530/sq.mi.), far greater than the country's density of 102 people per square kilometer. The most densely populated areas are the southwest, west and northwest of the city center as well as the European side, while the most densely populated district is the Asian side, Uskudar. The most rapid growth in the city came during the late 20th century, when its population increased from 983,000 in 1950 to 10,923,000 in 2000. The population has grown partly from expanding the city limits, especially in the 1980s, when Istanbulites doubled in number. Istanbul is growing at a rate of 3.45%, making it the fastest growing of the 78 largest metropolitans in the world (Istanbul Population 2016, n.d.) Istanbul is known for its long-term monocentric planning with a dominant central core and linear public transport and highway systems. However, construction of the peripheral highways in 1970s and 1980s as an answer to increasing highway travel demand for extensive transportation networks contributed to the polycentric development of the city. Moreover, decentralization of industry in Istanbul which contributed to a sustainable urban system transformed the city into a linear and polycentric form, which will have a positive impact on the various aspects of urban transportation systems as the regional planning policy has embraced a shift towards a polycentric planning policy since 1980s. Analysis of the travel patterns at the metropolitan level provided insights into the concentration of passenger flows at multiple levels as a potential mechanism in the identification of the multicentric development of the urban structure in Istanbul (Ozus et al., 2012). Districts of Istanbul are mapped in

Figure 1. Map of Istanbul with its districts including Gebze
ITMP, 2011.

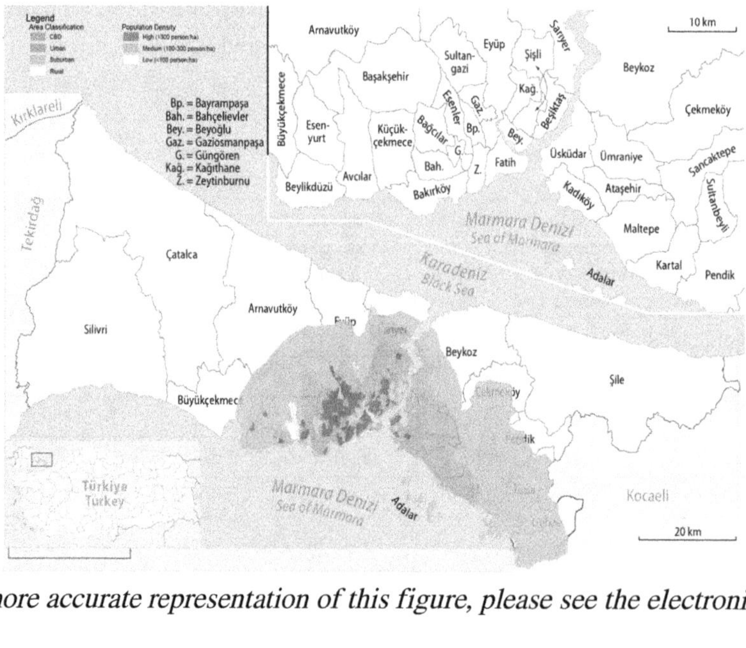

**For a more accurate representation of this figure, please see the electronic version.*

Figure 1. There are 25 districts located in the European and 14 districts in the Asian side of the city. Gebze, a very west district of the neighboring province, Kocaeli, is also included because of its strong interaction with the Asian districts of Istanbul.

BACKGROUND

Transportation systems have long been recognized as vital parts of city systems. They affect the quality of live (Vuchic, 1999) and have strong effects on shaping the form of settlements. Urban transport systems can mainly be grouped into public, private, and freight transport systems. In most cases, they can be complementary, but they may usually compete against each other under the lack of a comprehensive planning effort. Urban transportation system, i.e., types, capacity and characteristics of travel modes, infrastructures, and users, and the cost or time of travel, have a unique spatial imprint and shapes the urban form continuously at every major transportation investment. Thus, each network design and its characteristics create different spatial form. For instance, while some cities have an urban form shaped by mainly highway modes, and others have different spatial forms compared to the former because various public transport modes, including non-motorized modes and infrastructure developments are taken place in a more balanced transportation system. Urban spatial structure is affected by spatial interactions among different activity locations and land uses over the city's transport system since each city has its unique circulation pattern of passengers and goods. In that sense, it is claimed in this chapter per the authors that urban spatial structure can be modeled using interzonal (O/D) travel data. Sohn (2005) already proved that commuting patterns reflect the corresponding urban spatial structure. Thus, urban geography can be characterized by analyzing travel patterns of people and goods in and across urbanized areas. OD data are the outputs of the trip distribution step from the travel demand modeling (TDM) process. Thus, by comparing the spatial structures obtained from the models created by employing Hierarchical Cluster Analysis (HCA) for a base year and future scenarios, the effects of alternative transport network design of a metropolitan city can be modeled based on the level of interaction, represented by distance or similarity calculated based on trip interchanges, over the transportation network. For example, a network with substantial highways projects and the other having substantial public transportation investments can be compared with respect to the spatial structure impacted by the dominant elements of the transportation system such as highway, public transit modes, and major transport terminals.

This chapter presents a case study of the analyses of urban spatial structures of Istanbul Metropolitan City of Turkey (Figure 1) for future network design alternatives: i) rail-based, ii) highway-based, and iii) balanced or master plan scenario. Results of the analyses can be used to present the effects of different network alternatives on the sustainable urban transportation planning by confirming the expansion of urban city core as well as other urban and suburban developments and their emergence with their neighboring clusters. Decision makers will be able to develop their capacities to plan ahead in order to meet future needs better from the spatial restructuring in urban agglomerations, when they know the extent and location of the future growth ahead.

Istanbul's population and employment figures are given in Table 2. It is seen that over the course of 35 years' the population of Istanbul increased from 2.96 million to 11.59 million by 291.7%. Even the 5-year population increase from 2000 to 2005 is 16.4%, which results a yearly increase of 3.27% (ITMP, 2011). Employment increased from 2.52 million to 3.86 million by 53.4% in ten years from 1996 to 2006. Although 34.6% of the city's population live in the Eastern region, only 26% of the city employments are

Table 2. Population and employment increases in Istanbul

Region	Population (in 1000s)					Employment (in 1000s)	
	1970	1980	1990	2000	2005	1996	2006
East[1]	677.5	1383.5	2440.5	3449.4	4005.2	661.5	1081.5
West[2]	1485.5	2264.4	3662.0	5230.4	6251.1	1317.3	2119.9
North[3]	795.5	996.8	1280.6	1280.6	1333.5	538.2	660.5
Total	2958.5	4644.7	3662.3	9960.4	11589.8	2517.0	3861.9

Source: ITMP, 2011.

1. East: Adalar, Kadikoy, Uskudar, Maltepe, Umraniye, Beykoz, Kartal, Sultanbeyli (after 1990), Pendik, Tuzla, Sile, 2. West: Eminonu, Fatih, Eyup, Zeytinburnu, Bayrampasa, Gaziosmanpasa, Esenler, Gungoren, Bagcilar, Bahcelievler, Bakirkoy, Kucukcekmece, Buyukcekmece, Avcilar, Catalca, Silivri, 3. North: Beyoglu, Sisli, Besiktas, Kagithane, Sariyer

in the Eastern region. This creates a great amount of daily traffic flow between two sides of the city. In 2009, the number of vehicles passed over the two highway bridges was 386354 per day, which is 27596 vehicles per bridge-lane. This also indicates the lowest level of service (LOS) F over the bridges with a rough estimate of 2760 veh/h/ln, which is quite higher than the maximum service flow rate of 1900 pc/h/ln for LOS E, at fee-flow speed of 70 km/h (HCM, 2010).

There is a considerable amount of research in structural changes and the patterning of cities, but much of these have been focused on explaining urban structure and form at a single point in time, as if cities were in some sort of perpetual equilibrium. Clearly the absence of favoring data through time has been a major constraint on the ability to produce an appropriate science of urban dynamics and thus most of the thinking about urban change has been speculative and non-rigorous. However, this has already been changing. New data sets to control and manage urban sprawl, and new techniques such as urban remote sensing to process information from satellite and aerial photographic data, are becoming important tools for modeling the changes in urban form and structure (Besussi et al., 2010).

Urban geography can be characterized by analyzing travel patterns of people and goods in and across urbanized areas. Modeling urban structures is essential for supporting an evidence-based spatial planning policy (Cats et al., 2015) for examining the sustainability of urban growth. In recent years, there have been some efforts to model the urban growth and expansion using clustering techniques based upon the idea that commuting patterns reflect the corresponding urban spatial structure (Sohn, 2005). For example, the research by Akin and Alasalvar (2016), Cats et al. (2015), Akin (2014), Ozus et al. (2012), and Akın (2011) employed hierarchical clustering (HC) in modeling the change of urban structure. Hierarchical clustering algorithms (HCA) and partitioning algorithms are two major clustering methods. Partitioning algorithms divide a data set into a single partition. However, hierarchical algorithms divide a data set into a sequence of nested partitions. In HCA, the clusters are represented hierarchically through a dendrogram, at varying levels of granularity. Two major hierarchical algorithms can be named, such as agglomerative algorithms, and divisive algorithms. Depending upon whether this hierarchical representation is created in top-down or bottom-up fashion, these representations may be considered either agglomerative or divisive (Aggarwal & Reddy, 2013). Agglomerative algorithms start with every single object in a single cluster, while divisive ones start with all objects in one cluster and repeat splitting large clusters into small pieces (Gan et al., 2007). In general, clustering is an exploratory data mining, and a common technique for statistical data analysis, used in many fields, such as machine learning, pattern recognition, image processing, artificial intelligence, information technology, and biology, psychology, and marketing. Such analysis procedures help grouping a set of objects (zones in urban modeling) into clusters, where those objects are more similar based on defined measurements to each other than to those in other clusters (Jain et al., 1999). As such, researchers often find it very difficult to determine a proper algorithm for their applications and/or to compare novel ideas with existing results. The type of data is directly associated with data clustering, and it is a major factor to consider in choosing an appropriate clustering algorithm. There are many data types, and once researchers get familiar with data types, they should be able to adjust the algorithms accordingly. In many applications of cluster analysis, the raw data, or actual measurements, are not used directly unless a probabilistic model for pattern generation is available (Jain & Dubes, 1988). Preparing data for cluster analysis requires some sort of transformation, such as standardization or normalization. Data standardization makes data dimensionless. It is useful for defining standard indices. After standardization, all knowledge of the location and scale of the original data may be lost. It is necessary to standardize variables in cases where the dissimilarity measure, such as the Euclidean distance, is sensitive to the differences in the magnitudes or scales of the input variables (Milligan & Cooper, 1988; Aggarwal & Reddy, 2013). In many cases, raw data should be normalized and/or transformed before a cluster analysis is performed. A reason to do normalization is that objects in raw data may be described by variables measured with different scales, and another is to reduce the size of the data to improve the effectiveness of clustering algorithms (Gan et al., 2007).

A literature review revealed that studies with the subject of the multicenteric development course of cities were very few in developing countries due to the lack of quality data (Akin, 2014; Ozus et al., 2012; Akın, 2011; Zhao et al., 2011; Berkoz & Turk, 2010; Alpkokin et al., 2008; Bhandari et al., 2007; Bhandari et al., 2007; Alpkokin et al., 2005; Berkoz, 2000; Dokmeci & Berkoz, 1994). Thus, in this chapter the relationship between urban structure and travel patterns over alternative network designs was studied as a case study of the metropolitan city of Istanbul, Turkey, by using hierarchical cluster analyses.

DATA AND METHODOLOGY

Data used in the analyses were borrowed from the four-step travel demand model of the metropolitan city of Istanbul, calibrated for 2010 and validated for 2023. During 2006-2010, Istanbul's Transportation Master Plan for 2023 was prepared based on projected Landuse Development Plan including 39 districts of Istanbul and one neighboring district (Gebze) of the city of Kocaeli on the east of the city of Istanbul. The data included O/D travel data for four network design alternatives: i) base network (existing and current projects until 2014), ii) highway-based, iii) rail-based, and iv) balanced or master plan scenario for 2023.

Data

Total, automobile, company & school shuttles, and rail transit trips summarized in Table 3 include 40 districts of the study area, estimated for 2023.

As seen in Figure 2, person-trips per hour by rail modes were increased in the highway- and rail-based, and the master-plan networks by 0.25, 2.82, and 0.95%, respectively, compared to the base network. The highest demand was observed at the rail-based network by 2.82%, on the other hand the lowest was observed at the highway-based network by 0.25% as expected. The master-plan network resulted in about 0.95% increase. Person-trips per hour by company/ school shuttles were increased in the highway- and

Table 3. Morning peak-hour demand at 2023 alternative transport networks

Year 2023 Morning Peak-Hour Demand	2023 Alternative Transport Networks						
	Base (1)	Highway-Based (2)	Diff. (1-2)	Rail-Based (3)	Diff. (1-3)	Master-Plan (4)	Diff. (1-4)
Person-trips per hour by rail	1,499,566	1,503,286	+0.25%	1,541,786	+2.82%	1,513,770	+0.95%
Person-trips per hour by company & school shuttles	522,391	526,592	+0.80%	516,807	-1.07%	525,387	+0.57%
Auto-trips (vehicles/hour)	940,288	939,192	-0.12%	922,557	-1.89%	934,340	-0.63%
Total vehicles/hour (auto+shuttles)	651,149	650,871	-0.04%	639,297	-1.82%	647,660	-0.54%

Source: ITMP, 2011.

Figure 2. Changes in estimated travel activities for 2023 between base and alternative transport networks

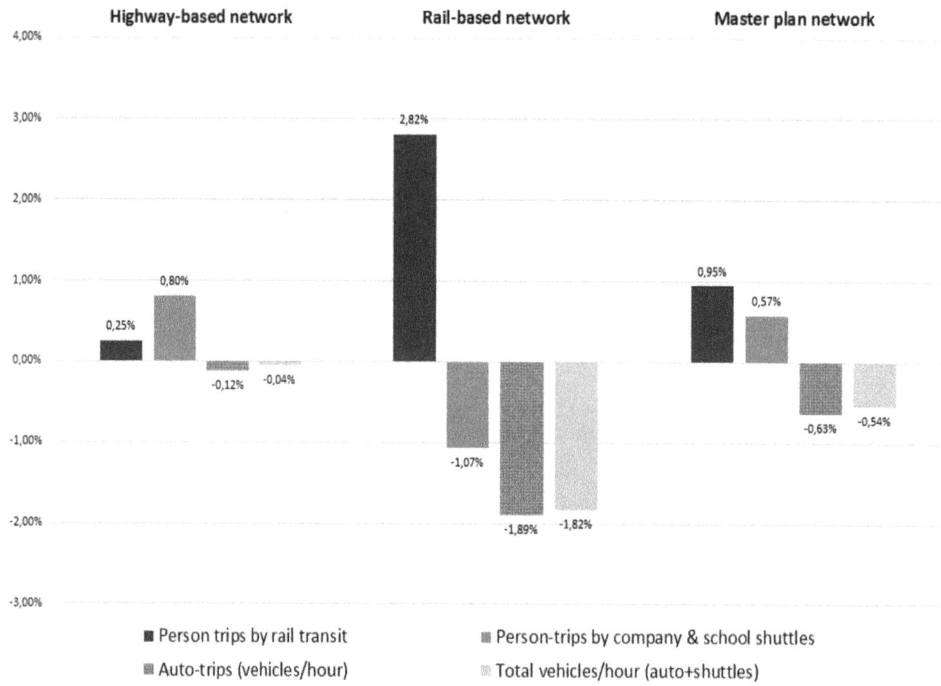

rail-based, and the master-plan networks by 0.80, -1.08 and 0.57, respectively, compared to the base network. Automobile-trips and total vehicle trips reduced by 1.89 and 1.82% by the rail-based network. Thus, it should be underlined that the rail-based network eliminates about two-percent of vehicle trips by highway modes.

Table 4 summarizes the characteristics of the 2023 alternative transport networks. Although the total cost of auto trips did not vary among alternative networks, travel time and cost by auto and public transit varied.

Table 4. Characteristics of 2023 alternative transport networks

Network Characteristics	For Trips among Districts during Morning-Peak Hour				
	Total Highway Length (km)	Total Travel Time by Auto Trips (min)	Total Travel Cost by Auto (TL in 2009)	Total Travel Time by Public Transit (min)	Total Travel Cost by Public Transit (TL, in 2009)
Highway-based network	37,183.18	113.282,32 (+0.78%)+	48,654.26	125,238.81 (+1.56%)+	3,704.42 (+2.60%)+
Rail-based network	37,183.18	115.776,83 (+3.00%)+	48,654.26	124,137.49 (+0.67%)+	3,606.15 (-0.12%)+
Master-plan network	37,183.18	112.406,95	48,654.26	123,312.48	3,610.55

Source: ITMP, 2011.
TL= Turkish Lira, + Changes with respect to the master plan network.

Methodology

In the HCA, the analysis starts with an input matrix *D (i, j)* with i measurements (observations) of j variables, denoting similarity or distance among the objects. In data clustering, a similarity measure or distance (dissimilarity measure) is used to quantitatively describe the similarity or dissimilarity of two data points or two clusters. Similarity and distance measures are fundamental elements of a clustering algorithm, without which any meaningful cluster analysis cannot be obtained (Gan et al., 2007). Distance-based methods are very popular in the literature, because they can be used with almost any data type, as long as an appropriate distance function is created for that data type. Thus, the problem of clustering can be reduced to the problem of finding a distance function for that data type. Therefore, distance function design has itself become an important area of research for data mining in its own right (Aggarwal, 2003; Wang & Sun, 2012; Aggarwal & Reddy, 2013). The greater the similarity measure d_{ij} is, the more the resemblance is between the objects. The greater the distance measure d_{ij} is, then the higher the dissimilarity is. In determining similarity measure to use in HCA, the nature of the observations plays a significant role (Hardle & Simar, 2007). In HCA, there are various measurements of distance. In this study, the "correlation distance" defined in terms of the correlation coefficient (r_{ij}) is seen in Equation 1:

$$d(i,j) = 1 - r_{ij}$$

where r_{ij} denotes the Pearson correlation coefficient which is used to measure the correlation among Istanbul's districts with respect to the number of trips attracted to. Each district has 40 trip values attracted to from other districts. It means that each district is represented by a column vector with 40 rows (i, j=1 to 40). Thus, the correlation coefficients were calculated among all districts and then, r_{ij} matrix is formed using a statistical software. The correlation distance d_{ij} yields values between 0 and 1 for positive correlations, and between 1 and 2 for negative correlations. In this approach, positively correlated measurements are to be less apart than negatively correlated ones in the clustering analysis. Thus, in this regard the choice of the similarity measure looks just fine for the purpose of analyses of this study.

In agglomerative methods, a bottom-up approach is used, in which starting points are individual data points and successively merge clusters in order to create a tree-like structure. A variety of choices are possible in terms of how these clusters may be merged, which provide different tradeoffs between quality and efficiency. Some examples of these choices are single-linkage, all-pairs linkage, centroid-linkage, and sampled-linkage clustering. In single-linkage clustering, the shortest distance between any pair of points in two clusters is used. At each step combining two clusters that contain the closest pair of elements not yet belonging to the same cluster as each other. In all-pairs linkage, the average over all pairs is used, whereas in sampled linkage, a sampling of the data points in the two clusters is used for calculating the average distance. In centroid-linkage, the distance between the centroids is used. Some variations of these methods have the disadvantage of chaining, in which larger clusters are naturally biased toward having closer distances to other points and will therefore attract a successively larger number of points. Single linkage clustering is particularly susceptible to this phenomenon. A number of data domains such as network clustering are also more susceptible to this behavior (Aggarwal & Reedy, 2013). In this study, single, complete, and average linkage rules are employed as the commonly used methods to calculate the interzonal distances when there are multiple zones that may be clustered together. Detailed definitions of the linkage rules can be seen in Tan et al. (2006), Johnson and Wichern (1992), and Milligan (1980).

Single linkage rule is based on the cluster proximity between the closest two points that are in different clusters. When points *x* and *y* are in cluster C_1, and *w* and *z* are in C_2, then the distance (*d*) between two clusters, C_1 and C_2, is given as follows:

$$d(C_1, C_2) = \min(d(x,w), d(x,z), d(y,w), d(y,z))$$

Complete linkage rule, however, alternatively takes the proximity between the farthest two points in different clusters as the cluster proximity. Therefore, the results can be sensitive to outliers. The distance between two clusters, C_1 and C_2, is then formulated in Equation 3:

$$d(C_1, C_2) = \max(d(x,w), d(x,z), d(y,w), d(y,z))$$

Another method as an alternative to the two techniques above, the group average rule defines the cluster proximity as the average pairwise proximities of all pairs of points among different clusters. In this case, the distance between clusters C_1 and C_2 with the number of observations n_1 and n_2, respectively, is defined in Equation 4.

$$d(C_1, C_2) = \frac{\sum_{\substack{x \in C_1 \\ y \in C_2}} d(x,y)}{n_1 * n_2}$$

RESULTS

Data visualization is the final step of data mining applications. The HCA was performed with both raw and normalized data, and it was observed that the clustering results did not change. The travel data was normalized as follows:

$$t_{ij} = \frac{T_{ij}}{T_j}$$

where T_{ij}= number of daily travel volumes between districts i and j presented in Tables 5, 6, 7 and 8, T_j= total number of trips attracted to zones j, and t_{ij}=normalized travel volumes between zones i and j.

HCA Results for 2023 Alternative Transport Networks

The HCA results for the alternative transport networks using the single, complete and average linkage rules are visualized through dendrograms in Figures 3, 4, 5 and 6. Clusters are differentiated by colors and totally eight distinct clusters are observed in the dendrograms. Districts numbered from 1 to 25 are European Districts and those from 26 to 40 are Asian Districts. The district of "40-Gebze" is also considered as an Asian District of Istanbul though it is the very west district of Kocaeli Province since it has a strong bound with Istanbul's Asian Districts.

Validation of Clustering Results

Cluster evaluation or validation should be a part of a cluster analysis to see how well any approach (distance definition and a linkage technique) fits the data well. The evaluation measures that can be applied to judge various aspects of the cluster validity are traditionally classified into the following three types: "internal or unsupervised", "external or supervised", and "relative or supervised or unsupervised" (Tan, 2006). We performed the cluster validation using the cophenetic correlation which is a popular evaluation measure for hierarchical clustering. The cophenetic distance between two districts is the proximity at which an agglomerative hierarchical clustering technique puts the districts in the same cluster for the first time (Tan et al., 2006).

The cophenetic distance matrices for the linkage techniques tested (single, complete, and average) were calculated based on the clustering results presented in the dendrograms for the alternative network scenarios. The cophenetic correlation coefficients (CPCCs) for the alternative clustering linkage techniques are calculated as the correlation between the original distance matrix (d_{ij}) and the cophenetic distances (read from Figures 3 to 6.). By looking at the CPCC values in Table 9, calculated for three major link approaches (i.e., single, complete, and average linkages), all techniques fit the data well ($p \leq 0.000$); however, the best clustering is yielded by the group average method with the highest correlation coefficient (0.916).

With this result, it is seemed to be redundant to calculate CPCCs for other networs scenarios and the best clustering technique is assumed to be the average linkage method for all cases.

Comparing Clustering Results for Alternative Transport Networks

Figures 7, 8 and 9 present the dendrograms by single, complete and average linkage techniques, respectively, for 2023 alternative transport networks. As seen there are eight distinct clusters denoted by different colors. In one hand, the dendrograms of the highway-based and master plan networks seem very identical in both sides of the city, and on the other hand, the dendrograms of the base and the rail-based networks are almost identical for the European districts, but somewhat different for the Asian districts.

Table 5a. 2023 average morning peak-hour trip interchange matrix (T_{ij}) by urban rail modes over the base network (1-20)

T_{ij}	1	2	3	4	5	6	7	8	9	10	11	12	13	14	15	16	17	18	19	20
1-Arnavutkoy	4112	389	699	517	370	806	423	517	353	439	650	146	264	446	414	1143	176	362	290	609
2-Avcılar	989	3665	1007	989	789	1964	535	613	1384	540	1788	111	362	2717	339	1532	191	530	338	1402
3-Bagcilar	2244	1949	7964	5774	3230	4726	2600	2180	1141	1830	1795	192	2919	1736	1580	4890	1064	2927	1273	5510
4-Bahcelievler	1486	1851	4206	5373	3184	2722	1536	1559	937	1283	1288	127	1361	1387	986	3804	589	1785	858	4055
5-Bakirkoy	248	402	620	1075	1482	464	429	443	186	404	244	21	330	280	242	1397	144	469	228	811
6-Basaksehir	1427	1582	2048	1725	1228	7187	919	914	1150	852	1955	163	772	2183	623	2098	419	781	577	3508
7-Bayrampasa	631	492	1352	1215	932	1185	2316	989	300	821	464	49	1153	453	1117	1891	1038	914	622	1145
8-Besiktas	126	89	199	185	150	170	196	2117	58	503	78	10	116	81	173	489	90	125	499	185
9-Beylikduzu	1144	986	697	607	459	874	377	587	1935	489	1254	97	258	1073	270	1380	145	392	285	704
10-Beyoglu	396	291	723	746	613	725	802	1810	172	3330	257	35	494	268	936	1706	393	476	1283	697
11-Buyukcekmece	1938	920	900	778	582	1114	497	701	1041	613	4336	268	309	1521	317	1777	166	531	359	912
12-Catalca	616	130	163	129	97	209	97	128	133	112	458	539	57	167	70	284	34	97	71	167
13-Esenler	1097	853	3508	2427	1537	2545	2141	1329	575	1121	914	90	3160	918	1077	2817	909	2133	804	2391
14-Esenyurt	1397	2759	1190	1055	818	2899	637	659	2548	601	3408	182	425	5969	380	1618	226	626	385	1445
15-Eyup	1410	604	1511	1347	1019	1592	1898	1679	377	1471	595	74	865	572	2867	2363	1703	865	1418	1454
16-Fatih	561	512	1120	1286	1410	981	1178	1819	300	2219	420	47	712	461	876	12037	479	881	820	1142
17-Gaziosmanpasa	1235	692	1984	1565	1146	1977	2715	1661	453	1479	712	87	1216	702	2597	2662	2825	1115	1319	1737
18-Gungoren	796	770	1764	1949	1378	1146	1059	965	423	775	608	60	1285	579	571	2366	363	1664	520	1320
19-Kagithane	728	414	1051	955	749	1157	1038	3564	272	1904	427	56	569	420	1377	1939	631	643	4634	1031
20-Kucukcekmece	1768	2743	4091	4799	3588	5321	1589	1450	1254	1268	1722	183	1369	1986	1096	3716	684	1485	894	10472
21-Sariyer	676	239	563	433	333	557	439	1828	172	725	274	38	229	240	416	1173	176	335	944	494
22-Silivri	8251	1517	2261	1663	1265	2385	1366	2001	1454	1717	4051	750	768	1695	849	4525	438	1452	1025	1970
23-Sultangazi	2037	902	2669	1845	1170	2755	1838	1779	612	1493	1000	137	1178	852	1521	2823	1204	1215	1177	2185
24-Sisli	347	239	570	537	434	553	571	2168	155	1675	221	30	340	227	641	1272	308	357	1428	553
25-Zeytinburnu	464	525	988	1497	1742	794	1007	986	274	1009	363	38	738	426	577	4366	325	989	500	1108
26-Adalar	27	12	26	24	24	25	24	53	8	53	12	2	13	11	16	157	8	20	22	24
27-Atasehir	531	171	343	284	267	310	272	534	111	521	173	25	145	143	165	1549	83	249	232	284
28-Beykoz	787	185	446	319	254	454	332	738	146	488	240	38	166	197	206	1238	111	280	397	368

continued on next page

Table 5a. Continued

T_{ij}	1	2	3	4	5	6	7	8	9	10	11	12	13	14	15	16	17	18	19	20
29-Cekmekoy	268	65	150	118	108	157	119	229	48	218	85	16	59	68	73	611	35	112	116	128
30-Kadikoy	782	234	448	367	341	391	340	653	158	629	241	34	187	189	212	1859	112	309	284	374
31-Kartal	590	171	372	303	284	354	305	535	118	555	187	31	157	157	179	1546	92	280	242	313
32-Maltepe	640	227	443	381	358	400	359	732	140	702	220	30	195	179	222	2003	111	326	297	373
33-Pendik	1449	329	744	577	532	744	590	990	240	1022	415	77	297	325	339	2811	170	557	468	619
34-Sancaktepe	611	131	316	242	223	349	253	468	102	453	189	37	123	146	153	1250	74	237	238	272
35-Sultanbeyli	463	90	225	168	154	256	178	306	72	312	134	27	84	105	103	851	49	169	155	194
36-Sile	133	27	64	44	38	67	42	70	23	64	40	7	22	30	27	176	14	39	38	53
37-Tuzla	241	61	137	105	95	131	103	156	48	164	82	15	56	63	61	453	35	96	81	114
38-Umraniye	1185	317	672	542	493	643	536	1052	224	976	360	64	279	290	334	2790	169	501	498	554
39-Uskudar	721	207	453	367	340	442	363	687	153	678	231	38	183	205	222	1936	113	332	342	393
40-Gebze	3697	537	1252	863	742	1291	870	1377	450	1310	821	165	445	564	512	3624	264	851	683	985

Table 5b. 2023 average morning peak-hour trip interchange matrix (T_{ij}) by urban rail modes over the base network (21-40)

T_{ij}	21	22	23	24	25	26	27	28	29	30	31	32	33	34	35	36	37	38	39	40
1-Arnavutkoy	242	1069	232	1029	421	6	237	142	55	682	348	264	595	88	48	55	515	333	336	1172
2-Avcılar	268	1384	226	1214	587	7	240	141	46	606	361	276	530	72	38	50	459	293	299	859
3-Bagcilar	1004	2105	1378	3992	1982	23	644	450	132	1533	878	732	1240	195	96	122	1037	805	831	1857
4-Bahcelievler	667	1445	690	2803	1532	17	479	291	88	1182	691	577	949	138	71	85	785	546	591	1366
5-Bakirkoy	164	297	127	715	593	5	150	67	24	376	222	174	289	41	21	19	224	157	170	409
6-Basaksehir	419	1819	681	1794	776	10	323	201	67	831	441	334	718	102	53	65	622	422	436	1247
7-Bayrampasa	426	581	687	1709	793	9	237	170	48	551	315	261	432	72	35	39	345	300	321	608
8-Besiktas	410	142	89	1370	149	3	98	76	20	214	112	85	150	30	15	12	101	127	144	202
9-Beylikduzu	300	1524	162	1157	484	6	250	156	58	804	392	310	621	88	48	68	569	341	368	1254
10-Beyoglu	553	361	375	3385	521	8	204	158	41	440	236	197	318	61	30	28	236	257	292	420
11-Buyukcekmece	327	5798	176	1482	656	9	365	195	78	1026	552	396	934	123	63	89	868	500	490	1849
12-Catalca	65	1194	41	281	113	1	76	38	17	203	108	75	188	26	13	17	170	104	100	369

continued on next page

Table 5b. Continued

T_{ij}	21	22	23	24	25	26	27	28	29	30	31	32	33	34	35	36	37	38	39	40
13-Esenler	560	1030	873	2396	1235	13	355	242	68	829	479	395	657	102	48	59	545	436	460	940
14-Esenyurt	303	2128	245	1382	667	8	265	169	56	645	372	283	593	88	45	62	534	347	340	1032
15-Eyup	814	844	1189	3143	923	10	385	281	81	911	449	361	670	117	56	65	560	507	536	1068
16-Fatih	548	572	355	2877	1732	30	476	197	70	981	628	565	765	127	68	45	524	452	470	852
17-Gaziosmanpasa	754	950	1896	3149	1105	12	423	311	86	920	487	388	695	123	56	67	596	555	547	1081
18-Gungoren	430	749	340	1670	1008	11	311	179	54	783	452	387	626	90	47	49	501	345	379	884
19-Kagithane	1836	598	632	5992	732	9	416	358	87	879	416	341	589	122	55	57	467	539	586	832
20-Kucukcekmece	679	1803	898	2740	1509	16	460	307	94	1109	654	530	923	141	73	92	774	571	577	1412
21-Sariyer	3953	524	190	2807	415	7	328	243	64	764	389	288	572	94	44	52	487	433	435	950
22-Silivri	1052	28594	465	4567	1794	25	1296	696	279	3767	1917	1384	3440	442	233	361	3346	1786	1700	7671
23-Sultangazi	940	1460	1972	3342	1165	14	576	432	127	1297	635	533	1001	180	87	110	871	765	740	1694
24-Sisli	828	372	303	4540	406	7	222	173	46	479	244	193	337	70	35	30	238	289	324	467
25-Zeytinburnu	324	442	248	1601	2799	15	294	127	44	651	402	348	499	77	40	31	376	290	305	608
26-Adalar	19	20	8	83	30	12	116	17	11	238	138	202	114	22	11	4	63	72	97	88
27-Atasehir	203	363	82	929	357	47	4052	452	361	3804	1763	1911	1964	561	285	137	1233	2283	2801	1946
28-Beykoz	395	481	118	1376	367	22	1282	4636	339	2375	1124	983	1585	414	182	193	1183	1865	2014	2117
29-Cekmekoy	103	183	35	447	158	17	1061	379	760	1409	638	654	935	658	201	152	569	1599	1247	831
30-Kadikoy	263	540	113	1139	439	73	2171	392	248	6371	1673	1624	1882	366	200	121	1228	1586	3011	2342
31-Kartal	198	392	90	992	395	42	1796	354	217	3018	7951	3172	5943	645	467	118	2251	1394	1670	2850
32-Maltepe	265	429	109	1208	470	61	2470	407	258	3842	3594	3556	3282	543	304	118	1736	1612	2190	2625
33-Pendik	386	937	165	1968	777	53	2955	724	519	5073	9354	4155	21045	1266	2083	309	6333	2687	2769	7431
34-Sancaktepe	208	404	71	938	333	31	2336	644	1059	2668	2212	1797	3127	2436	966	259	1674	2925	2169	2317
35-Sultanbeyli	123	284	46	637	237	20	1573	331	388	1796	2150	1313	5108	1137	2181	170	1717	1445	1208	2034
36-Sile	37	93	15	148	54	2	161	84	63	286	150	123	264	80	44	517	172	219	184	290
37-Tuzla	69	208	37	314	131	9	489	127	87	796	1251	616	2762	212	220	43	3445	442	450	1870
38-Umraniye	500	776	168	1978	705	59	3864	1308	1163	5518	2609	2600	3346	1375	517	363	2238	6251	4772	3596
39-Uskudar	296	470	112	1314	473	64	3166	1087	584	5737	2100	2025	2328	688	295	181	1563	3748	8340	2454
40-Gebze	697	2009	272	2992	1162	43	2307	924	457	5568	4220	2841	8079	942	644	395	7901	2610	2656	32722

Table 6a. 2023 average morning peak-hour trip interchange matrix (T_{ij}) by urban rail modes over the highway-based network (1-20)

T_{ij}	1	2	3	4	5	6	7	8	9	10	11	12	13	14	15	16	17	18	19	20
1-Arnavutkoy	4268	391	711	521	380	833	434	558	351	466	587	136	279	446	426	1149	185	372	307	634
2-Avcılar	1085	3663	1053	1029	813	2038	554	620	1401	559	1690	107	377	2777	351	1598	200	553	345	1437
3-Bagcilar	2394	1980	7984	5896	3297	4720	2643	2266	1138	1867	1801	179	2935	1756	1712	5036	1176	2955	1362	5742
4-Bahcelievler	1563	1870	4251	5469	3184	2751	1567	1575	929	1375	1196	120	1399	1379	1014	4155	624	1814	875	4198
5-Bakirkoy	262	397	628	1075	1478	465	429	458	183	425	223	20	328	277	244	1475	146	473	232	811
6-Basaksehir	1565	1657	2088	1790	1264	7213	955	974	1155	869	1950	153	813	2217	720	2136	484	814	639	3474
7-Bayrampasa	665	491	1364	1225	936	1190	2285	1006	299	834	449	46	1149	457	1116	1966	1025	915	635	1156
8-Besiktas	135	88	209	188	154	175	199	2057	57	522	75	10	120	81	190	525	102	126	523	187
9-Beylikduzu	1261	1005	720	642	486	876	391	588	1923	501	1128	91	269	1063	280	1430	152	411	289	735
10-Beyoglu	412	289	720	752	619	711	795	1899	171	3289	241	33	497	268	940	1867	396	477	1289	691
11-Buyukcekmece	1944	1031	986	846	624	1277	538	687	1186	616	4294	283	341	1709	346	1746	184	564	368	1017
12-Catalca	694	145	177	140	107	227	106	136	150	120	408	526	63	181	78	303	39	106	76	183
13-Esenler	1178	863	3510	2473	1562	2536	2152	1395	580	1152	916	86	3126	932	1177	2887	972	2127	861	2411
14-Esenyurt	1536	2777	1258	1106	859	2999	666	672	2519	621	3240	178	448	5893	400	1704	241	657	397	1526
15-Eyup	1475	609	1636	1390	1039	1713	1886	1846	380	1570	592	70	913	583	2835	2585	1698	886	1472	1520
16-Fatih	604	515	1132	1318	1382	991	1197	2043	300	2310	407	46	727	465	918	12042	495	893	898	1142
17-Gaziosmanpasa	1300	694	2154	1639	1163	2130	2703	1837	452	1580	720	81	1277	712	2612	2815	2805	1149	1407	1838
18-Gungoren	842	779	1796	2000	1415	1177	1053	977	420	818	575	57	1274	578	586	2477	375	1666	532	1381
19-Kagithane	760	410	1073	959	759	1152	1044	3665	266	1991	401	52	584	415	1403	2117	663	649	4537	1018
20-Kucukcekmece	1905	2772	4338	4996	3617	5219	1653	1478	1262	1314	1665	174	1428	2022	1164	3938	747	1569	926	10446
21-Sariyer	642	237	567	444	344	548	452	1915	167	798	246	33	238	235	454	1242	198	337	998	494
22-Silivri	5069	1363	1832	1429	1073	2128	1104	1435	1376	1296	4498	682	627	1768	675	3306	344	1168	785	1781
23-Sultangazi	2116	891	2727	1860	1177	2781	1838	1870	600	1535	972	126	1214	847	1525	2838	1207	1232	1225	2211
24-Sisli	366	240	579	549	444	548	576	2244	154	1702	211	29	348	227	658	1392	322	362	1433	550
25-Zeytinburnu	488	531	1009	1559	1822	805	1000	1067	274	1087	337	36	727	430	595	4570	327	987	529	1138
26-Adalar	29	12	27	24	24	26	25	54	8	52	11	2	13	11	17	154	9	20	23	24
27-Atasehir	609	172	352	291	268	313	282	550	113	517	165	27	151	147	180	1497	92	258	264	286
28-Beykoz	863	195	474	344	275	476	359	770	151	503	225	36	180	206	228	1179	124	300	425	392

continued on next page

Table 6a. Continued

T_{ij}	1	2	3	4	5	6	7	8	9	10	11	12	13	14	15	16	17	18	19	20
29-Cekmekoy	302	65	160	120	108	170	121	241	51	210	82	16	61	72	77	584	39	111	131	137
30-Kadikoy	857	230	452	369	341	390	345	641	157	605	222	32	190	189	218	1784	116	311	292	372
31-Kartal	666	168	376	303	284	355	309	531	117	534	178	31	160	157	184	1483	96	281	252	311
32-Maltepe	708	223	450	384	359	402	366	731	140	679	207	30	200	178	231	1929	116	329	312	372
33-Pendik	1695	331	783	592	535	803	604	995	255	992	423	80	305	350	360	2708	180	567	496	653
34-Sancaktepe	702	135	347	253	223	385	260	494	111	439	189	39	127	161	166	1200	82	239	271	298
35-Sultanbeyli	551	97	261	186	159	297	189	327	83	316	142	30	92	123	117	842	55	178	175	225
36-Sile	150	28	69	47	40	71	44	72	24	62	38	7	24	32	28	168	15	40	42	57
37-Tuzla	279	63	146	109	95	144	105	159	53	159	85	16	57	69	66	436	37	98	88	123
38-Umraniye	1325	305	679	529	486	665	537	1083	217	941	334	63	280	288	346	2651	180	494	558	555
39-Uskudar	796	202	465	365	338	457	370	720	152	651	218	37	187	206	231	1856	122	330	378	399
40-Gebze	4195	548	1320	887	739	1341	868	1336	472	1235	769	158	449	589	521	3413	271	846	704	1038

Table 6b. 2023 average morning peak-hour trip interchange matrix (T_{ij}) by urban rail modes over the highway-based network (21-40)

T_{ij}	21	22	23	24	25	26	27	28	29	30	31	32	33	34	35	36	37	38	39	40
1-Arnavutkoy	269	609	239	1094	434	7	262	153	59	735	365	284	663	100	61	57	594	358	364	1256
2-Avcılar	276	953	232	1238	615	8	256	149	50	649	379	292	571	81	45	54	498	315	331	939
3-Bagcilar	1056	1347	1401	4091	2054	24	683	478	142	1625	893	746	1276	212	108	129	1073	865	916	1944
4-Bahcelievler	693	905	698	2846	1617	18	500	304	95	1241	717	602	989	149	79	89	817	584	657	1426
5-Bakirkoy	168	188	126	729	615	5	152	70	24	381	226	178	295	43	24	19	230	161	181	417
6-Basaksehir	449	1250	729	1873	813	10	350	214	73	891	449	352	774	117	66	69	676	452	471	1317
7-Bayrampasa	442	379	674	1728	805	9	260	179	52	596	322	269	449	78	40	41	361	326	359	643
8-Besiktas	449	95	98	1385	155	3	112	78	22	234	121	95	164	34	18	13	113	139	160	209
9-Beylikduzu	309	874	168	1171	508	7	272	165	62	877	419	337	701	99	59	72	651	364	403	1392
10-Beyoglu	601	243	377	3368	541	8	230	165	44	483	244	208	333	67	34	30	252	279	325	437
11-Buyukcekmece	327	5231	194	1472	676	9	380	196	80	993	551	391	967	132	73	89	913	503	490	1766
12-Catalca	73	821	46	299	122	2	89	43	19	235	124	90	224	32	18	19	206	117	114	425

continued on next page

Table 6b. Continued

T_{ij}	21	22	23	24	25	26	27	28	29	30	31	32	33	34	35	36	37	38	39	40
13-Esenler	600	685	902	2475	1260	14	382	260	74	890	489	405	680	113	55	62	567	476	516	994
14-Esenyurt	317	1606	258	1421	705	9	294	181	62	707	398	306	668	101	55	68	604	379	378	1145
15-Eyup	906	538	1172	3297	954	12	433	301	88	1001	479	395	733	132	69	69	627	551	597	1138
16-Fatih	663	392	361	3075	1761	30	470	214	70	970	623	562	769	128	71	46	531	453	486	861
17-Gaziosmanpasa	867	614	1856	3354	1124	13	480	340	94	1012	514	417	756	140	68	72	651	608	612	1149
18-Gungoren	450	447	343	1697	1041	12	316	188	58	801	457	393	635	95	51	51	509	368	418	898
19-Kagithane	1908	385	636	6042	757	10	477	377	97	974	454	382	648	138	66	62	517	597	656	874
20-Kucukcekmece	699	1231	931	2796	1586	18	494	325	102	1184	682	556	986	157	85	97	831	616	641	1522
21-Sariyer	3887	286	210	3004	423	7	360	244	70	801	399	307	591	104	52	54	501	459	466	920
22-Silivri	689	29654	357	3315	1382	18	982	448	194	2491	1417	943	2480	323	174	219	2392	1297	1217	4624
23-Sultangazi	1016	878	1978	3437	1174	14	626	452	133	1393	668	565	1073	197	101	114	940	811	798	1775
24-Sisli	895	254	312	4439	424	7	258	182	50	531	263	214	370	79	43	32	267	319	363	488
25-Zeytinburnu	363	291	250	1696	2780	15	295	136	44	650	403	350	504	79	42	32	379	291	312	613
26-Adalar	22	12	8	85	31	12	119	19	13	237	141	204	118	24	13	5	64	78	94	90
27-Atasehir	256	223	89	1000	370	47	4087	503	375	3748	1796	1962	2036	596	340	141	1378	2398	2809	2014
28-Beykoz	431	279	129	1442	390	23	1333	4462	336	2455	1150	1020	1654	417	195	189	1263	1847	2044	2137
29-Cekmekoy	125	116	40	472	158	18	1083	389	741	1435	666	689	960	652	204	149	610	1572	1227	885
30-Kadikoy	279	284	115	1136	446	72	2161	417	260	6347	1682	1621	1932	379	218	128	1275	1649	3007	2386
31-Kartal	219	244	91	1000	399	41	1810	378	235	2971	7956	3205	6014	660	470	128	2313	1489	1635	2907
32-Maltepe	299	244	112	1225	478	61	2460	444	285	3801	3615	3590	3338	565	315	131	1775	1728	2136	2697
33-Pendik	439	602	174	2024	793	54	3010	785	539	5044	9342	4182	21177	1310	2104	322	6548	2830	2777	7605
34-Sancaktepe	251	266	83	997	336	32	2378	675	1047	2696	2229	1854	3166	2412	993	260	1794	2910	2166	2363
35-Sultanbeyli	151	201	55	695	248	21	1693	377	398	1868	2139	1371	5084	1184	2162	172	1802	1520	1273	2042
36-Sile	43	54	17	154	54	3	164	83	61	294	159	132	273	80	44	506	187	217	182	323
37-Tuzla	78	140	40	327	133	9	510	138	95	794	1248	624	2813	225	227	47	3463	478	456	1879
38-Umraniye	590	459	186	2065	702	65	4018	1344	1153	5666	2732	2763	3515	1394	563	363	2446	6200	4820	3775
39-Uskudar	342	294	120	1374	473	64	3249	1113	579	5770	2108	2075	2392	705	334	183	1670	3737	8371	2474
40-Gebze	744	1076	287	2985	1159	42	2296	931	481	5517	4167	2779	8114	934	629	434	7810	2672	2596	32921

Table 7a. 2023 average morning peak-hour trip interchange matrix (T_{ij}) by urban rail modes over the rail-based network (1-20)

T_{ij}	1	2	3	4	5	6	7	8	9	10	11	12	13	14	15	16	17	18	19	20
1-Arnavutkoy	4012	499	872	662	473	1290	490	638	368	525	670	164	328	483	476	1276	223	423	360	878
2-Avcılar	1072	3604	1061	1029	809	1986	547	621	1372	548	1811	117	376	2678	349	1561	200	542	346	1435
3-Bagcilar	2485	2003	7965	5876	3276	4844	2635	2301	1173	1859	1993	231	2931	1822	1719	4931	1176	2933	1376	5763
4-Bahcelievler	1617	1896	4284	5466	3196	2829	1559	1575	956	1333	1309	141	1397	1411	1019	4036	626	1806	872	4187
5-Bakirkoy	286	405	636	1080	1472	476	432	447	189	410	250	24	331	284	247	1440	151	475	232	807
6-Basaksehir	1828	1576	2214	1862	1277	7319	968	1029	1163	893	2203	212	834	2254	760	2113	501	813	675	3626
7-Bayrampasa	686	498	1370	1220	931	1216	2288	1035	305	825	494	56	1150	465	1131	1891	1037	908	664	1172
8-Besiktas	151	91	213	188	153	184	204	2098	60	506	85	12	121	84	186	496	102	127	512	193
9-Beylikduzu	1137	985	727	626	473	909	382	589	1895	497	1229	98	265	1054	277	1412	151	398	289	725
10-Beyoglu	442	296	731	750	613	736	799	1867	177	3307	268	39	495	275	931	1712	397	478	1290	701
11-Buyukcekmece	2139	1068	1118	882	650	1524	558	750	1263	651	4709	308	375	1813	384	1866	212	579	404	1110
12-Catalca	700	146	214	152	114	307	113	149	144	128	498	531	73	195	84	316	46	109	86	218
13-Esenler	1230	866	3508	2452	1536	2574	2145	1415	589	1141	1004	109	3119	953	1181	2832	975	2116	872	2430
14-Esenyurt	1470	2735	1289	1097	851	3035	656	676	2496	615	3433	189	452	5818	406	1661	247	641	400	1528
15-Eyup	1436	618	1650	1393	1036	1820	1921	1862	390	1537	670	91	924	608	2856	2438	1710	885	1500	1565
16-Fatih	625	530	1136	1331	1425	993	1183	1868	311	2252	442	53	713	480	883	12171	485	896	842	1156
17-Gaziosmanpasa	1348	707	2168	1642	1168	2199	2724	1879	468	1556	814	108	1289	750	2623	2716	2816	1145	1452	1888
18-Gungoren	873	789	1787	1986	1396	1190	1055	980	432	802	628	68	1278	596	590	2482	381	1662	530	1374
19-Kagithane	820	423	1104	965	754	1225	1099	3675	280	1931	459	65	591	437	1415	1959	691	650	4650	1066
20-Kucukcekmece	2066	2741	4382	4989	3580	5393	1656	1501	1273	1296	1824	216	1445	2041	1191	3785	771	1540	946	10370
21-Sariyer	750	243	572	437	334	561	455	1810	176	721	282	43	234	243	426	1173	193	336	940	499
22-Silivri	8390	1678	2675	1852	1383	3002	1498	2135	1648	1821	4403	763	895	1973	985	4763	530	1554	1133	2331
23-Sultangazi	2114	892	2712	1841	1159	2869	1814	1898	610	1518	1104	168	1203	889	1523	2750	1207	1208	1242	2249
24-Sisli	400	244	586	545	437	570	578	2218	159	1672	233	34	345	233	654	1278	325	360	1432	563
25-Zeytinburnu	510	538	1013	1549	1746	811	1027	1012	283	1040	373	42	733	441	594	4562	348	996	521	1117
26-Adalar	29	13	26	24	25	25	24	53	9	53	12	2	13	12	17	158	8	21	23	24
27-Atasehir	588	177	348	294	271	311	282	546	117	532	181	30	149	152	181	1564	95	258	256	289
28-Beykoz	838	189	449	322	257	452	335	740	150	488	242	41	169	201	213	1247	117	282	402	371
29-Cekmekoy	287	66	151	118	109	156	121	231	50	219	86	18	60	69	75	612	38	112	120	129

continued on next page

Table 7a. Continued

T_{ij}	1	2	3	4	5	6	7	8	9	10	11	12	13	14	15	16	17	18	19	20
30-Kadikoy	854	240	448	377	346	389	342	656	166	629	247	39	189	199	219	1869	117	314	293	376
31-Kartal	637	175	373	310	289	353	306	542	124	556	192	35	159	166	186	1557	97	284	255	315
32-Maltepe	707	232	444	390	363	398	360	741	147	702	227	35	197	189	233	2010	118	330	317	375
33-Pendik	1566	339	751	591	542	749	600	1009	253	1031	432	88	303	345	357	2843	182	568	492	629
34-Sancaktepe	652	134	322	246	226	351	259	475	107	458	194	42	126	152	160	1261	79	241	249	277
35-Sultanbeyli	497	94	233	174	159	263	186	316	77	322	140	31	88	112	111	876	54	175	162	201
36-Sile	142	27	65	45	39	67	44	71	24	65	41	8	23	31	28	177	15	40	41	54
37-Tuzla	268	63	139	108	97	133	105	162	51	167	87	17	57	68	65	461	37	99	87	117
38-Umraniye	1236	318	670	540	493	635	545	1058	225	980	359	70	281	293	343	2775	180	500	522	552
39-Uskudar	777	211	458	372	344	444	369	693	158	680	235	42	186	210	228	1949	119	336	352	397
40-Gebze	3778	556	1266	885	758	1295	883	1397	477	1322	835	181	455	601	528	3678	272	867	708	1007

Table 7b. 2023 average morning peak-hour trip interchange matrix (T_{ij}) by urban rail modes over the rail-based network (21-40)

T_{ij}	21	22	23	24	25	26	27	28	29	30	31	32	33	34	35	36	37	38	39	40
1-Arnavutkoy	321	1094	288	1204	474	7	273	166	67	699	371	290	667	109	61	68	564	370	360	1182
2-Avcılar	272	1404	230	1222	597	7	242	145	48	603	358	275	537	77	41	51	462	298	308	880
3-Bagcilar	1046	2256	1397	4099	2019	23	660	470	140	1529	862	717	1226	206	104	129	1025	833	858	1871
4-Bahcelievler	675	1485	699	2823	1581	18	481	300	92	1176	693	577	948	144	74	88	777	555	612	1371
5-Bakirkoy	165	310	128	721	600	5	150	69	24	373	222	174	289	42	22	19	222	156	173	406
6-Basaksehir	476	1964	765	1937	795	10	346	217	74	843	437	342	747	116	62	73	636	450	453	1256
7-Bayrampasa	446	608	673	1738	799	9	249	177	51	561	311	259	431	76	37	41	342	315	336	616
8-Besiktas	410	152	98	1377	153	3	108	80	21	226	119	92	162	33	17	13	108	135	153	210
9-Beylikduzu	301	1550	165	1152	490	6	259	159	60	809	398	317	659	94	53	69	595	344	376	1295
10-Beyoglu	563	378	377	3396	528	8	219	166	44	460	237	200	326	66	33	30	242	271	309	432
11-Buyukcekmece	355	5631	234	1541	691	10	388	203	84	1016	573	410	1001	135	71	96	914	515	501	1846
12-Catalca	78	1127	54	310	121	2	88	43	20	222	123	89	222	32	17	20	195	116	112	404

continued on next page

Table 7b. Continued

T_{ij}	21	22	23	24	25	26	27	28	29	30	31	32	33	34	35	36	37	38	39	40
13-Esenler	590	1102	899	2473	1238	13	368	256	73	842	476	393	655	109	52	62	540	457	483	954
14-Esenyurt	310	2156	268	1399	682	8	277	174	59	651	375	287	627	96	50	64	555	356	351	1064
15-Eyup	911	919	1186	3299	952	11	423	299	90	948	472	386	717	132	64	72	592	544	564	1103
16-Fatih	561	606	353	2939	1766	31	478	206	70	985	634	572	773	131	71	46	526	451	476	857
17-Gaziosmanpasa	877	1038	1872	3373	1128	13	470	339	96	956	502	407	730	138	64	74	615	601	578	1101
18-Gungoren	433	777	340	1681	1025	12	308	183	56	771	448	383	615	92	48	50	489	347	389	865
19-Kagithane	1868	634	666	6038	753	10	465	376	96	922	446	373	637	137	63	63	500	586	618	866
20-Kucukcekmece	705	1891	971	2796	1536	17	472	319	100	1104	647	526	936	152	80	95	782	590	597	1446
21-Sariyer	3902	546	199	2783	420	7	361	251	70	791	411	310	608	105	50	56	510	460	450	971
22-Silivri	1136	28307	579	4743	1876	27	1384	727	300	3846	2024	1457	3667	483	258	387	3509	1856	1767	7807
23-Sultangazi	1000	1558	1972	3414	1149	14	605	442	133	1303	643	541	1028	192	95	117	884	788	751	1681
24-Sisli	825	389	313	4507	414	7	245	180	49	503	255	204	359	76	39	31	250	305	342	479
25-Zeytinburnu	335	463	255	1642	2811	15	297	133	44	654	406	352	504	79	41	32	376	289	308	610
26-Adalar	20	21	8	84	30	12	118	17	13	237	141	206	118	25	13	5	64	76	96	88
27-Atasehir	229	389	90	972	368	47	4118	474	369	3795	1795	1948	2018	592	313	137	1297	2379	2896	1999
28-Beykoz	398	501	120	1371	371	23	1345	4558	339	2400	1160	1031	1666	427	194	191	1232	1871	2009	2135
29-Cekmekoy	109	190	36	451	159	18	1083	376	747	1423	672	683	1000	657	208	149	611	1590	1237	886
30-Kadikoy	269	565	114	1142	447	73	2165	394	253	6336	1697	1643	1925	382	212	125	1257	1618	2998	2353
31-Kartal	209	413	91	1005	401	42	1814	362	234	3023	7921	3156	5984	648	468	125	2297	1475	1690	2881
32-Maltepe	282	451	111	1224	477	61	2459	417	277	3834	3598	3556	3297	565	311	127	1774	1697	2201	2641
33-Pendik	408	996	171	2007	792	55	3031	753	569	5111	9343	4155	21177	1362	2142	327	6549	2839	2850	7569
34-Sancaktepe	220	425	75	955	338	33	2393	653	1053	2720	2244	1862	3274	2452	982	254	1769	2939	2198	2373
35-Sultanbeyli	131	305	50	658	244	22	1663	349	401	1889	2150	1332	5194	1170	2154	170	1777	1506	1286	2075
36-Sile	40	97	16	151	55	3	165	83	62	292	160	128	282	80	44	508	185	217	183	321
37-Tuzla	74	221	38	324	134	9	516	134	103	813	1265	629	2863	238	238	48	3442	485	469	1896
38-Umraniye	521	801	173	1985	706	64	4036	1306	1161	5639	2742	2759	3522	1406	546	358	2362	6246	4778	3747
39-Uskudar	305	493	115	1326	478	64	3269	1082	583	5758	2145	2094	2424	716	324	181	1633	3747	8268	2493
40-Gebze	718	2121	274	3020	1184	43	2356	939	503	5620	4230	2818	8179	975	669	441	7852	2725	2678	32686

Table 8a. 2023 average morning peak-hour trip interchange matrix (T_{ij}) by urban rail modes over the balanced or master plan network (1-20)

T_{ij}	1	2	3	4	5	6	7	8	9	10	11	12	13	14	15	16	17	18	19	20
1-Arnavutkoy	4171	504	874	660	477	1286	503	629	367	526	623	153	336	482	480	1286	229	434	360	876
2-Avcılar	1164	3637	1071	1032	814	1978	556	622	1373	560	1694	111	380	2686	353	1602	203	555	348	1425
3-Bagcilar	2556	2009	7979	5897	3302	4791	2640	2277	1161	1868	1847	212	2931	1813	1714	5028	1179	2948	1369	5794
4-Bahcelievler	1665	1876	4287	5467	3177	2795	1565	1575	932	1375	1199	128	1400	1381	1018	4157	629	1813	877	4192
5-Bakirkoy	294	398	630	1075	1474	466	429	458	184	425	225	22	327	276	245	1477	148	472	232	803
6-Basaksehir	1901	1593	2182	1853	1286	7349	985	1022	1163	902	2065	199	844	2267	770	2158	509	825	673	3575
7-Bayrampasa	707	493	1363	1223	934	1202	2280	1032	301	834	456	51	1146	461	1122	1962	1032	912	661	1160
8-Besiktas	153	89	211	189	154	180	205	2049	58	522	77	11	122	81	192	525	105	127	520	191
9-Beylikduzu	1253	999	744	642	488	913	391	588	1915	501	1131	91	272	1052	283	1433	155	412	291	740
10-Beyoglu	442	291	723	754	620	717	798	1909	172	3280	244	35	497	268	930	1870	393	477	1292	691
11-Buyukcekmece	2117	999	1083	855	618	1473	538	696	1173	615	4422	288	360	1722	365	1740	202	562	379	1076
12-Catalca	778	154	225	158	119	319	121	155	154	133	450	537	78	199	90	329	50	116	89	224
13-Esenler	1269	867	3507	2466	1556	2559	2146	1405	587	1152	936	100	3113	956	1179	2882	978	2118	869	2420
14-Esenyurt	1606	2747	1303	1108	862	3058	669	676	2497	623	3248	181	458	5826	412	1707	251	659	402	1538
15-Eyup	1477	613	1644	1397	1045	1800	1913	1864	385	1563	616	83	923	600	2826	2579	1699	889	1493	1549
16-Fatih	646	518	1133	1317	1380	991	1197	2043	302	2309	409	50	723	466	916	12035	498	890	897	1137
17-Gaziosmanpasa	1385	702	2162	1647	1173	2171	2726	1871	462	1581	752	98	1292	743	2607	2826	2801	1154	1442	1872
18-Gungoren	906	783	1794	2000	1413	1178	1051	978	422	817	573	62	1272	584	588	2474	380	1664	535	1384
19-Kagithane	820	413	1081	963	762	1178	1098	3674	269	1992	412	58	593	421	1418	2115	691	654	4585	1034
20-Kucukcekmece	2130	2770	4394	4991	3592	5282	1664	1497	1268	1322	1695	201	1444	2037	1186	3936	770	1565	943	10405
21-Sariyer	699	238	569	446	344	552	468	1911	168	797	249	36	240	236	457	1242	207	338	997	498
22-Silivri	5378	1301	1988	1420	1043	2394	1080	1436	1334	1276	4574	677	648	1731	694	3254	369	1143	797	1861
23-Sultangazi	2189	890	2722	1853	1174	2848	1831	1873	601	1525	1009	152	1214	875	1521	2814	1208	1227	1229	2247
24-Sisli	404	241	581	551	445	556	581	2238	155	1701	213	32	348	228	657	1390	325	362	1427	554
25-Zeytinburnu	520	534	1009	1553	1812	803	1022	1064	276	1086	340	38	725	432	605	4573	347	985	536	1130
26-Adalar	31	12	27	24	24	26	25	54	8	52	11	2	13	11	17	154	9	20	24	24
27-Atasehir	639	171	351	289	269	312	286	547	113	515	166	29	151	146	182	1505	96	258	269	286
28-Beykoz	897	196	476	345	276	474	361	769	153	504	229	39	181	205	230	1180	127	301	426	392

continued on next page

Table 8a. Continued

T_{ij}	1	2	3	4	5	6	7	8	9	10	11	12	13	14	15	16	17	18	19	20
29-Cekmekoy	317	66	161	120	109	170	123	242	51	210	84	17	61	72	80	585	41	111	134	138
30-Kadikoy	912	230	449	369	341	387	345	640	158	606	221	35	190	190	219	1786	118	312	293	370
31-Kartal	696	168	374	303	284	352	307	530	118	534	178	34	160	158	186	1486	98	281	256	310
32-Maltepe	756	224	447	383	359	399	364	728	141	679	206	33	199	181	235	1930	120	329	318	370
33-Pendik	1740	329	773	585	532	793	598	987	254	983	419	85	302	349	361	2698	182	562	500	648
34-Sancaktepe	722	135	347	253	224	384	262	493	113	439	191	41	127	160	168	1205	85	238	275	300
35-Sultanbeyli	554	96	256	182	158	289	186	322	82	307	139	30	89	119	114	834	55	173	174	221
36-Sile	156	29	69	47	40	71	45	72	25	62	38	8	24	31	29	168	16	40	43	58
37-Tuzla	295	62	145	108	95	143	104	158	53	159	84	17	57	70	66	436	38	97	89	122
38-Umraniye	1369	306	683	529	486	668	549	1084	220	942	341	67	283	288	354	2654	188	496	567	558
39-Uskudar	833	203	467	366	338	457	373	722	154	653	219	40	188	206	233	1861	123	331	380	399
40-Gebze	4126	553	1315	891	742	1333	866	1337	478	1237	768	165	450	599	523	3424	271	850	710	1040

Table 8b. 2023 average morning peak-hour trip interchange matrix (T_{ij}) by urban rail modes over the balanced or master plan network (21-40)

T_{ij}	21	22	23	24	25	26	27	28	29	30	31	32	33	34	35	36	37	38	39	40
1-Arnavutkoy	323	635	296	1216	487	7	288	172	70	753	387	307	700	114	64	71	596	391	384	1261
2-Avcılar	277	953	233	1240	617	8	256	150	50	650	378	292	569	81	43	54	492	316	333	942
3-Bagcilar	1058	1422	1395	4092	2053	24	691	480	144	1625	891	746	1268	214	107	132	1066	876	920	1947
4-Bahcelievler	693	912	697	2846	1616	18	499	306	95	1241	716	601	981	149	76	90	809	586	660	1427
5-Bakirkoy	168	188	126	729	613	5	152	71	24	381	225	178	293	43	23	19	227	161	182	416
6-Basaksehir	481	1338	779	1953	816	10	366	224	78	910	458	363	789	123	65	76	676	476	485	1344
7-Bayrampasa	457	383	669	1744	810	9	264	180	53	597	320	269	445	79	39	42	356	332	361	643
8-Besiktas	447	96	98	1380	156	3	112	79	22	235	122	95	164	34	17	13	109	140	161	210
9-Beylikduzu	310	869	169	1172	510	7	273	166	63	877	419	338	698	99	56	72	635	365	405	1395
10-Beyoglu	606	245	372	3370	542	8	231	166	45	486	244	208	332	68	33	30	247	281	328	439
11-Buyukcekmece	326	5236	221	1460	667	9	377	191	80	994	558	394	972	129	68	90	891	500	488	1785
12-Catalca	87	844	58	329	129	2	96	47	22	250	134	99	244	35	18	22	218	128	123	455

continued on next page

Table 8b. Continued

T_{ij}	21	22	23	24	25	26	27	28	29	30	31	32	33	34	35	36	37	38	39	40
13-Esenler	602	711	896	2476	1256	14	386	262	76	892	489	405	675	113	54	64	561	483	519	995
14-Esenyurt	318	1596	272	1423	707	9	295	182	63	708	398	307	667	101	53	67	595	380	380	1148
15-Eyup	921	563	1173	3313	968	12	446	305	93	1007	488	404	744	137	67	74	615	570	600	1149
16-Fatih	662	395	359	3072	1758	30	470	214	70	970	624	562	765	129	69	46	527	453	488	862
17-Gaziosmanpasa	898	653	1849	3398	1143	13	495	346	99	1019	521	426	761	144	67	76	642	630	616	1152
18-Gungoren	449	451	340	1694	1040	12	315	188	59	800	456	392	629	94	49	51	504	368	419	895
19-Kagithane	1929	395	640	6045	768	10	487	378	99	977	460	389	653	141	65	64	511	609	657	883
20-Kucukcekmece	711	1266	961	2815	1580	18	497	327	104	1182	679	555	979	158	83	99	823	623	643	1523
21-Sariyer	3868	288	210	2987	425	7	362	244	70	803	400	309	589	105	50	54	487	460	467	920
22-Silivri	671	29698	397	3239	1339	19	959	427	188	2482	1427	939	2469	308	156	215	2308	1269	1200	4641
23-Sultangazi	1012	925	1979	3420	1171	14	636	451	138	1389	670	569	1067	200	99	120	922	827	797	1758
24-Sisli	888	254	311	4419	424	7	259	182	50	533	264	214	369	79	41	32	257	320	364	488
25-Zeytinburnu	363	292	256	1694	2763	15	295	137	44	651	404	351	502	79	41	32	378	291	314	614
26-Adalar	22	12	8	85	31	12	119	19	13	237	140	207	118	25	13	5	65	78	94	90
27-Atasehir	259	223	91	996	371	47	4110	506	378	3766	1795	1951	2026	598	314	142	1313	2414	2849	2033
28-Beykoz	431	281	130	1438	392	23	1345	4450	334	2453	1147	1026	1650	421	192	189	1226	1841	2039	2137
29-Cekmekoy	125	117	42	472	158	18	1091	388	739	1436	667	683	993	649	206	149	609	1572	1227	889
30-Kadikoy	278	285	115	1134	447	72	2162	416	260	6346	1685	1633	1920	384	210	128	1268	1651	3011	2383
31-Kartal	220	246	91	998	401	41	1802	378	237	2973	7970	3159	6013	658	471	127	2322	1486	1642	2922
32-Maltepe	298	246	112	1222	480	61	2460	444	285	3800	3621	3588	3325	575	315	131	1805	1726	2142	2697
33-Pendik	436	598	174	2002	789	54	2994	776	574	5023	9342	4136	21261	1374	2150	331	6584	2837	2772	7628
34-Sancaktepe	252	267	85	994	337	33	2393	674	1048	2722	2232	1860	3253	2439	977	257	1766	2916	2175	2377
35-Sultanbeyli	149	195	54	677	243	21	1639	363	408	1852	2145	1333	5179	1180	2159	173	1778	1504	1253	2081
36-Sile	43	55	17	154	55	3	165	83	61	294	158	128	280	79	44	504	186	217	182	323
37-Tuzla	78	138	40	324	133	9	507	137	103	795	1256	621	2854	238	237	48	3460	480	455	1890
38-Umraniye	590	463	192	2064	706	65	4033	1342	1150	5679	2730	2759	3515	1394	543	363	2373	6193	4818	3784
39-Uskudar	344	297	121	1377	476	65	3264	1112	578	5788	2114	2072	2392	708	319	183	1621	3731	8365	2484
40-Gebze	745	1087	284	2988	1166	42	2302	932	491	5524	4190	2775	8141	958	654	434	7879	2685	2603	32896

Figure 3. Dendrograms for the base network by the linkage techniques

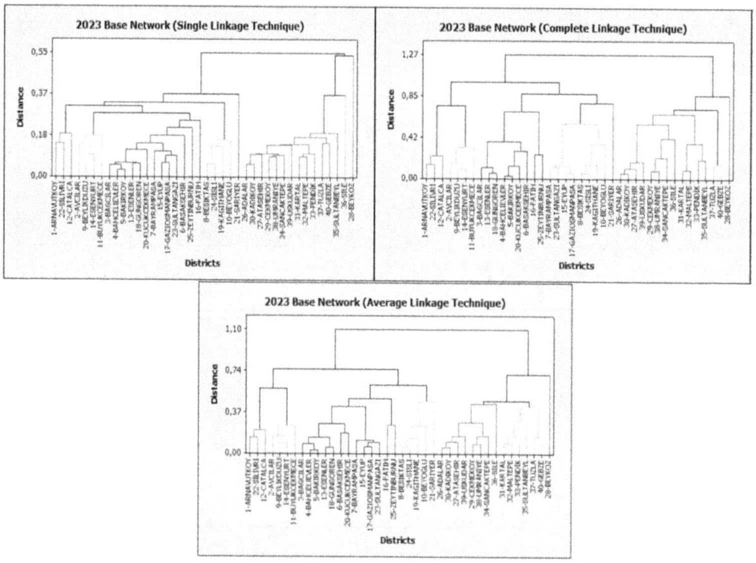

Figure 4. Dendrograms for the highway-based network by the linkage techniques

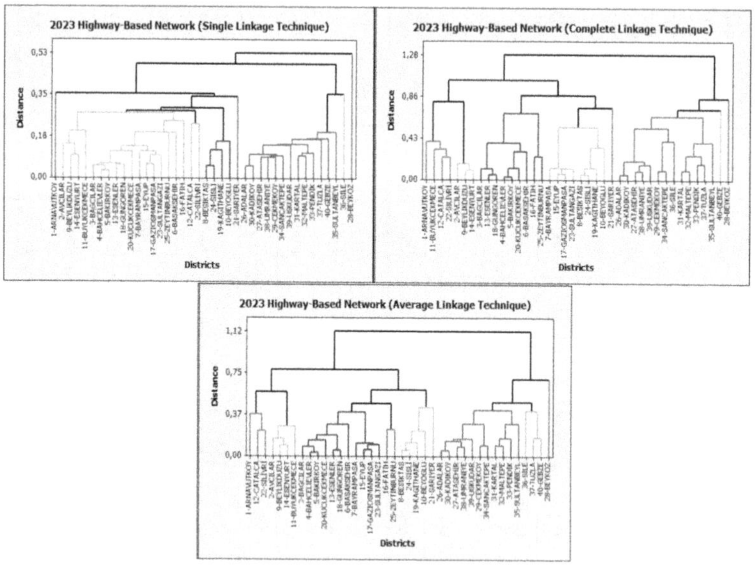

As seen the clustering results by the single linkage in Figure 7, in the Asian side of the city there is one major cluster formed with the addition of two other districts; “36-SILE” and “28-BEYKOZ”. This did not change for all alternative networks. In the European side of the city, there are four major clusters observed and on top of those, “21-SARIYER” joined them in the base and rail-based networks. In the highway-based network, “1-Arnavutkoy” district joined the European clusters after “21-SARIYER” district. The reverse is observed in the master plan network. At this point, it needs to be mentioned that a drawback of single clustering method is that it tends to produce long thin clusters in which nearby ele-

Figure 5. Dendrograms for the rail-based network by the linkage techniques

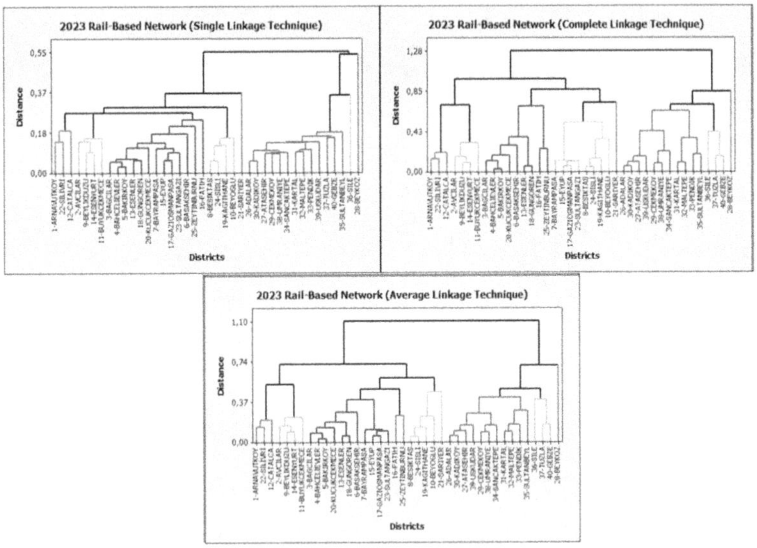

Figure 6. Dendrograms for the master plan network by the linkage techniques

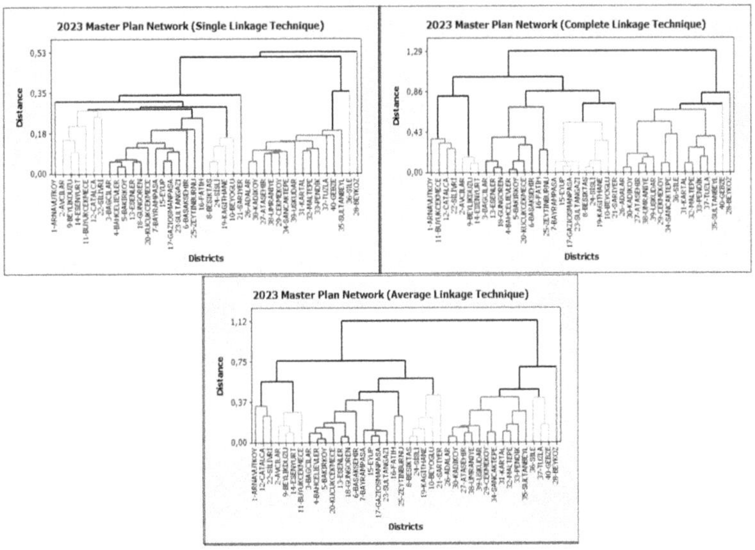

Table 9. Cophenetic Correlation Coefficients (CPCCs) for the 2023 base network scenario

Linkage Techniques	N+	CPCC	*p*, Significance (2-Tailed)
Original Distance (d_{ij})	780	1.000	N/A
Single Linkage	780	0.877	0.000*
Complete Linkage	780	0.908	0.000*
Average Linkage	780	0.916	0.000*

+N: (n x m)/2-(n/2); N/A: Not applicable; *: Significant at level 0.001.

Figure 7. Dendrograms for the 2023 transport networks by the single linkage technique

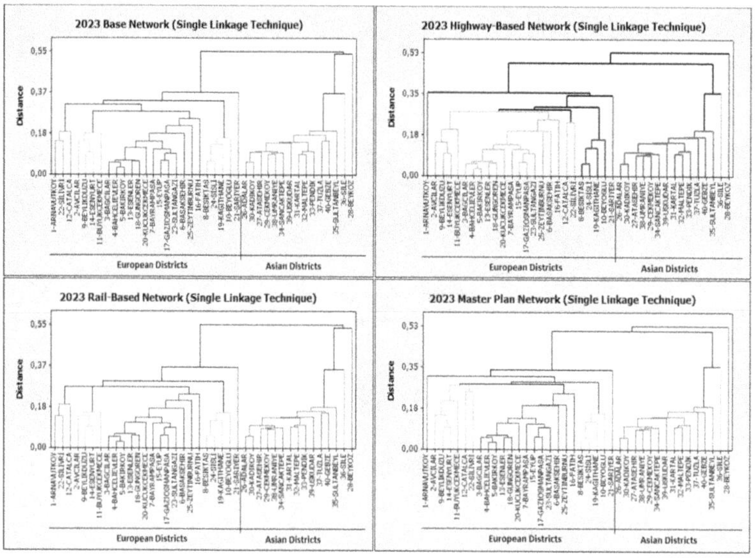

Figure 8. Dendrograms for the 2023 transport networks by the complete linkage technique

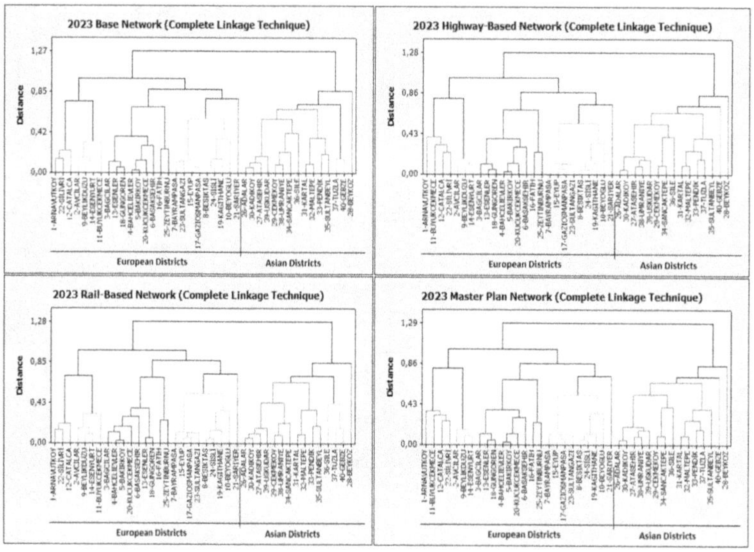

ments of the same cluster have small distances, but elements at opposite ends of a cluster may be much farther from each other than to elements of other clusters. This may lead to difficulties in defining clusters that could usefully subdivide the data. Single linkage clustering is sensitive to outliers (Everitt, 2011).

In the highway-based network, Asian and European clusters (except the district of "28-BEYKOZ") are linked at correlation distance of 0.48 by the single linkage. If we include "28-BEYKOZ" to the Asian side of the city, then the distance between the two sides of the city increases to 0.53 correlation distance. This distance is 0.55, 0.55, 0.51 for the base, rail-based, and master plan networks, respectively. As seen, the highway-based and master plan networks yield shorter distances between European and Asian sides

Figure 9. Dendrograms for the 2023 transport networks by the average linkage technique

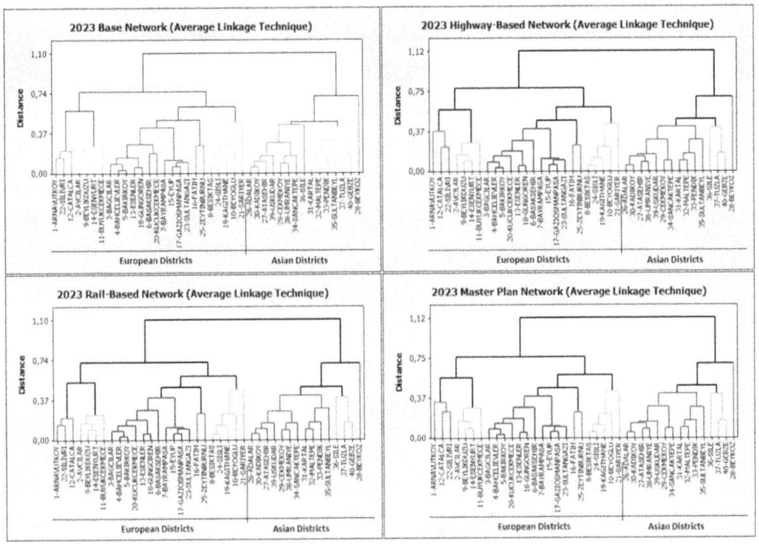

of the city that means that they yield higher interactions between the two sides of the city because of better transportation links for highway modes.

Figure 8 presents the dendrograms by the complete linkage technique. In the complete-linkage clustering, the link between two clusters contains all element pairs, and the distance between clusters equals the distance between those two elements (one in each cluster) that are farthest away from each other. The shortest of these links that remains at any step causes the fusion of the two clusters whose elements are involved. The method is also known as farthest neighbor clustering. Since clusters are formed at farthest distances between districts and clusters, the correlation distances are greater in the dendrograms of the complete linkage methods in Figure 8 than those by the single linkage methods in Figure 7. Again, the dendrograms for the base and rail-based networks, and those for the highway-based and master plan networks look quite similar.

Single and complete linkage techniques can have some practical problems. Single linkage suffers from chaining. In order to merge two groups, only need one pair of points to be close, irrespective of all others. Therefore, clusters can be too spread out, and not compact enough. Complete linkage avoids chaining, but suffers from crowding. Because its score is based on the worst-case dissimilarity pairs, a point can be closer to points in other clusters than to points in its own cluster. Clusters are compact, but not far enough apart.

On the other hand, average linkage technique tries to strike a balance. It uses average pairwise dissimilarity, so clusters tend to be relatively compact and relatively far apart. Figure 9 presents the dendrograms by the average linkage technique. Similarly, the dendrograms of the highway-based, rail-based and master plan networks are very much alike. The dendrogram of the base network in the European side is very much similar to the one of the rail-based network.

CONCLUSION

This chapter presented the case study of the analyses of urban spatial structures of Istanbul Metropolitan City of Turkey for future (2023) network design alternatives: i) rail-based, ii) highway-based, and iii) balanced or master plan scenario. Results of the analyses can be used to present the effects of different network alternatives on the sustainable urban transportation planning by confirming the expansion of urban city core as well as other urban and suburban developments and their emergence with their neighboring clusters. Decision makers will be able to develop their capacities to plan ahead in order to meet future needs better from the spatial restructuring in urban agglomerations, when they know the extent and location of the future growth ahead. Such analyses provide a tool to compare urban structure models representing possible snapshots of human settlements during the course of growth and expansion under several alternatives of transportation network developments.

Results of the analyses led to the following conclusions:

- Cluster analyses can be used to determine urban centers and subcenters using trip interchange data among the districts of a metropolitan city. However, for better results and more detailed analysis, smaller analysis units such as travel analysis zones should be used.
- In parallel to small differences among the outcomes of the alternative transport networks with respect to the sustainability indicators given in Table 1, the clustering results show only small differences. Perhaps for better results, instead of peak-hour, all day travel volumes should be used as input for the cluster analyses.
- Last but not least, both highway and rail investments increase trip interchanges among the districts and that's why the clustering results for the alternative networks look very much alike.

The results of this case study are expected to be used by planners, administrators and policy makers to comprehend the interactions over a spatial structure, and to plan for creating efficient subcenters with a more balanced distribution of travel patterns over urban agglomerations. Consequently, by this approach the impact of changes in land use, and population and employment densities on the hierarchical pattern of movements and hubs, which may indicate potential areas for future developments, can be modeled.

FUTURE RESEARCH DIRECTIONS

Modeling urban growth and expansion is a very complex issue since it includes many factors such as socioeconomic factors, physical and environmental factors, and emerging market and development trends. Future research should include other important variables in modeling urban growth and expansion.

REFERENCES

Aggarwal, C. C. (2003). Towards systematic design of distance functions for data mining applications. In *Proceedings of the ninth ACM SIGKDD international conference on Knowledge discovery and data mining*. ACM. doi:10.1145/956750.956756

Aggarwal, C. C., & Reddy, C. K. (Eds.). (2013). *Data Clustering: Algorithms and Applications*. CRC Press.

Akın, D. (2011). *Kentsel Yapının Değişiminin Kentsel Yolculukların Hiyerarşik Kümeleme Analizi ile İncelenmesiyle Ortaya Konması: İstanbul Metropoliten Alanı 1996-2006 Yılları* [Studying the Change of Urban Structure by Hierarchical Cluster Analysis: Istanbul Metropolitan Region between 1996 and 2006]. *2nd Symposium of Urban and Regional Studies*, Ankara, Turkey.

Akin, D. (2014). Is Istanbul's growth monocentric or multicentric? Analyses of intraurban travel data between 1996 and 2006. In *OPT-i,1st International Conference on Engineering and Applied Sciences Optimization* (pp. 2773-2799). Institute of Structural Analysis and Antiseismic Research, School of Civil Engineering National Technical University of Athens (NTUA), Greece.

Akin, D., & Alasalvar, S. (2016). Estimate Urban Growth and Expansion by Modeling Urban Spatial Structure Using Hierarchical Cluster Analyses of Interzonal Travel Data. *International Journal of System Dynamics Applications*, *5*(4), 16–40. doi:10.4018/IJSDA.2016100102

Alpkokin, P., Cheung, C., Black, J., & Hayashi, Y. (2008). Dynamics of clustered employment growth and its impacts on commuting patterns in rapidly developing cities. *Transportation Research Part A, Policy and Practice*, *42*(3), 427–444. doi:10.1016/j.tra.2007.11.002

Alpkokin, P., Hayashi, Y., Black, J., & Gercek, H. (2005). Polycentric employment growth and impacts on urban commuting patterns: Case study of Istanbul. *Journal of the Eastern Asia Society for Transportation Studies*, *6*(1), 3835–3851.

Berköz, L. (2000). Location of financial, insurance, and real estate firms in Istanbul. *Journal of Urban Planning and Development*, *126*(2), 75–88. doi:10.1061/(ASCE)0733-9488(2000)126:2(75)

Berkoz, L., & Turk, S. S. (2010). Intrametropolitan determinants of foreign investment firms in Istanbul. *Journal of Urban Planning and Development*, *136*(3), 234–242. doi:10.1061/(ASCE)UP.1943-5444.0000025

Besussi, E., Chin, N., Batty, M., & Longley, P. (2010). The structure and form of urban settlements. In *Remote Sensing of Urban and Suburban Areas* (pp. 13–31). Springer Netherlands. doi:10.1007/978-1-4020-4385-7_2

Bhandari, K., Alpkokin, P., Kachi, N., Jia, P., & Kato, H. (2007). *Dynamics of employment distribution and accessibility index: Case of Delhi. Proc., Infrastructure Planning*. JSCE.

Bhandari, K., Jia, P., Alpkokin, P., Mukhopadhyay, D., & Gangopadhyay, S. (2007). Policies, commuting patterns and accessibility in a non-monocentric city: Case study of Delhi. In *Proceedings of the Eastern Asia Society for Transportation Studies* (pp. 192-192). Eastern Asia Society for Transportation Studies.

Cats, O., Wang, Q., & Zhao, Y. (2015). Identification and classification of public transport activity centres in Stockholm using passenger flows data. *Journal of Transport Geography, 48*, 10–22. doi:10.1016/j.jtrangeo.2015.08.005

Dökmeci, V., & Berköz, L. (1994). Transformation of Istanbul from a monocentric to a polycentric city. *European Planning Studies, 2*(2), 193–205. doi:10.1080/09654319408720259

Everitt, B. (2011). *Cluster analysis*. Chichester, UK: Wiley. doi:10.1002/9780470977811

Gan, G., Ma, C., & Wu, J. (2007). *Data clustering: theory, algorithms, and applications* (Vol. 20). Siam. doi:10.1137/1.9780898718348

Halkidi, M., Batistakis, Y., & Vazirgiannis, M. (2002). Cluster validity methods: Part I. *SIGMOD Record, 31*(2), 40–45. doi:10.1145/565117.565124

Härdle, W., & Simar, L. (2007). *Applied multivariate statistical analysis*. Berlin: Springer.

Manual, H. C. (2010). (*HCM*). Washington, D.C.: Transportation Research Board.

ITMP. (2011). *2023 Istanbul Metropolitan Area Transport Master Plan*. Istanbul: Istanbul Metropolitan Municipality.

Istanbul Population 2016. (n.d.). Retrieved December 23, 2016, from http://worldpopulationreview.com/world-cities/istanbul-population/

Jain, A. K., & Dubes, R. C. (1988). *Algorithms for clustering data*. Prentice-Hall, Inc.

Jain, A. K., Murty, M. N., & Flynn, P. J. (1999). Data clustering: A review. *ACM Computing Surveys, 31*(3), 264–323. doi:10.1145/331499.331504

Johnson, R. A., & Wichern, D. W. (1992). *Applied multivariate statistical analysis* (Vol. 4). Englewood Cliffs, NJ: Prentice hall.

Milligan, G. W. (1980). An examination of the effect of six types of error perturbation on fifteen clustering algorithms. *Psychometrika, 45*(3), 325–342. doi:10.1007/BF02293907

Ozus, E., Akın, D., & Çiftçi, M. (2012). Hierarchical cluster analysis of multicenter development and travel patterns in Istanbul. *Journal of Urban Planning and Development, 138*(4), 303–318. doi:10.1061/(ASCE)UP.1943-5444.0000121

Rodrigue, J. P., Comtois, C., & Slack, B. (2013). *The geography of transport systems*. Routledge.

Sohn, J. (2005). Are commuting patterns a good indicator of urban spatial structure? *Journal of Transport Geography, 13*(4), 306–317. doi:10.1016/j.jtrangeo.2004.07.005

Tan, P., Steinbach, M., & Kumar, V. (2006). Introduction to data mining. Addison-Wesley.

Theodoridis, S., & Koutroubas, K. (1999). *Pattern Recognition*. London: Academic Press.

Tsompanoglou, S., & Photis, Y. N. (2013). Measuring urban concentration: A spatial cluster typology based on public and private sector service patterns. *World Review of Science. Technology and Sustainable Development, 10*(4), 185–202.

Vliet, J. V., White, R., & Dragicevic, S. (2009). Modeling urban growth using a variable grid cellular automaton. *Computers, Environment and Urban Systems*, *33*(1), 35–43. doi:10.1016/j.compenvurbsys.2008.06.006

Vuchic, V. R. (1999). *Transportation for livable cities*. Rutgers Center for Urban Policy Research.

Wang, F., & Sun, J. (2012). *Distance metric learning in data mining.SDM Conference (Tutorial).*

Zhao, P., Lu, B., & de Roo, G. (2011). The impact of urban growth on commuting patterns in a restructuring city: Evidence from Beijing. *Papers in Regional Science, 90*(4), 735-754.

Chapter 11
A Comparison of Transport Modes in Terms of Energy Consumption

Zafer Yilmaz
Turkish Army, Turkey

Serpil Erol
Gazi University, Turkey

Ebru Vesile Öcalir-Akunal
Gazi University, Turkey

ABSTRACT

Energy consumption of transport modes are partly dependent on the chosen travel mode. For specific OD pairs a comparison of travel modes in terms of energy consumption gives an idea of the possibilities to change the energy budget of a city. In this study a comparison of transport modes in terms of energy consumption is given for two chosen routes in the case study of Ankara, Turkey. The results show that there are differences with respect to chosen travel mode in terms of energy use and travel cost reflected to traveler for a certain OD couple.

INTRODUCTION

During the evolution of settlements, "an optimum" in space organisations with "a minimum" energy for mobility (adopted to the human dimension) had emerged until periods of rapid industrialisation. During the 20th century, cities have followed a disordered and limitless growth, which can be called as "sprawl", "metropolisation", "urban diffusion", etc. This process has created a mobility and energy consumption problem because it has resulted in a settlement model wholly dependent on private car. With the introduction of motorised modes, urban development began to change. Energy, which had been restricted to the use of body energy of a person, has found new possibilities to create new orders and therewith to use the resources differently. The new technologies have created a new form of mobility,

DOI: 10.4018/978-1-5225-2116-7.ch011

which is dependent on external energy. The use of external energy for mobility has changed the natural structure of the settlements and has created dispersed structures (Yildiz, 2003, p. 32).

Energy use in transport is determined by the demand for movement of people and goods. It is influenced by the spatial environment in which people live and by the activities that create movement. Any trip done by an individual by means of a transport mode generates an energy consumption. The level of this consumption and the type of energy required depend on the length of the journey and on the mode of transport (the characteristics of the travel modes) used. The structure of cities and their geographical constraints will influence the length of trips, thus energy consumption for passenger transport (Yildiz, 2003, p.32).

The (body) energy requirement of pedestrian, which can be measured in pulse frequency, is 50-60% higher than the car driver who sits still in his car (for unit distance). In inclined roads, this consumption raises to an additional 20%- 75%. Car drivers allocate their (body) energy savings to longer distances of travel. In addition, the cars they drive consume extreme amounts of external energy, which can be obtained from natural resources. Public transport modes can be considered to be more efficient compared to car due to their high occupancy levels but they are also dependent on external energy. They follow the same trend with car in supply of mobility needs. They also offer travelers the chance to save the body energy for unit distance and let them travel longer distances (Yildiz, 2003, p.33).

The main transportation modes in urban public transport are buses, trams, trolleybuses, trains and metro, and in some cities also ferries (Teodorović & Janić, 2017). Cities can have one or many of these systems according to their specific needs and opportunities. In some cities para-transit is also an important component of the transportation system.

Mode choice is a fundamental problem in transportation planning process and is dependent on many factors. There can be several travel alternatives to travel between a certain O (Origin) and a D (Destination). Public transport users try to make the best decision among these alternatives and sometimes they cannot choose the most rational one. This study presents a set of the possible alternatives for a certain OD couple with a comparison in terms of energy use.

Energy use is a basic indicator of the sustainability in transport systems. Public transport modes should follow a greener tendency to be more sustainable. There are a number of beneficial forms of green transportation that support and improve urban systems. The green transportation options make people's lives easier, reduce congestion, and reduce dependence on cars and foreign oil. Green transportation is also safer and less costly which helps save the earth.

It is always much more comfortable and convenient to drive one's own private vehicle to go to office, home or market every day. Citizen sensitivity lies on using green transportation that is easily accessible to everyone. Green transport's sorting begins with pedestrians and bicycles and scooters come after them. Public transportation opportunities are followed by private cars.

Transportation is being faster and cheaper in today's world. The differences among costs of transport get more negligible for transport users. However the important differences in energy use still exist and they can be overestimated sometimes. The cities all around the world are getting more crowded and that results many problems in urban life. As the cities are being more crowded, the need of energy increases. The dependency to external energy forces countries pay enormous amounts of money to import energy. The need of people for energy forced the big countries to get some precautions to control the region which has the energy (petroleum, natural gas, etc.). The need for energy has initiated many wars during the last century. So, it is a main duty of the city authorities to find more economic and energy saving solutions for public transportation.

In literature there are some studies dealing with the methodologies to chose the best alternative public transport system in a city or in a particular route. Some of the most relevant are listed below:

Buwana et al. (2016) discuss the most important criteria and choose the appropriate alternative to develop sustainable transportation systems in the Kasongan City using Analytical Hierarchy Process (AHP). Their analysis results show the most appropriate alternative is the optimization of the integrated road—river transportation systems. Social acceptance is supposed to be the main aspect that must be met in order to increase economic activity. Implementation of this alternative is realized by developing the integrated transit locations in Kasongan City. This program could increase public transport users and reducing CO_2 emissions.

The main aim of Sebhatu & Enquist (2016) in their paper is to assess and analyze how sustainable public transit (SPT) services contribute to building living cities in emerging economies through value network and resource integration/value co-creation, which is embedded on societal and environmental perspectives. The study identified the challenges of transforming emergent and fast urbanized cities, and understanding the role of public transport services from the cities perspective in an international context by embedding social and environmental perspectives. The challenges of transforming today's cities have created many new opportunities for prosperity through new ways of living and working in urban areas. The paper makes a contribution to the study of public transit services by explicitly linking the adoption of SPT value network in emerging economies based on sustainability, and resource integration/value co-creation for building resilient living cities.

Beaudoin et al. (2015) provide a framework for evaluating public transit investment, taking account of the effects of transit investment on traffic congestion and air quality. They discuss how to assess the sustainability of transit investment and the mechanisms through which public transit investment can affect equilibrium auto travel volumes and the associated congestion and air quality outcomes.

A dynamic model for planning and development of sustainable transportation systems is presented by Maheshwari et al.(2015). This is given by a system of three nonlinear differential equations representing the dynamics of the three independent states, namely, transportation, activity, and environmental systems. A policy scenario considering investment in energy efficient technologies and its effects on the states is discussed to assist making investment decisions. Optimal control techniques are used to design the controls. The results of their study show that it is possible to formulate an optimal control to achieve the desired target. They proposed that the methodology discussed in the paper will be helpful to decision makers in making optimal decisions. The contribution of this research work is the introduction of a systems and controls methodology to develop optimal policies for the design of sustainable systems.

Ştefănescu et al. (2014) analyze the importance of trip planners for public transportation operators and for passengers who use public transportation. The authors present trip planners used today in different cities from different countries and information provided to passengers. The paper shows how a trip planner can be built based on a certain algorithm, the inputs of data that are required and the generated outputs. The trip planner presented in the paper is designed for the city of Timişoara (Romania) and links the city's transportation ways such as: air, rail and road with all the other transportation modes.

Roy (2009) discusses the climate change adaptation and mitigation issues that can be addressed through such a method in Dhaka, using a scenario-based approach. Two alternative development options (constrained and unconstrained) are constructed and analyzed. A common set of indicators are used: first to evaluate which alternative is more sustainable; and then the mitigation and adaptation issues addressed.

Li (2016) identifies the problems that green transportation of Beijing is facing, and proposes countermeasures and suggestions on improving the green transportation system of Beijing and advocating green travelling.

In their study, Ahmad and Oliveria (2016) quantify the influence of city features and socio-economic and socio-cultural variables on mobility patterns to identify evidence-based policy interventions for promoting more inclusive and sustainable transportation paths. Their research estimates the amount of public and private transport and modal choices, employing multivariate analyses, using a nationally representative household survey from the 98 largest Indian cities. Findings of the study reveal that densification in Indian cities reduces the amount of transport as well as enhances the probability of using public transport. Small and medium sized cities predominantly use private transport, whereas large cities prefer public transport but lack non-motorized transportation. Moreover, income is the most important determinant of the amount of transport and the use of motorized and private transport. The top quintile is the only with positive income elasticity of transport demand. Results show that public and non-motorized transport infrastructures will provide sustainable and inclusive development, besides other co-benefits, such as energy security. Based on these findings, several policy recommendations are proposed to improve the sustainability and inclusivity of urban mobility in Indian cities.

The literature survey shows that the tendency is towards chosing the most sustainable transport mode. Conservation of energy, enhancing the air quality and keeping social acceptance are common objectives of the contemporary studies about public transport. Sustainable and green modes of transport are getting more and more important.

In this study, the focus will be on the comparative analysis of transport modes for public transportation with respect to energy and travel cost in chosen routes in the case study Ankara.

DETERMINANTS OF ENERGY USE IN TRANSPORT

Energy use in transport is determined by the demand for movement of people and goods. It is influenced by the spatial environment, in which people live and by the activities that create movement, such as work, shopping, social interaction and distribution of products. Any trip done by an individual by means of a transport mode generates an energy consumption. The level of this consumption and the type of energy required depend on the length of the journey and on the mode of transport (the characteristics of the travel modes) used. The structure of cities and their geographical constraints will influence the length of trips, thus energy consumption for passenger transport (Yildiz, 2003, p. 32). A schematic representation of the factors influencing the total expenditure for passenger traffic is presented in Figure 1.

Total energy expenditures are dependent on passenger traffic (in pass-km) and energy expenses of the chosen travel modes. Many factors influence travel demand, thus passenger traffic, but the modal split plays an important role in calculating total energy consumption as different modes have specific energy consumption levels.

Energy Consumption of Travel Modes

All animals, human beings and goods need energy for their movements. Human and animals supply the energy they need for movement (without any appliances) through their nutrition. Human can move and transport by vehicles some of which also need just energy of body and not any form of external energy

Figure 1. Determinants of energy expenditures for passenger traffic
Source: Yildiz, 2003, p. 33.

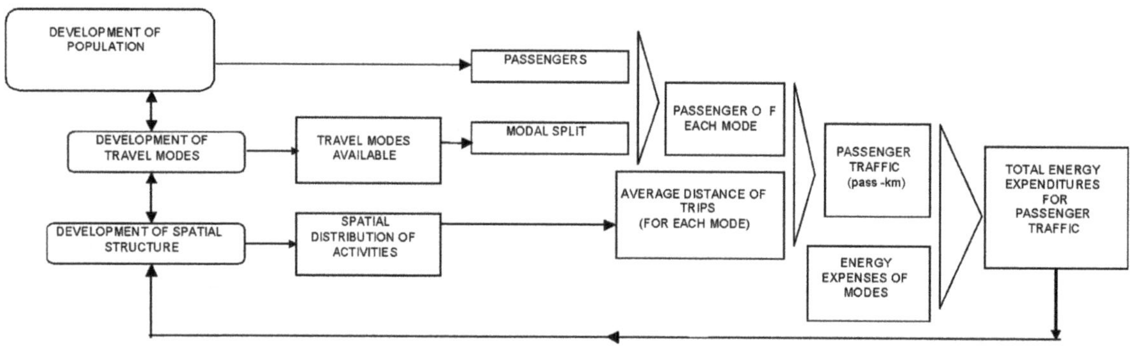

(e.g. bicycle and boat). Other modes of transport, such as car, train, underground systems, trolleys, bus, etc.) need external energy, mostly fossil-based energy. Traffic expenses are the product of the number of trips and the distances travelled (person-km for people and ton-km for goods) (Yildiz, 2003, p. 33).

A distinction is made between primary, secondary and final energy due to the different stages of their manufacture: Primary energy represents both energy sources which have been taken from nature and which may be used as petroleum or natural gas and those which have no economic value before being processed. Secondary energy represents any form of energy which has been processed once or several times. Final energy is measured directly observing the quantities supplied on the market by the producers to the consumers (Yildiz, 2003, p. 33).

Non-Motorised Modes

Walking and bicycle are the only modes of travel which use renewable energy and have no emission.

Knoflacher (1987; pp. 68-70) accepts heart pulse's frequency rate as the indicator for subjective energy consumption of people (Figure 2). The (body) energy requirement of pedestrian is 50-60% higher than the car driver, who sits still in his car. In inclined roads, energy consumption of pedestrians raises to an additional 20%- 75% which can be measured in pulse frequency. For pedestrians (m=70 kg) the efficiency and energy need dependent on the speed is given in Table 1 (Yildiz, 2003, p. 34).

The (body) energy requirement for different kinds of bicycles is given in Table 2.

Motorised Modes

Car

The specific energy consumption of motorised modes is supplied from necessary fuels (Table 3).

Car use requires natural resources to gain energy that it needs for its movement. Unfortunately just the 24% of the energy (31% in diesel motors) obtained from fuel is able to reach the wheels in a combusted motor. All the rest is lost during the operation of the motor (Figure 3) (Yildiz, 2003).

Growing car use has an important effect on energy consumption because cars tend to be an inefficient mode of transport relative to available alternatives in the urban setting. Increasing concentration of vehicles in urban areas leads to more congestion and thus less efficient vehicle use than before (Yildiz, 2003, p. 34).

Figure 2. The changing pulse frequencies and energy consumption by certain step lengths and speeds
Source: Knoflacher, 1987, p.68, as cited in Yildiz, 2003.

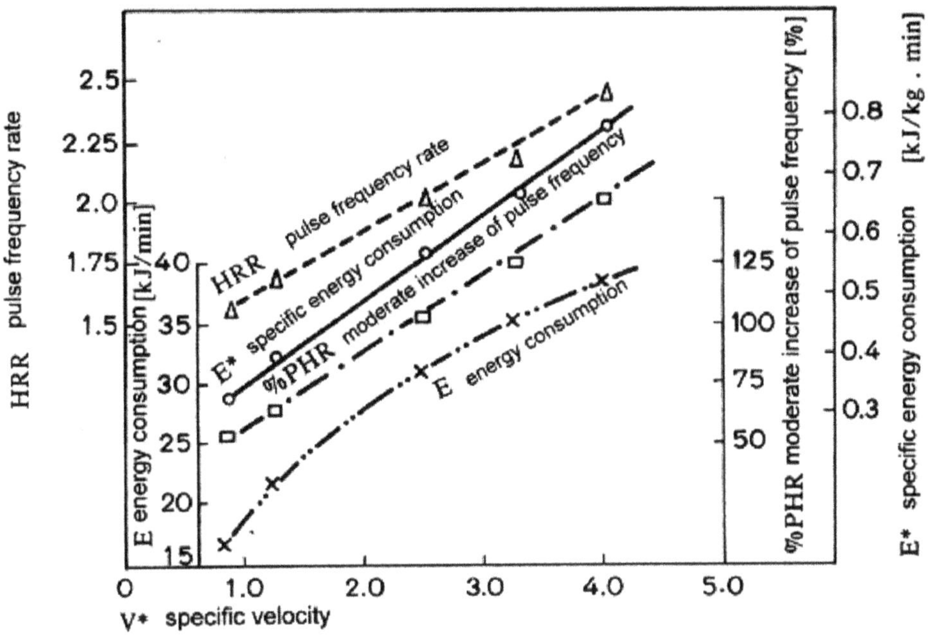

Table 1. Energy need for walking

Speed		Energy Need
km/h	m/s	(kJ/km)
4	1,11	188
5	1,39	205 +3,5. *h (incline factor)*
6	1,67	250

Source: Schnabel & Lohse, 1997; p.581.

Table 2. Energy need for cycling

Speed (km/h)	Energy Need (kJ/km)		
	Tour	Sports	Race
10	43	22	16
15	58	32	23
20	79	46	34
25	106	64	47
30	139	86	63
35	-	111	82
40	-	-	104

Source: Schnabel & Lohse, 1997; p.579.

Table 3. Energy consumption values of car (in Germany) 1995

Occupancy Level	Consumption/Specific Final Energy Consumption			Specific Primary Energy Consumption	
(P/vehicle)	(1/100 km)	(1/P.100 km)	(kJ/P.km)	(kJ/P.km)	(kWh/P.km)
1,4	9,50	6,79	2188	2432	0,675
2	9,68	4,84	1561	1734	0,482

Source: Schnabel & Lohse, 1997; p.581.

Figure 3. Energy flow in a car
Source: Stadtwerke-Generaldirektion, 1984,cited in Knoflacher &Macoun, 1989, p. 37; Yildiz, 2003.

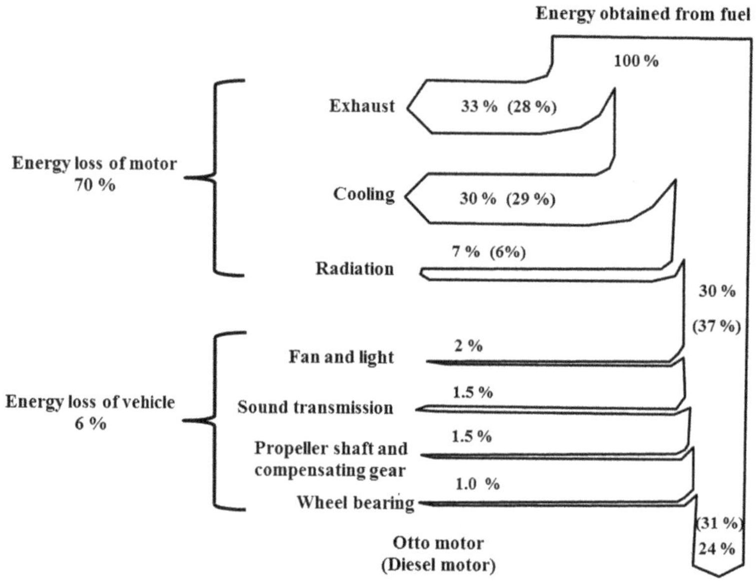

Mass Transport

The energy consumption of mass transport modes is given in Table 4. The values are subject to change due to the characteristics of the traffic network (Yildiz, 2003, p. 34).

Public road transport relies upon the road infrastructure which it shares with private cars and goods vehicles. There exists differences in the annual distances per vehicle which may be explained by the heterogeneity of the services ensured by public road transport modes and the relative share of their various uses (Yildiz, 2003, p. 35).

The indicator which can most clearly reflect the supply of public modes of transport is defined by either vehicles-km or seats-km available for the public. This indicator also be used to determine the energy consumption of public modes of transport. It is necessary to complete this view of the supply by the situation of the infrastructure (network and availability of vehicles) (Yildiz, 2003, p. 35).

Energy efficiency is dependent, in part, on the load factor of the vehicle, in other words, on how well its capacity is utilized. Congestion leads to slower travel and less efficient operation of fuel-powered vehicles. Traffic congestion may be the major deterrent to further increases in car ownership in developing country cities (Yildiz, 2003, p. 35).

Table 4. Specific final and primary energy consumption values of mass transport modes

Transport Mode		Final Energy Consumption	Primary Energy Consumption
		(kJ/ P. km)	
Bus	Diesel-Standard bus (Dieselomnibus) 41,5 l/100 km, 45% occupancy	330	367
	Diesel bus, 35% occupancy	485	540
Tram	21% occupancy	428	1223
Metro	21% occupancy	302	863
Plane	Medium distance flights, 65% occupancy, 850 km/h	1296	1440

Source: Schnabel & Lohse, 1997, p. 583.

A Comparison of Modes

In Figure 4, a comparison of the travel modes with respect to their energy requirement for movement and produced trip value in unit distance is given. The bicycle, which requires just 1/3 of the human's body energy to produce 2.1 times higher trip value, seems to be the most efficient travel mode. Bicycle and walking are the only travel modes that do not require any external energy.

In a comparison of motorised modes, public transport has a relative advantage that can be explained by their higher occupancy levels. Car use consumes the most energy per person km and has a low level of energy efficiency (Yildiz, 2003, p. 34).

A CASE STUDY IN ANKARA

Turkey is divided into 7 geographical regions. Kumbaralioglu and Arikan (2009) makes a comparison of CO_2 emissions by transport modes in these regions. Ankara is located in Ic Anadolu Region, where car use results about 75% of the CO_2 emissions (Figure 5).

The population of Ankara is 4,965,542 people and 4,630,735 of them live in the urban areas. Annual population increase is 15.15%. Car ownership level is 0.209.

Parallel to the increasing population and growing city, many modes for mass transportation have emerged in Ankara after 1923. Some of these modes were lost and some new ones were added due to increasing demand, changing city structure and characteristics of the passengers. The share of pedestrian trips decreased while the share of mass transport modes increased. The share of private car however, could not increase as expected due to low-income profiles. Today, the increasing travel demand for motorised trips, which was created by changing spatial character of the city, is met mainly by mass transport modes (Yildiz, 2003, p. 67) (Table 5).

Parallel to the increasing population and growing city, many modes for mass transportation have emerged in Ankara after 1923. Some of these modes were lost and some new ones were added due to increasing demand, changing city structure and characteristics of the passengers. The share of pedestrian trips decreased while the share of mass transport modes increased. The share of private car however, could not increase as expected due to low-income profiles. Today, the increasing travel demand for motorised trips, which was created by changing spatial character of the city, is met mainly by mass transport modes

Figure 4. Comparison of energy for movement and produced trip value in unit distance
Source: Knoflacher & Macoun, 1989; p. 41, cited in Yildiz, 2003.

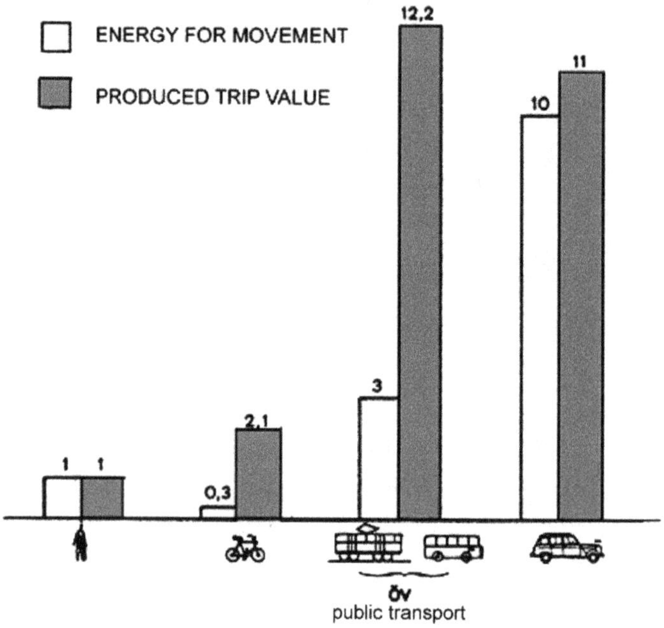

The inner city transportation depends on road in Ankara which causes more greenhouse gas emission and environmental pollution than rail transportation. Since most of the cars in Ankara uses LPG rather than oil, and buses use natural gas, the level of environmental pollution decreases. 23% of energy used for transportation depends on LPG in Ankara which makes Ankara more environmental friendly city than the other cities of Turkey.

The most environmental friendly way of public transportation is using railway which are Ankara Metro and Ankaray in Ankara. The information of rail public transportation can be found in Table 6.

Comparison of Travel Alternatives

Panora-Ulus Route

Panora-Ulus route is a case study example for this study. A traveller has many alternatives to go from Panora to Ulus (Figure 6):

Alternative 1: Bus 185-1 line: Goes directly from Panora to Ulus.
Alternative 2: Bus+Metro. Bus from Panora to Kızılay and Metro from Kızılay to Ulus.
Alternative 3: Dolmuş+Metro. Dolmuş: The person can use Dolmuş from Panora to Kızılay and Metro from Kızılay to Ulus.
Alternative 4: Dolmuş+Bus. Dolmuş: The person can use Dolmuş from Panora to Kızılay and a bus from Kızılay to Ulus.

Figure 5. CO2 emission percentage per person for inner city transportation of 7 regions of Turkey
Data Source: Kumbaralioglu & Arikan, 2009.

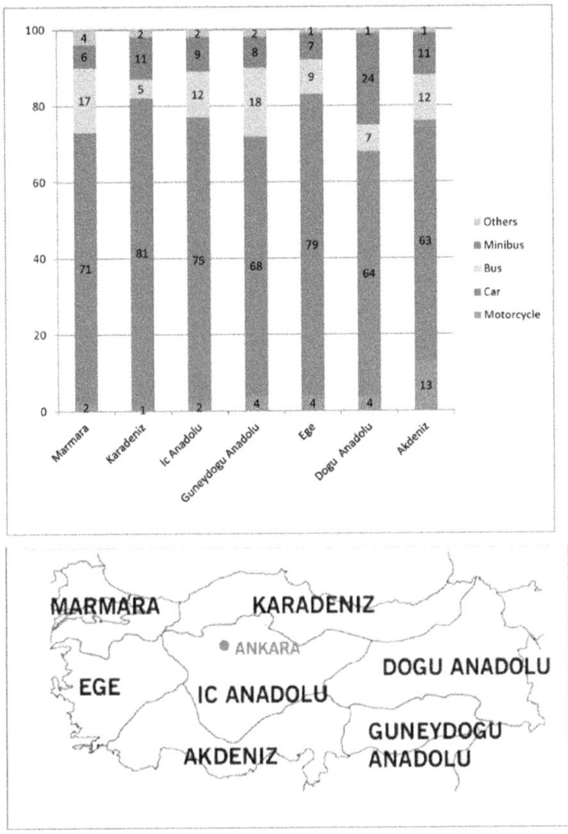

Table 5. Mode choice in Ankara (May-2013)

Modes	Passenger Carried	%	Share of Public Transport in Total%
EGO Buses (by municipality operated)	710,400	13.9	22.7
LRT (Ankaray) (by municipality operated)	105,450	2.1	3.4
HRT (Metro) (by municipality operated)	170,940	3.3	5.5
Suburban Train (by state operated	65,000	1.3	2.1
Minibuses (Privately operated)	1,152,900	22.5	36.8
Service Vehicles	610,000	11.9	19.5
Buses (Privately Operated)	190,000	3.7	6.1
Other buses (Privately Operated)	88,800	1.7	2.8
Town public vehicles	40,320	0.8	1.3
Public Transportation Total	3,133,810	61.2	100
Taxi	290,000	5.7	
Car	1,700,000	33.2	
Private Transportation Total	1,990,000	38.8	
General Total	5,123,810		100

Source: Ankara Kalkınma Ajansı, 2013.

Table 6. Railway public transportation in Ankara (2012)

Metro	Passenger Carried	Energy Spent (Kwh)
Ankara Metro	56,631,842	39,958,763
Ankaray	37,862,314	16,906,073
Total	94,494,156	56,864,836

Source: Ankara Kalkınma Ajansı, 2013.

Figure 6. Travel alternatives in Panora-Ulus Route
Source:Google Maps and ArcGIS.
a. Dolmuş Line from Panora to Kızılay Güvenpark
b. Bus Line from Panora to Uus
c. Private Car from Panora to Ulus (Shortest Route)
d. Metro Line (fromKızılay to Ulus)
e. Bus Line from Panora to Kızılay Güvenpark

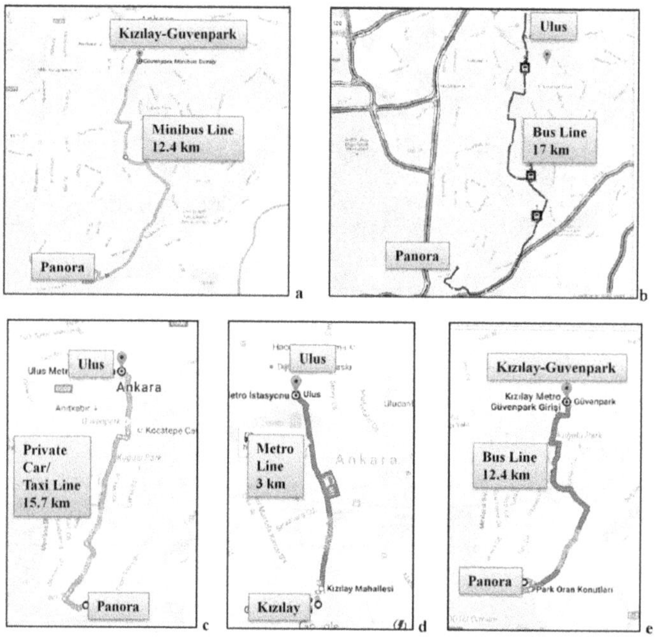

Alternative 5: Dolmuş+Taxi. Dolmuş: The person can use Dolmuş from Panora to Kızılay and taxi Kızılay to Ulus.
Alternative 6: Taxi: The person can use a taxi.
Alternative 7: Private Car: The person can use his or her private car.

For 7 alternatives, the calculated total costs and transport energy consumption can be seen in Table 7. The best alternative with respect to cost is Alternative-1 with the value of 2.25 TRY and the Alternative-2 with respect to total energy consumption with the value of 1,157 Kcal. It is obvious that as the energy being used to travel increases total CO_2 emissions will also increase. So, to be more eco-friendly people should prefer the transportation mode which spends less energy.

Table 7. Cost and energy used for 7 alternatives to travel from panora to ulus centrum

Alternative	Total Distance (km)	Total Time (Minutes)	Total Cost (TRY)	Transport Energy Consumption (Kcal)	Explanations
1	17	50	2.25	17x105=1,785	Walk(100m)- Bus 185-1 line and Walk(200m)
2	15.4	40-45	4.5	12.4x105+3x85= 1,557	Walk(100m)- Bus 185-1 line-Walk (70m)- M3 and Walk(150m)
3	15.4	35-40	4.75	12.4x275+3x85=3,665	Walk(100m)- Dolmuş-Walk(80m)- M3 and Walk(150m)
4	15.8	55-60	4.75	12.4x275+3.4x105=3,767	Walk(100m)- Dolmuş-Walk(100m)- Bus and Walk(150m)
5	15.8	35-40	15-20	12.4x275+3.4x550=5,280	Walk(100m)- Dolmuş- Taxi
6	15.7	30-35	39.68	15.7x550=8,635	Taxi
7	15.7	30-35	8-10	15.7x550=8,635	Private Car

The urban transport energy consumption per passenger-km; for rail 85 Kcal, for buses 105 Kcal, for minibuses 275 Kcal and for private cars 550 Kcal.

The focus can be on many different alternative routes to travel in Ankara, but it is supposed that the results will not change. People living in Ankara should use public transportation modes to travel more economic and nature friendly.

People always concern about their comfort thus they use private cars to travel inside cities. They do not want to walk long distances especially at the days with uncomfortable weathers. So, Ankara municipality should focus on placing huge parking areas (not inside the city center but near to the highly populated areas) next to big bus and metro stations so that people can drive short distances to the station. They can leave their cars at station parking areas and use the public transportation modes.

Eryaman-Besevler Route

Eryaman-Beşevler line is the second example for this study. A person has many alternatives to go from Eryaman to Beşevler. Since there is no Dolmuş line between Eryaman and Beşevler the focus will be on 4 alternatives in this example (Figure 7):

Alternative 1: Bus 543-1 line: Goes directly from Eryaman to Beşevler.
Alternative 2: Metro line (M3 and M1): M3 goes from Eryaman to Kızılay and M1 goes from Kızılay to Beşevler.
Alternative 3: Taxi: The person can use a taxi. 22.4 km, 20-26 minutes.
Alternative 4: Private Car: The person can use his or her private car by using the best path.

For four alternatives the calculated total costs, time and transport energy consumption can be seen in Table 8.

The best alternative with respect to cost is Alternative-1 with the value of 2.25 TRY and the Alternative-2 with respect to total energy consumption with the value of 1,870 Kcal. It is obvious that as the energy being used to travel increases total CO_2 emissions will also increase. So, to be more eco-friendly people should prefer the transportation mode which spends less energy.

Figure 7. Travel alternatives in Eryaman-Besevler route
Source: Google Maps and ArcGIS.
a. Traveling between Eryaman and Beşevler by bus
b. Traveling between Eryaman and Beşevler by metro rail
c. Traveling between Eryaman and Beşevler by taxi.

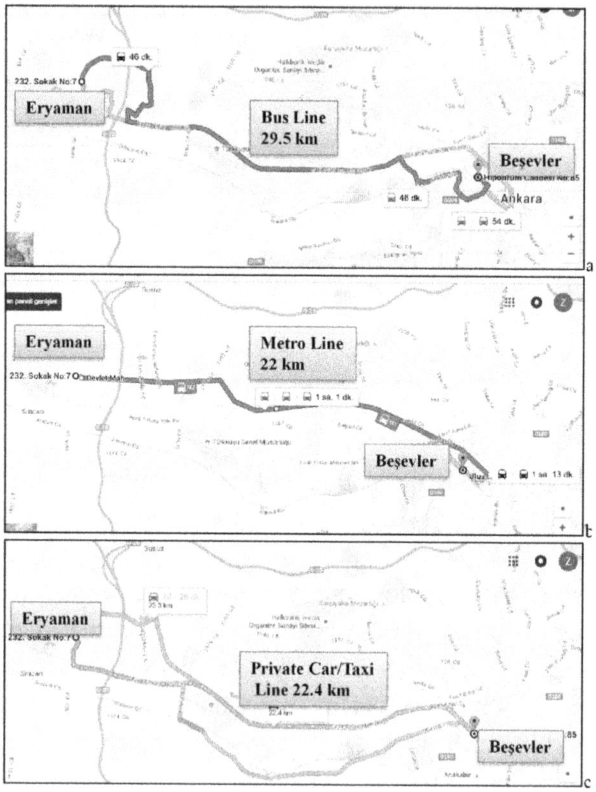

Table 8. Cost and energy used for 7 alternatives to travel from panora to ulus centrum

Alternative	Total Distance (km)	Total Time (Minutes)	Total Cost (TRY)	Transport Energy Consumption (Kcal)	Explanations
1	29.5	45-50	2.25	**29.5x105=3,097.5**	Walk(100m)- Bus 543-1 line and Walk(200m)
2	22	70-75	3.9	**22x85=1,870**	Walk(300m)-Metro(M3)-Metro(M1) and Walk(1700m)
3	22.4	25-30	56.46	**22.4x550=12,320**	Taxi
4	22.4	25-30	12-15	**22.4x550=12,320**	Private Car

Urban transport energy consumption per passenger-km; for rail 85 Kcal, for buses 105 Kcal, for minibuses 275 Kcal and for private cars 550 Kcal.

The number of examples can be increased but the results will not change. The sort of alternatives with respect to total energy used will be Metro line-Bus Line-Private Car-Taxi while the sort of alternatives will be Bus Line-Metro line- Private Car-Taxi with respect to total cost. When the focus is on the total travel time, the best alternatives will change with respect to the period of day traveled. If the passenger travels in rush hour (in the morning while people going their works of in the evening people returning from work to their houses) the best alternative may be the Metro line. But during ordinary period, the best alternative may be using the private car or taxi.

CONCLUSION

The creation of new developments with a balanced number of workplaces and housing opportunities does not always result in a liveable environment. If the dependency on a big city still continues and if distances can be overtaken by motorised modes, then high energy consumption for mobility will be inevitable.

In the interaction of big city (system) and the satellite settlements (sub-systems) the energy flux is through motorised traffic. The heavier the traffic is, the less chance there is for high-level organizations within a zone. Only the city parts, which have a good pedestrian circulation, can preserve their energy levels.

Mobility is the result of potential differences in space (Knoflacher, 1998; p.20). These differences emerge as a result of mistakes in location decisions of work places, shopping centres, etc. The higher the dispersion, the more energy that is required for mobility.

In urban spaces, the structures that minimize the need for motorised travel should be formed. Mobility is the number of daily trips per person. Traffic avoidance means the reduction of time spent in traffic and aims to reduce trip distances.

"The city of shorter routes" should be aimed for when designing city structures. A mixture of activities (housing, working) can help to reduce the travel distances and increase the share of non-motorised modes. In big cities like Ankara, this can be achieved at the zone level.

Modal split cannot be considered as a simple "choice". Instead, it is a very important determinant of the urban region. If managed perfectly, it can be a very useful tool in creating livable environments. It should be remembered that what is important is the desire to reach certain activities. There are activities for which movement of people and coverage of distance is desired. The choice of how the movement is to be accomplished creates the modal split. The distance to be covered, time constraints (constant travel time), personal pereference, economic resources and availability of different transport modes shape this choice. The amount of energy used for the movement is dependent on the tranport mode chosen.

The way we travel, our mode choice, how we design our streets and the mobility form we accept as a base for our transport policy are reflections of our culture. What we understand from "culture" is of course not to keep people in narrow side-walks and to design the streets for machines. Traffic culture emerges when the whole traffic system is modified to have an harmony with pedestrians.

The car is an obstacle for the population growth as it uses limited space required for public use. Slower and less car traffic means, fewer victims and fewer car accidents, less external energy dependency, less air and noise pollution, less land consumption and less dispersion of activities.

The results of the case studies show that there are differences with respect to chosen travel mode in terms of energy use and travel cost reflected to traveler for a certain OD couple.

The energy budget of a city profits from even very small changes in modal split in favour of non-motorised modes. The very high energy consumption of motorised modes, multiplied by increasing travel distances, absorb the energy of the city-system which should in fact rather be used for the emergence of more sustainable structures. An efficient pedestrian network is only possible in a compact city form with a harmonised mixture of activities.

REFERENCES

Ahmad, S., & de Oliveira, J. A. P. (2016). Determinants of urban mobility in India: Lessons for promoting sustainable and inclusive urban transportation in developing countries. *Transport Policy*, *50*, 106–114. doi:10.1016/j.tranpol.2016.04.014

Alan, A. W. (1986). *Urban transit systems: guidelines for examining options*. Word Bank Technical Paper.

Ankara Kalkınma Ajansı. (2013). *İstatistiklerle Ankara 2013, 36*. Retrieved from www.ankaraka.org.tr/tr/files/yayinlar/istatistiklerle-ankara-2013.pdf)

Beaudoin, J., Farzin, Y. H., & Lawell, C. Y. C. L. (2015). Public transit investment and sustainable transportation: A review of studies of transits impact on traffic congestion and air quality. *Research in Transportation Economics*, *52*, 15–22. doi:10.1016/j.retrec.2015.10.004

Buwana, E., Hasibuan, H. S., & Abdini, C. (2016). Alternatives selection for sustainable transportation system in Kasongan City. *Procedia: Social and Behavioral Sciences*, *227*, 11–18. doi:10.1016/j.sbspro.2016.06.037

Knoflacher, H., & Macoun, T. (1989). *Ökologie und Strassenverkehr, Reports UBA-89-035*. Wien: Umweltbundesamt.

Knoflacher. (1987). *Verkehrsplanung für den Menschen Band 1 Grundstrukturen*. Wien: Verlag Orac.

Li, H. R. (2016). Study on Green Transportation System of International Metropolises. *Procedia Engineering*, *137*, 762–771. doi:10.1016/j.proeng.2016.01.314

Luke, S., & MacDonald, M. (2006). Public transport mode selection: a review of international practice. *European Transport Conference*.

Maheshwari, P., Kachroo, P., Paz, A., & Khaddar, R. (2015). Development of control models for the planning of sustainable transportation systems. *Transportation Research Part C, Emerging Technologies*, *55*, 474–485. doi:10.1016/j.trc.2015.03.024

Roy, M. (2009). Planning for sustainable urbanisation in fast growing cities: Mitigation and adaptation issues addressed in Dhaka, Bangladesh. *Habitat International*, *33*(3), 276–286. doi:10.1016/j.habitatint.2008.10.022

Schnabel, W., & Lohse, D. (1997). Article.*Grundlagen der Strassenverkehrstechnik und der Verkehrsplanung*, *1*, 2.

Sebhatu, S. P., & Enquist, B. (2016). Sustainable Public Transit Service Value Network for Building Living Cities in Emerging Economies: Multiple Case Studies from Public Transit Services. *Procedia: Social and Behavioral Sciences*, *224*, 263–268. doi:10.1016/j.sbspro.2016.05.458

Stadtwerke-Generaldirektion. (1984). *Energie für Wien, Energiekonzept der Stadt Wien 1, Fortschreibung*. Wien: Stadtwerke-Generaldirektion.

Stefanescu, P., Mocan, M., Stefanescu, W., & Neculai, P. V. (2014). Trip Planners Used in Public Transportation. Case Study on the City of Timisoara. *Procedia: Social and Behavioral Sciences*, 124.

Teodorović, D., & Janić, M. (2017). *Public Transportation Systems*. Transportation Engineering.

Wilson, S. S. (1973). Bicycle Technology. *Scientific American*, *228*(March), 81–90. doi:10.1038/scientificamerican0373-81

Yildiz, E. V. (2003). *The Analysis of Spatial Structure and Mobility in Ankara* (Unpublished Ph.D. Thesis). TU Wien Institut für Verkehrsplanung und Verkehrstechnik.

KEY TERMS AND DEFINITIONS

CO_2 Emissions by Transport Modes: Transportation is a major contributor to global warming through emission of CO2. Transport modes show differences in their levels of emissions.

The Energy Budget of a City: Total energy available for needs of a city. Energy is limited and in order to fulfill some demands some others should be sacrificed.

Travel Mode: Available travel alternatives. Travelers freely chose one of them for their travels and try to maximize their individual profits.

Chapter 12
Effect of a Motorway on Development of Accidents in a Big City

Hermann Knoflacher
Vienna University of Technology, Austria

ABSTRACT

Traditional transport practice and science call motorways as safe roads since the accident rate (accidents per 1 million vehicle kilometres) is lower than on ordinary roads. This is based on the scientific inadmissible comparison. Similar, but not comparable things, are compared: the accident situation of two different kinds of transport systems, which are basically different not only in its construction elements, but also in operation condition, their environments and the composition of their users. All of them are different. To compare, comparability is a necessary precondition. If we take this into consideration we can recognize, that the existing contradiction between accident-research-results and physics can be removed. Systems with a higher speed have a higher risk. But these facts are continuously neglected by traditional educated practitioners: they cling to their myths. One of them is "Since the number of accidents per kilometer driven on motorways is less than on other roads, motorways are safer roads."

INTRODUCTION

In 1960ies, during the period of fast motorization in West-European countries, the number of traffic accidents and victims in the road transport sector increased with increasing car ownership and reached a peak in the early 1970ies. In Austria the number of traffic accident victims exceeded the figure or 2100 people killed in traffic accidents in 1972. In Germany more than 21.300 people were killed in traffic accidents in 1970. Since then the number of traffic accidents victims is decreased continuously in this countries. There were many speculations, what might have been the cause for this change. Ministries and traffic safety organizations claimed the positive effect for her traffic safety measures. But at this time not very much of this kind of measures were implemented. Speed limits outside built-up areas were not implemented, no mandatory safety belt use and many other measures implemented today,

DOI: 10.4018/978-1-5225-2116-7.ch012

were introduces mainly since the first energy-crisis in 1973. The cause of this change in the time-series of traffic accidents was found by analyzing the "internal dynamic" of the transport system during this transition period from not motorized to motorized society. In the beginning inexperienced drivers are coming into the transport system. The risk of an accident of this group is more than seven times higher than the risk of experiences drivers. At the same time pedestrians, cyclists and even public transport users are becoming car users. After some years the surviving cohort of beginners are becoming safe transport users. When his group start to dominate the transport system behavior, the "Eigendynamic" of the transport system, concerning "accident-production", change into as "safe mode". By carrying out a dynamic analysis of the transport system introducing these internal structures of the system users, it could be shown that the accident figures observed in the statistics can be reproduces nearly perfectly and explain the change of the system dynamic before and after the peak. After 1975 the effects of safety belt law, speed limits, better enforcement had to be taken into account to reproduce the observed data. (Knoflacher 1981, 1985) But one variable has no effect on the development of traffic accidents: the motorways. At this time this conclusion of the analysis was in contradiction to the general professional belief that motorways contribute always to the safety of the whole system.

This wishful thinking can be found in reports, for example in Eurostat (2016) till today. The authors have made some reservations, that other factors might have influenced the development of traffic accidents in some regions. In general most of research is done to improve traffic safety on motorways, by implementing speed limits or traffic safety devices on the existing infrastructure. Although some studies indicate that the linear extrapolation of "more motorways – less accidents" might not be valid (Germany 2014), no in-depth studies have been made. The question: "What are the effects of a new motorway on the traffic safety of a region or a city?", was never asked or a study published on this issue till recently. The broad system-view, concerning traffic safety, in this chapter, was not common in the profession of transport engineering, when the study if this chapter was carried out in 2001. Just recently as study "Effect of new urban motorway infrastructure on road traffic accidents in the local area: a retrospective longitudinal study in Scotland" war published (Olsen et. al. 2016). But this study was carried out by an Institute for Health and Wellbeing, University of Glasgow in Scotland and not by an institute of transport engineering. This underlines the necessity for new tools in transport engineering for the future. The study presented here is so far the only one from the transport engineering perspective.

Motorways are built in Austria since the 1950ies. But the traffic accident situation was not affected by this important, expensive and powerful - concerning capacity – infrastructure. The dynamic simulation had even the contrary effect: if Austria would not have built motorway, the country would have less traffic accidents. (Knoflacher 1985) At this time the findings were not recognized by the professionals, since it was no direct proof available till 1999. The city if Vienna asked the author to develop measures for further enhancement of traffic safety in the city. (Knoflacher 2001)

COMPARISON IS ONLY POSSIBLE BETWEEN COMPARABLE THINGS

Traditional transport practice and science call motorways as safe roads since the accident rate (accidents per 1 million vehicle kilometers) is lower than on ordinary roads. So the argument goes to build motorways for safety reasons. This seems logical, but the arguments are based on a scientific inadmissible comparison. To compare, comparability has to be executed. If somebody compares things or affairs, which are not comparable, it´s is a popular saying that he is "comparing apples with pears." The popular saying shows the sensitivity of the society concerning preconditions of comparability: apples and pears are both pomaceous fruits, they differ only in form and taste. But in the transport field, similar, but not comparable things, are compared, if we use the traditional accident rate to assess the traffic safety of motorways and ordinary roads. The difference in design, environment and conditions of operation between these roads is much greater then between apples and pears (Table 1):

There are fundamental differences in design elements, operation condition and road traffic rules – and totally different conflict-situations between transport system users between motorways and roads..

From the point of view of physics, the system with the higher speed, hat also the higher accident risk. But the accident rate shows us the contrary figures. The fester system has a lower "risk-rate". Since physics is a very well developed science the bias must be in the transport indicators used so far. If we want to compare the risk of accidents between different roads, we have to remove all elements, which are not comparable. This has been done already decades ago, by comparing accident rates of different roads in a scientific manner (Knoflacher, 1970). This systematic analysis show: the traffic-accident-rate, between motorways and urban roads is absolutely in agreement with the laws of physics if basic principles of comparability are observed. Traffic on roads with lower speed limits has less and less severe accidents than traffic on high-speed roads – under comparable circumstances.

Each change of transport structures in one sector has effects not only on all other sectors of the transport system, but also on the urban and economic system too. A system approach is therefore necessary to prevent misinterpretations of indicators in transport and to prevent momentous undesirable effects. Such examples are detour roads. Decisions to build a detour road are very often justified with traffic safety arguments. Sound research show, that this is not the case. Scholas (1988) analyzed the effects of detour roads in the national and country road network of Germany (not motorways) in as systematic way concerning traffic safety. The number of accidents on the urban road decreased, but the number and severity of accidents on the detour road overcompensate the decline. The overall number and severity of accidents was enhances in the system with detour roads. An analysis on the national level in Germany, carried out by Heusch/Boesefeldt (1997), furnished these effects also.

To compare, comparability is a necessary precondition. If we take this into consideration we can recognize, that the existing contradiction between accident-research-results and physics disappear. Systems with a higher speed have a higher risk. (Knoflacher 1972). But these facts are continuously neglected by traditional educated practitioners: they cling to their myths. One of them is "Since the number of accidents per kilometer driven on motorways is less than on other roads, motorways are safer roads." Is this valid also for the system of a big city? So far such an analysis was not made so far. One of the reasons is the fact, that ministries or City administrations are mainly engaged in new projects and have

Table 1. Comparison of design, environment, and condition of operation between motorways and roads

Motorways	Roads
Separated opposite lanes	No separation
No level crossings	All kinds of level crossings
Separated from surrounded environment	No separation from environment
No parking along the carriageway	Parking along road edges
Cars, trucks and busses only	All motorized and not motorized vehicles: pedestrian, cyclists, animals, etc.

neither the time, nor the money, to analyze the effects of project realized. Politicians are in general not interested about the past, but much more about the future. The city of Vienna might be an exception, since the administration is supporting research to provide the decision makers with a sound scientific material. One part of a general traffic safety analysis of the City were time-series analyses of accident statistics. This gave the opportunity to study the effect of an new build motorway as a kind of detour road of the city in a more systematic and broader view than usual.

EFFECTS OF A MOTORWAY ON THE TRANSPORT SYSTEM AND ON URBAN STRUCTURES INSIDE AND OUTSIDE OF A CITY

If roads with many attractive lanes are opened in an urban environment, car traffic moves from urban roads to the motorway. Urban streets are relieved from heavy traffic volume – at least for the first moment. Urban streets are therefore becoming more attractive for potential car traffic demand, which could not be satisfied before. Car users are being "activated" to use now the cars not only on the relieved urban roads but also on the new motorway.

This relationship is expressed in the fundamental relationship between transport volume, density and speed: transport volume M, the number of transport system users D in the system and the speed of the system V are related by the formula

$$M = D \times V.$$

Since time for mobility is a constant, every increase of speed has the effect of an increase of transport volume in the sector of the transport system. If the sector is additionally increasing its attraction by providing new space for users, also the number of users D is increased. This has been proved in many cases, one of them is shown for the A23 motorway in Vienna.

Figure 1 shows the location motorway in the Vienna urban transport structure. This motorway has been opened in 1978 to relieve the parallel urban road called Schlachthausgasse from its heavy traffic volume. In the first year after opening the motorway, the volume on the urban road decreased to about 30% of the volume before. 10 years after this road had again 20% more traffic than it had before opening the motorway and in addition about 60.000 to 80.000 cars were "produced" by the motorway.

The expected relieve of the urban roads was only temporary. The induced traffic through the high attraction of conditions for car driving had not only compensated the relieve-effect but has created higher traffic volume in the urban system than before. Ten years after opening the motorway the traffic volume on the secondary road was even higher than before (see Figure 2). The effect of the motorway was not limited to the traffic system itself. In the following years new, unwanted and unplanned effects occurred in the city and outside of the administrative City-borders. Contrary to the transport planning goals, the share of car trips (modal split) increased, as well as car ownership was stimulated. Automobile Association became a more powerful lobby group, air pollution in the city and noise along the motorway became problems, which have not been solved till today. Commuters were encouraged to use their cars, instead public transport, intensifying parking problems in most of the districts of the city.

Beside this direct effect on the environment and the transport system, the motorway changed land use in and around the city. Big shopping centers were built in the municipalities close to the city at motorway exits and intersections, draining money out of the city into their pockets. This amount of money was now

Figure 1. Motorway A 23 in the Vienna network

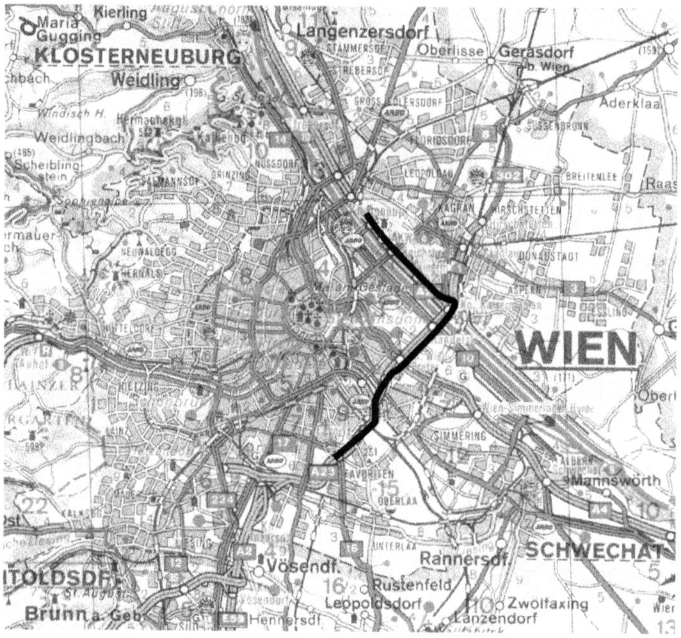

Figure 2. Effects of the motorway on the traffic volume of parallel road

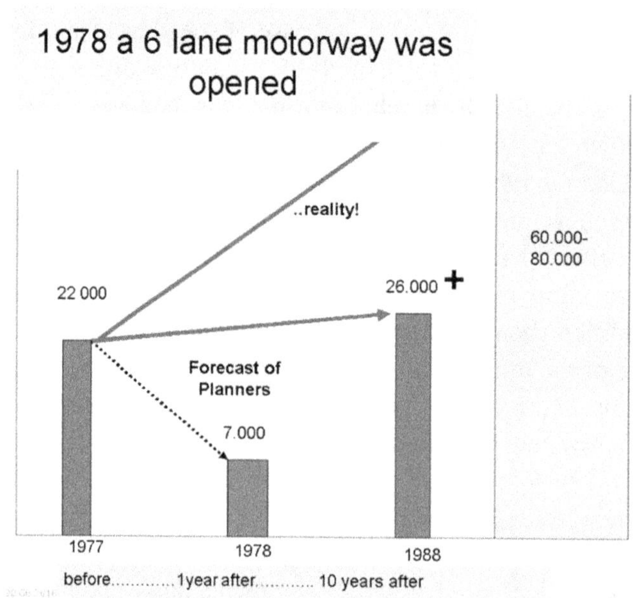

lost for the smaller urban shops and many of them went bankrupt. The motorway has changed the urban structure into an unwanted direction for many years instead of improving the situation of the city. More than 20 years of transport management and traffic calming measures were necessary to compensate the negative effects of the motorway on the urban transport, urban economic and urban social system. This would not happen if the system-behavior would have been taken into account and new tools, developed for the "paradigm change in transport", would have been applied. One of the interesting empirical results was the interruption of the development of traffic accidents in Vienna, after opening of the motorway.

ACCIDENT SITUATION OF THE CITY OF VIENNA

The city of Vienna is compared to other similar big cities (over 1.5 million habitants) a relative safe city. If we analyze the development of the number of accidents we recognize a decreasing tendency after 1972. The number of accidents with injured people decreased continuously. But in 1980 a kind of "bump" occurred, enhancing the number of accidents till 1998 as Figure 3 shows. If we look to the number of killed people in traffic accidents, we see a similar development in Figure 4. Something must have happened in the transport system of the city during this period. The existing tendency of decrease of the number of killed people in traffic accidents was interrupted in 1979: the pace of decrease of the number of killed people in traffic accidents became slower after this period.

The only fundamental change in the transport system during this period was the opening of the motorway in 1978 and the continuous sections in the following years.

Structural Changes, Physical Effects, Empirical Observation

Since a motorway is enhancing the speeds in the system, the kinetic energy in a system must increase. If the kinetic energy in the system is increasing the probability of more and heavy accidents must increase also. The empirical data are therefore consistent to the physical expectations and the expected system

Figure 3. Development of traffic accidents in Vienna 1970 – 1998

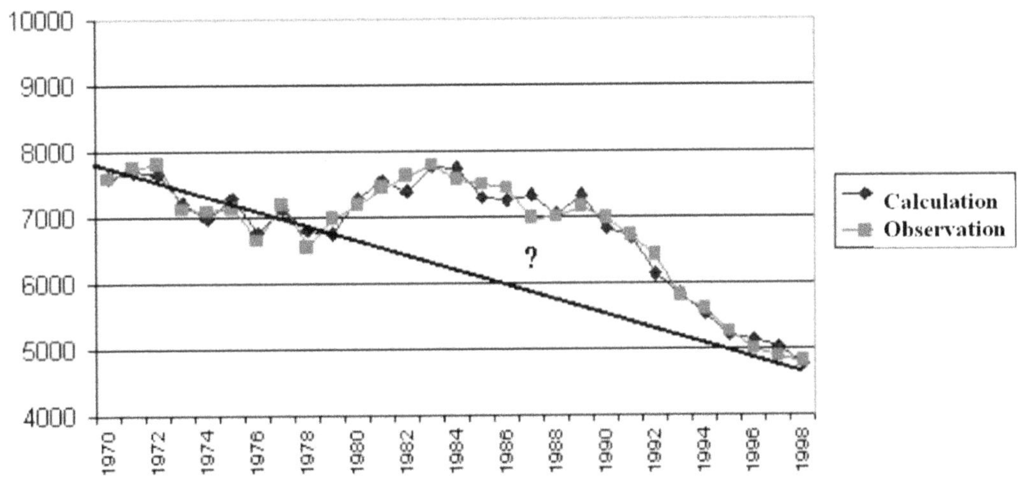

Figure 4. Development of killed people in traffic accidents in Vienna

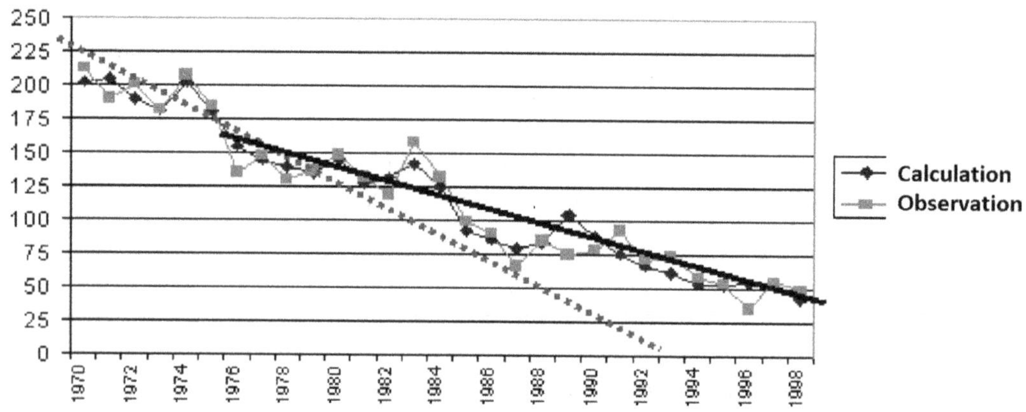

effects. Driving cars on Vienna urban roads became more attractive when urban street became more free capacity. The speed enhanced which encouraged car and truck traffic. In a short term view, the "traffic situation" improved (for motorized system users) from the traditional professional point of view. Traffic at this time was "car traffic". From the narrow traffic engineering perspective the motorway was a success. Speed of car traffic went up, a better level of service was achieved, congestion declined for the moment.

From the system behavior perspective, which has in view not only the automobile as a mode, but also walking, cycling and public transport use, the situation might look totally different. Laws of physics tell us, that the risk increases with the number of risk-producing system users and with increasing speed. This theoretical approach has to be empirical falsified or validated.

To prove this hypothesis an analysis of the accident situation in Vienna was carried out analyzing the development of traffic accidents for different sectors.

Analysis of the Trends of Accidents for Different Transport Modes

The unexpected increase of traffic accidents in the transport system of the city has to be analyzed in details. Is the observed trend the same for all modes, or are differences in the dynamics of risk between the modes?

The accident situation from 1970 to 1979 followed the downward trend of previous years. If we extrapolate the trend into the 1990ies, the real number of accidents in the system is again in the former trend line. It was therefore interesting to analyze the trends within this period.

Accidents with Person Cars from 1970 to 1989

In order to show the change in the "dynamic of the system" the time period between 1970 to 1989 is highlighted in Figure 5. We see two different trends in the development of the number of accidents in this period. In the first period from 1972 to 1978 we see a continuous tendency of the decrease of accidents with person cars.

Between 1978 and 1983 as steep increase of the number of accidents with person cars occurred followed by a continuous decrease of the number of accidents, as it happened before. But the "risk level" between the first and the third period is about 50 to 20% higher. The risk level was increased by some

Figure 5. Accidents with person cars

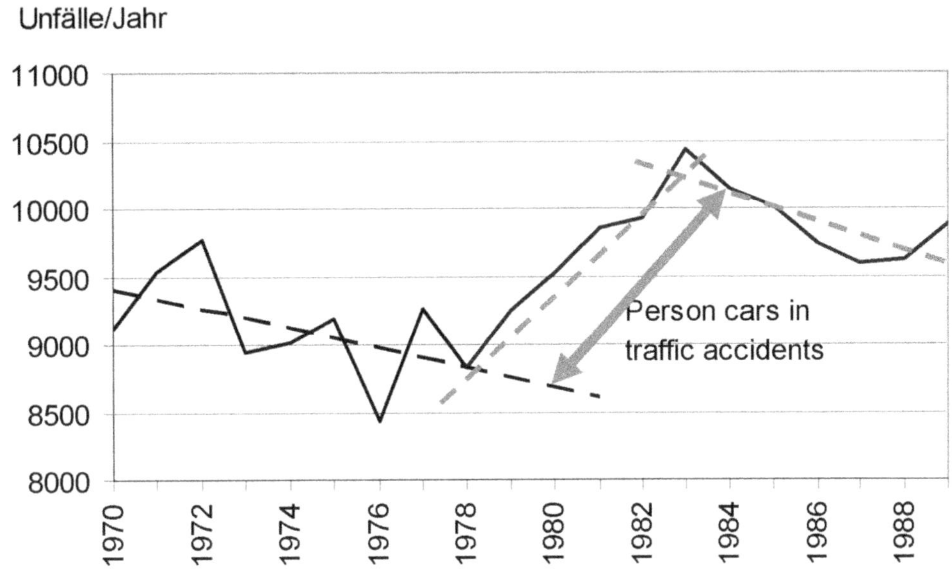

effects. If we observe such a kind of change in the trend of indicators of a system, the system dynamic must have changed.

The period of about 5 years between 1978 and 1983 show a totally different dynamic in the transport system of the city concerning traffic accidents with cars. Contrary to the time periods before and after, the system produced every year more accidents in this traffic mode. In 1983 the system stabilized again and switched into a mode, similar like before 1978. But the level of risk was much higher than before for the next decade. The number of accidents decreases asymptotic into the extrapolated line from the periode before 1978. The motorway has increased the number of accidents in the whole system by relieving traffic volume for urban roads and enhancing the speed for at least 5 years. This structural change in the urban transport system had a lasting negative effect on traffic safety for more than 15 years.

Accidents with Trucks

The accident statistic as well as the transport statistic distinguished between light and heavy trucks: light trucks with a gross weight less than 3.5 tons without trailers and heavy trucks with 3.5 tons and more with and without trailers.

A fundamental change in the development of the number of accidents with this two modes can be observed between 1978 and 1979 (see Figure 6). The tendency of decreasing the number of accidents with light trucks changed dramatically in 1978. An increase of the number of accidents with this kind of trucks can be observed in a similar form like for person cars. Accidents with heavy trucks were on decline also. But the existing tendency of decrease of heavy trucks also stopped in 1978, since this time the number of accidents with heavy trucks remain constant. The motorway is attracting truck traffic to use this route more than before. Higher speed and additional capacity as well as a convenient cross section with wide lanes reroute trucks through the City.

Figure 6. Accidents with heavy and light trucks

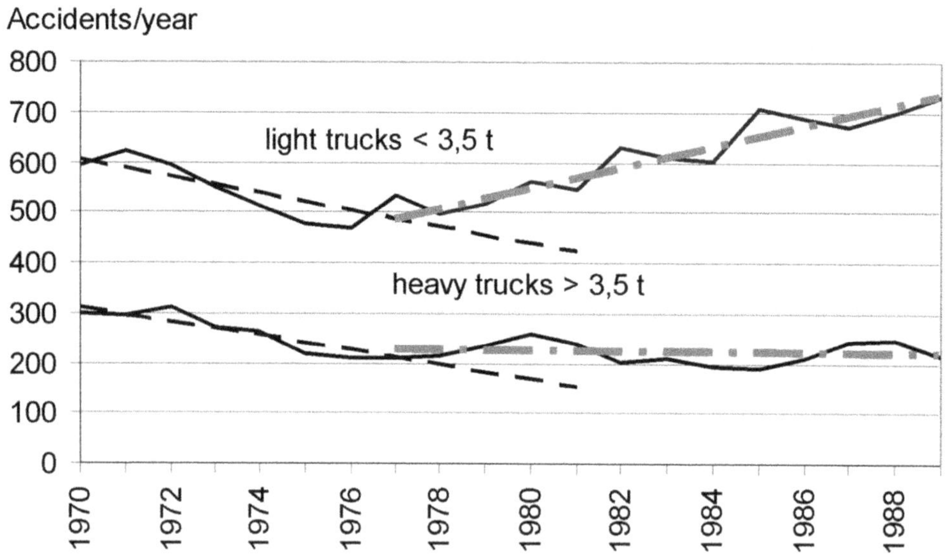

Hypothetical Explanation

The motorway has not only effects on the traffic volume, it has also dramatic effects on the economic structures of the city and around the city. South of Vienna new shopping centers around the motorway appeared, pulling out money from inner urban shops and destroying continuously the inner urban economic structure (SCS extension 1988) The same happened in the north of the Danube River when the section of the motorway was extended. More and more inner urban functions were transferred to the outer ring. The number of kilometers driven, especially for economic purposes, were enhanced by the motorway. This can explain the dramatic change of tendency in the development of accidents.

To prevent overlapping effects from opening the eastern border, the comparison is made only for the period before 1989. This interruption of the decreasing number of accidents happened in a period of economic continuity. No external economic or other influences during this period can be found.

Accidents with Motorcycles and Mopeds

During this period the number of motorcycles increased especially as a kind of sports mode.

Mopeds are not allowed to use the motorway. No significant change in the development of the number of accidents with mopeds can be therefore observed during this period as Figure 7 shows.

Accidents with Street Cars (Tram)

No external effects can be observed during the observation period.

Figure 7. Accidents with motorcycles and mopeds

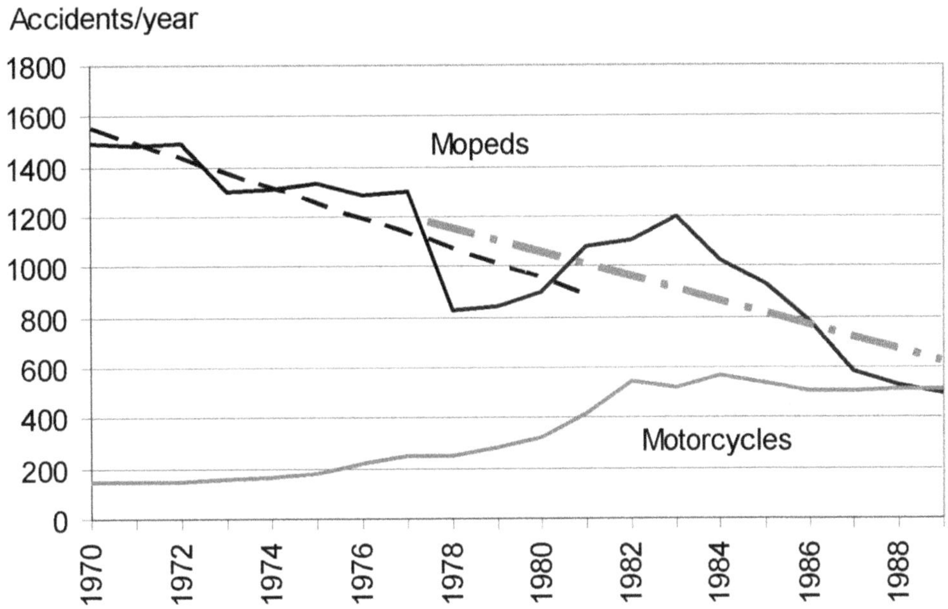

Accidents with Cyclists

Vienna had no cycling tradition till the end of the 1970ies. Then the cities started to build cycle tracks, which enhanced the amount of cyclists from close to 0 to more than 4% of all trips. Figure 8 shows that the number of cycle accidents increased with the number of cyclists during that period.

The number of cycle accidents is representing the increasing number of cyclists in this time period.

Figure 8. Accidents with cyclists

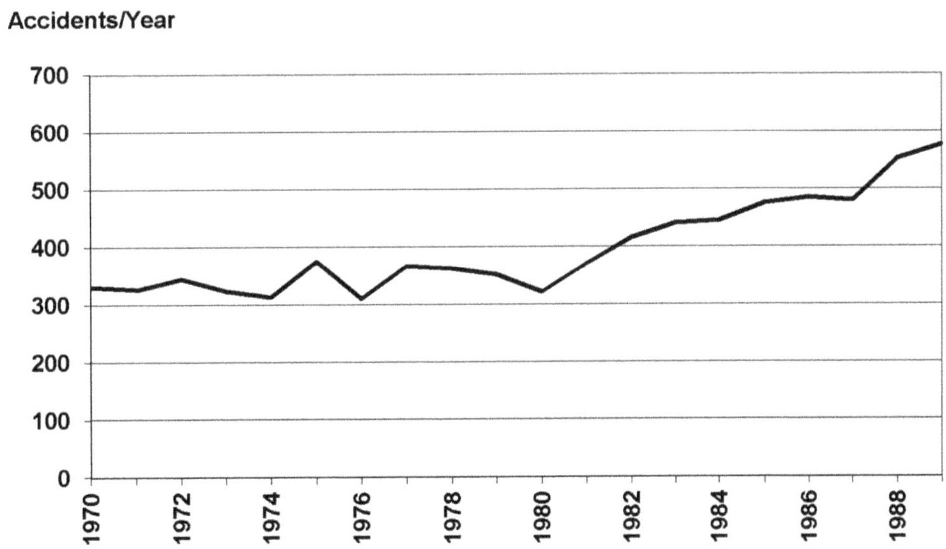

Pedestrian Accidents

It seems surprising, that the number of pedestrian accidents could be influenced by the opening of a motorway as it looks like in this diagram. But it is obviously, that after opening the motorway the positive tendency of the development of the pedestrian accidents in Vienna has been changed distinctively. The numbers are still declining continuously but significantly slower than before. One could argue since more car drivers are now in the system, we might have less pedestrians than before and the motorway will decrease the number of pedestrian accidents even more. But this diagram in Figure 9 shows obviously, the environment for pedestrian has become more risky and more unsafe. The relief of car traffic from urban roads has enhanced travel speeds in the existing road system. This could explain why the number of accidents is still decreasing but not with the same pace like before.

Results of Traffic Counts

There are only very few counting stations in the city of Vienna, which can be used to analyze the development of traffic flows, which might be affected by the new motorway. One of them is shown in the following figure. Even this counting station is 5 to 6 kilometers away from the this motorway, it can be seen, that after 1978 till 1988 traffic volume at this counting station remain more or less the same than it started to grow again (see Figure10). Therefore it can expected, that there was a relief even on this distant counting stations due to the effects of the motorway. A similar effect can be found also on a counting station close to the motorway, which was opened with the motorway opening. We can see that for a time period of about 4 to 5 years the traffic flow remain the same, but then it increased till the 90s till the full capacity was used again. The traffic counts show the time period between the initial effect and the time, when the effect of the motorway is compensated by reaching the limits of capacity again. If we understand the system behavior, we can use this effect for traffic management measures.

Figure 9. Pedestrian accidents

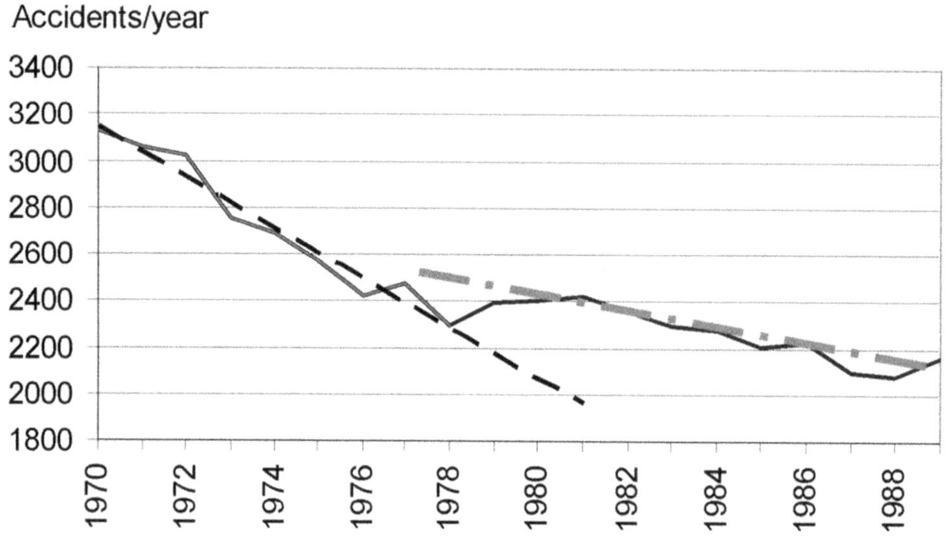

Figure 10. Traffic counts on a parallel secondary road show the relieve effect for about 8 years

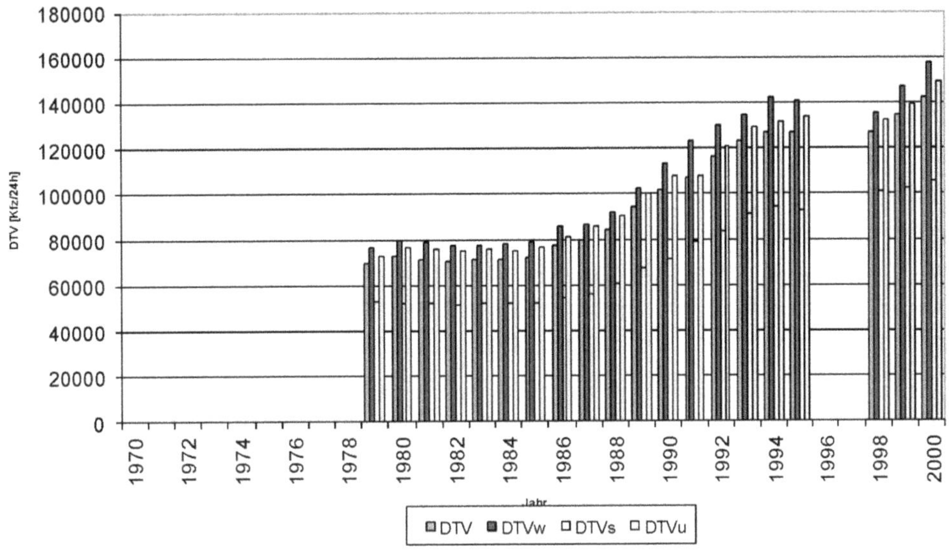

CONCLUSION

By analyzing time series of the development of traffic accidents in Vienna the period of continuous decrease of the accidents was interrupted at the end of the 70s by a rather fast increase of the number of accidents over a period of five years, till the system stabilized again on a much higher accident level than before. After this period the number of accidents started to decline again. On the macro scale traffic accidents are effects from the system dynamic of the transport system and therefore dependent on the structures. These structures are the build environment, the legal and the social system, the kind of transport management and the enforcement. A new motorway is a built structure with effects on the whole city and the city neighborhood as well. It change the modal split toward more car traffic, but it change also inner urban structures, like housing and location of economic activities as well as structures outside the city but with effects on the city itself. Vienna had no motorway bypass till 1978, when the A23 and A22 were opened for the traffic. The number of accidents with injured people declined continuously since 1970 due to traffic calming measures in some districts of the city. The expectation followed the extrapolation of this trend, with the promise, that the motorway, as a "safe road" will contribute to this positive development and bring more safety to the urban streets also, as well as less congestion, air pollution and noise.

After opening the motorway bypass, the car traffic system became more convenient, the speed was enhanced also in some urban roads and congestion was reduced temporary. Traditional educates engineers were satisfied with the effects – at least for some years. But the change of the system behavior is not visible with the instrument of traditional transport engineering tools, which use indicators like "time saving" or "growth of (car) mobility". The observation of increase of traffic accidents seem to have some "external causes" and cannot be related to the effects of the new motorway and its splendid traffic performance.

From the system point of view and knowing the real system behavior – as well as simple laws of physics – we have to doubt whether the promise and forecast, based on the traditional tools of transport engineering are in agreement with the real system – and human – behavior. The new motorway change the structure of the urban transport system, followed by behavioral change of residents, commuters, investors and as a conse quence, also land use plans. One of the effects was the interruption of the smooth steadily decrease of traffic accidents in the city.

If we extrapolate the development of accidents in the city before opening of the motorway bypass, the number of accidents came not back into the extrapolated development of the time period before, until 1994. The measure, which has increased the number of accidents needed about 16 years to be compensated and "digested" in the system. If we compare the difference between a continuous development of the number of accidents and this interruption, we can calculate about 15.000 accidents with injured people where caused by the effects of the measure. The only substantial change in the transport infrastructure of the city of Vienna was the opening of a part of the motorway in 1978. Since no other substantial measures have been introduced in the transport system during this period it can be suspected that a motorway had this adverse effect on the whole traffic safety system of a big city. If we analyze the development of the number of accidents of the different transport system users it can be seen that especially accidents with person car, light and heavy trucks have contributed to this increase of the accident figures. But there were also negative effects on pedestrian accidents. The analysis shows that:

- Traffic safety indicators based on driven kilometers, are nor useful to compare roads with different speeds, environments and operational conditions.
- Motorways as detour roads to relieve car traffic from urban streets can reduce the traffic safety of the city – even a big one.
- Traditional traffic engineering tools are not applicable for planning and political decisions in the system level.

Due to the fact that transport was relieved from the urban system the speeds in the urban roads has gone up also kinetic energy increased and the probability of the risk increased also. This is the first time that the effect of a urban motorway was analyzed on the whole system over a long time period.

The empirical data are consistent with a analysis of effects for the hypothesis based on physical effects, which changed with the opening of the motorway. If the increase of the speed by opening new road sections especially motorways is not compensated by an effective reduction of the speed in the secondary urban system, a ring motorway with the effects of relieving the inner urban transport system can reduce a traffic safety of the whole city substantially. It needs nearly two decades till the transport system has recovered from this lack of the transport safety. Very often transport system is seen in a much to narrow view. If we change the transport system we change also structures in the midterm view. This happens always by opening the motorway and can be proved not only in the city of Vienna but in all other cities, which have done the same decision, based on the biased models of traditional transport planning.

New economic activities as well as new housing activities occur along the new motorway interfering with the urban structure and destroying the sensitive inner urban structure economically and culturally. Traditional transport engineers and consultants having no idea about system effects of the elements they are dealing with, oversee these adverse effects because they are not in their model. They are modelling things, which occur immediately after opening such a technical infrastructure but not the effects of the changing urban structure, on the other modes and on non-motorized system users. Their results are – if we compare them with the scientific background of physics and logics absolutely contradiction to them. They use wrong indicators for example the accident rate of a section and compare them with system elements, which can't be compared because they have not comparable structures. This was the case during the last 50 years. If politicians are informed in such a way, they make wrong decisions creating the negative effects they want to prevent.

This empirical study has been carried out recently but there are some empirical evidences from other analysis, which support the findings of this study. The result of this study and the hypothesis, which is proved by the empirical data is in some contradiction to traditional transport engineering education and experience and their expectations. The results of this study are coherent with the laws of physics and logics: motorways in urban areas can enhance the risk of accidents and will enhance the risk of accidents if they are not compensated by risk reduction measures of the same amount as motorways increase the risk.

What can we learn from this example?

- Planning of detour roads, especially motorway bypasses must be accompanied by a set of measures which can compensate the speed enhancing effects of this new structure.
- The results of this long term study underestimate the negative effect of the motorway because the spatial borders of the system analyzed were kept constant, although the system speed enhanced. If this is the case, the system borders must be extended in accordance with the change of the speed of the transport system.
- As a simple rule of thumb for practitioners: We can keep the system in balance it the degree of enhanced attractiveness for car users by some measures is equal or less, compared to measures enhancing attractiveness for walking, cycling and public transport. This can be done by using transport models and tools based on knowledge about real system and human behavior.
- Any project in transport is always only a part of a system and cannot be treated in a responsible manner without analyzing its effect on the whole.
- A motorway or road project has effects on all other kind of modes in the system. A cross section and time-longitudinal modeling is necessary to reproduce the dynamic effects on traffic and spatial structures of the new system behavior.
- The tools we use so far to implement such a project are not appropriate to solve transport problems without unwanted effects.

To deal in a responsible way with transport problems we need:

- A multimodal approach in the computer-model structure.
- Public money, which is used for project is earmarked today to different sectors of the transport system. This is one of the causes, which destabilize the equilibrium of the transport system continuously.
- Budget has to be allocated to solve transport problems *of the system* and not only in one or the other sector.
- To solve transport problems, with effects across administrative borders, budgets have to be provided and allocated in accordance with the overall goals of the society before decisions are made. If this is omitted to do, one or more victimization is inevitable. In our case at least 15.000 in-

jured people were the victims of this narrow sighted view – and inappropriate tools in transport engineering.

- Budget allocation not only for the project, but for the whole transports system, e.g. budget for the motorway project *plus* budget for all measures to compensate adverse effects in the system, caused by the new infrastructure. In the example the ministry, which pay for the motorway construction, has to provide also the cost for the traffic calming and traffic management measures in the city, to prevent negative effects caused by the motorway.
- Models have to be used (which are available today) to provide a sound data base for these decisions.
- To understand and interpret the model-results, the transport engineers have to understand the real system behavior. The precondition to understand transport system behavior is to understand human and societal behavior, which should not be mixed-up with individual and/or societal wishes or aspiration of some lobbies.

REFERENCES

Automatische Straßenverkehrszählung – Jahresauswertung 2000. (2001). Bundesministerium für Verkehr, Innovation und Technologie, Wien.

Eurostat. Statistics explained. (2016). *Road safety statistics on regional level.* Retrieved from http://ec.europa.eu/eurostat/statistics-explained/index.php/Road_safety_statistics_at_regional_level

Germany. (2014). *Unfallentwicklung auf den Bundesautobahnen der Region im Fünfjahresvergleich.* Retrieved from www.pd-bs.polizei-nds.de/.../Verkehrsunfallstatistik_BAB_im_5-Jahresvergleich_bis_

Heusch/Boesefeldt. (1997). *Überprüfung der Bewertungsmodelle für die Bundesfernstraßen beim Bundesverkehrswegeplan 1992 anhand beobachteter Verhältnisse.* Aachen: Author.

Knoflacher, H. (1970). *Methoden der Unfallauswertung als Beitrag zur Erhöhung der Verkehrssicherheit auf Stadtstraßen.* Kleine Fachbuchreihe des Kuratoriums für Verkehrssicherheit (KfV), Band 9/1970.

Knoflacher H. (1981). *Ungenutzte Möglichkeiten zur Erhöhung der Verkehrssicherheit.* ARBÖ-Verkehrsjurist, Nr 55/56, 1981

Knoflacher, H. (1985). Ungenutzte Möglichkeiten zur Erhöhung der Verkehrssicherheit. Österreichische Gemeindezeitung, Wien 1985/2

Knoflacher, H. (1989). A 23 Südost-Tangente. Strukturelle Auswirkungen 1978 – 1988. Durchgeführt im Auftrag des Magistrats der Stadt Wien, MA 18 – Stadtstrukturplanung. Wien.

Knoflacher, H. (1990): Wie sicher sind Autobahnen? *Verkehr & Umwelt, 4*(2), 22-26.

Knoflacher, H. (1995). *Fußgeher- und Fahrradverkehr.* Planungsprinzipien. Böhlau Verlag Wien – Köln – Weimar.

Knoflacher, H. (1996). *Zur Harmonie von Stadt und Verkehr. Freiheit vom Zwang zum Autofahren. 2., verbesserte und erweiterte Auflage.* Böhlau Verlag Wien – Köln - Weimar.

Knoflacher, H. (1997). *Landschaft ohne Autobahnen. Für eine zukunftsorientierte Verkehrsplanung.* Böhlau Verlag Wien – Köln – Weimar.

Knoflacher, H. (2001). *Verkehrssicherheit Wien – Analyse des Bestands und Vorschläge für die Zukunft.* Durchgeführt im Auftrag des Magistrats der Stadt Wien – MA 46.

Olsen, J. R. (2016). *Effects of new urban motorway infrastructure on road traffic accidents on the local ares: a retrospective longitudinal study in Scotland. Journal on Epidemiology and Community Health.* doi:10.1136/jech-2016-207378

Scholas, M. (1988). *Auswirkungen von Ortsumgehungen im Zuge von Bundes- und Landesstraßen in Nordrhein-Westfalen – unter besonderer Berücksichtigung des Unfallgeschehens.* Dissertation Universität Dortmund 1988.

SCS Extension. (1988). Retrieved from https://de.wikipedia.org/wiki/Shopping_City_S%C3%BCd

Straßenverkehrsunfälle im Jahr 1970 – Straßenverkehrssicherheit im Jahre. (1989). Beiträge zur Österreichischen Statistik. Herausgegeben vom Österreichischen Statistischen Zentralamt, Wien.

Chapter 13
Multi-Modal Assessment of Highway Performance

Markus Mailer
Universität Innsbruck, Austria

ABSTRACT

This chapter presents a multi-modal method for the assessment of highway performance. It is derived by extending a traditional assessment concept step by step taking into account the capacity and quality of different modes on the road as well as in the corridor. It defines an appropriate performance target and explains why a multi-modal concept has to consider transport demand in persons and goods rather than traffic volumes in vehicle units. It is shown that the concept allows for different options and measures to improve traffic quality and so supports the efficient use of existing infrastructure and the effective allocation of limited funds.

INTRODUCTION

This paper discusses methods for the assessment of road design. Its starting point is the traditional approach which is essentially based on the method established in the Highway Capacity Manual (HCM), the first guideline for the assessment of road design and operation, published in the United States of America in 1950. This method and its updates formed the basis of many guidelines for the assessment of road design in different countries; one of them is the Austrian Standard. This example is used to discuss the method and show how a method to assess whether an existing or planned road can fulfil its function can be extended by a multi-modal approach taking into account the capacity and quality of different means of transport on the road as well as in a corridor. Finally, it shows that a multi-modal concept allows a more efficient, cost-effective design and use of traffic infrastructure and supports an efficient allocation of limited funds by considering different options to meet road traffic quality targets. In addition to changing road design parameters, the potentials of all modes of transport as well as measures of mobility and traffic management can be considered in the assessment and planning process.

DOI: 10.4018/978-1-5225-2116-7.ch013

BACKGROUND

The basis of the multi-modal concept described in the following chapter was laid when the Austrian Standard for the Assessment of Road Design RVS 3.7 (FSV, 1994) had to be revised in 2001. Austria, like many other countries, had to deal with growing road traffic volumes, the increasing financial problems of rail, and limited budgets. Moreover, there was evidence that the traditional approach of road standards assessing highway performance might not be useful to solve these problems but that tools which basically adjust road capacity to traffic demand might even be part of the problem.

In 1994, Knoflacher published considerations based on feedback loops that the level-of-service concept introduced by the HCM has repercussions on traffic growth (Knoflacher, 1994). The short-term satisfaction of transport demand, in conjunction with the assessment procedures and criteria used, forms a reinforcing feedback loop that generates ever greater demand for transport in the long term (see Figure 1).

The left side of the figure shows the logic of the HCM and of similar standards focused on road sections. When traffic volumes grow, service quality decreases. This inverse interaction is marked with a minus in the figure. Based on the standards, measures to increase road capacity have to been taken when service quality falls below a benchmark in order to raise it above the desired level again. The relation between travel quality and capacity seems to form a negative and therefore self-stabilizing feedback loop. But, in the long term, service quality, especially if it is directly related to travel speed, affects land use. This relationship is shown on the right side of the figure. Increasing travel speeds resulting in longer travel distances have caused developments such as urban sprawl and economic concentration in the last few decades. Longer travel distances led to more mileage hence to more road traffic in the road network which finally results in a higher traffic volume on the road section being the starting point of the loop. These proportional relations are marked with a plus. The inverse interaction between traffic volume and service quality again forms a negative, self-stabilizing feedback loop at the top. In combination, the two loops, however, form a positive and therefore reinforcing feedback circle. So, responding to traffic growth by increasing road capacity is generating even more traffic in the long term. Since there is a substantial time lag in the relations between increased travel speeds resulting from infrastructure improvement, land use changes, and traffic growth, this vicious circle is difficult to realize. However,

Figure 1. Reinforcing feedback loop between the level-of-service approach and traffic volume
Mailer, 2004.

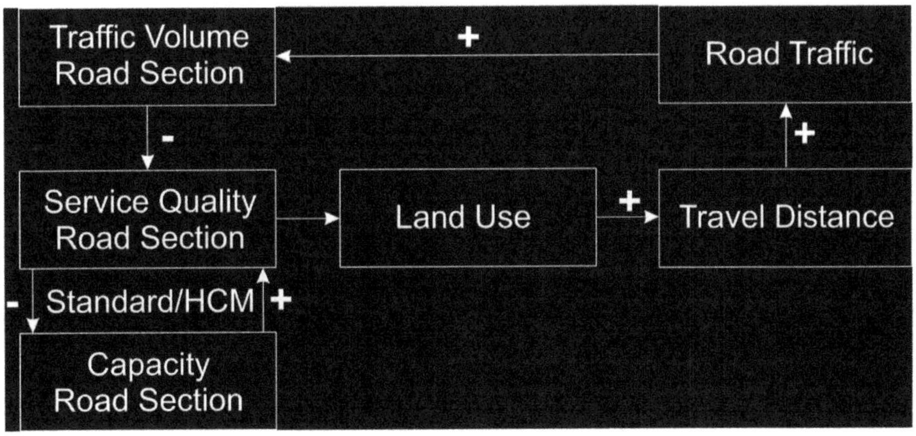

the growth of road infrastructure and traffic has reached several limits already, not at least a financing limit for infrastructure construction, operation, and maintenance (Mailer, 2004).

Considering this, road design should actually be derived from a general concept of space and urban planning taking into account social, economic and ecological requirements, boundary conditions, and limitations defining a "permissible traffic volume" (Schnabel & Lohse, 1997). So, some experts like Lohse state that this is the way road design has to be carried out under present conditions. A design according to service levels, which are only based on traffic developments and consequential concepts, would no longer be reasonable or only reasonable in partial systems.

Thus, the planned revision of the Austrian guideline should not only update the existing method based on recent findings concerning the influence of road design parameters on traffic flow and travel speed, but extend it using a multi-modal approach allowing the consideration of traffic management measures (Mailer, 2004).

At that time, traffic management and multi-modality did not yet have the attention they have today. Nowadays, they are seen in the light of the efficient use of existing transport infrastructure and services in order to cope with limited resources on the one hand and increasing costs for infrastructure maintenance on the other as well as in respect to sustainable mobility.

In Germany, initial findings about multi-modal behavior and travel patterns were published by Chlond and Lipps in 2000. Franke (2004) described multi-modal travelers as a new target group for mobility services and Beckmann et al. (2006) expected that the number of people with multi-modal behavior would increase in the next few years. Today, multi-modality is seen in the context of a trendy sharing lifestyle, a change in mobility culture, the increasing number of car-sharing services and users, and a reduced perception of cars being status symbols (VCÖ, 2014). However, when it comes to the multi-modal assessment of infrastructure design and performance, the focus is rather on the potential of the cooperation of transport systems to meet transport demand, i.e. the transport of people and goods in a given spatial and temporal frame, than on individual behavior, which of course still is the substratum of multi-modal strategies.

DEVELOPING A MULTI-MODAL CONCEPT

To present and explain the multi-modal approach for the assessment of highway performance, a traditional mono-modal concept is used as the starting point for a step-by-step extension of the method. In addition, it has to be clarified as to which parameter describing traffic quality is the suitable primary measure; hence, the target dimension of the assessment.

Traditional Mono-Modal Concept

In a traditional demand-oriented approach, the assessment of highway performance follows the procedure shown in Figure 2, which will subsequently be referred to as the existing method. A given traffic volume represented by the number of vehicles that are counted or predicted on the one hand and the design parameters of an existing or planned road on the other hand are used to calculate a traffic quality measure. This measure is compared to a benchmark dependent on the road function or category. If the quality requirement is not met, the design parameters have to be changed, i.e. the road design has to be improved.

Figure 2. Traditional mono-modal concept for the assessment of highway performance
Mailer, 2004.

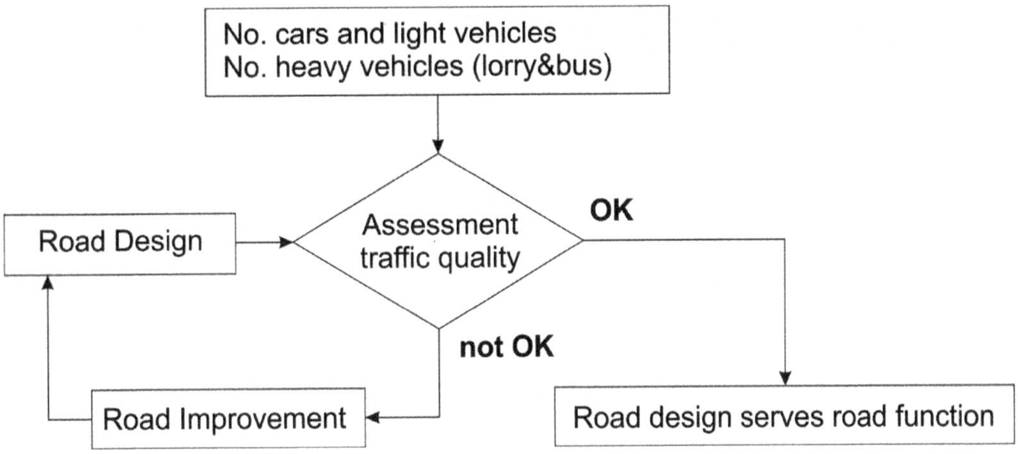

In the Austrian Standard RVS 3.7 and also in its update RVS 03.01.11 (FSV, 2012), the average travel speed of private cars, which is called operational speed (RVS 3.7) or traffic speed (RVS 03.01.11), is defined as the primary quality measure. These speeds are dependent on traffic volume and road design. The traffic volume - travel speed algorithm used in RVS 3.7 to calculate the average car travel speed for road sections in non-built-up areas - considered design parameters such as the number of lanes, lane width, and gradient as well as the load factor, i.e. the ratio of design volume to capacity. However, the capacity value did not reflect capacity in its true sense. An algorithm similar to that defined in the third edition of the HCM (TRB, 1985, 1994) for computing (service) flow rates was used to calculate this value for a rural road segment. Hence, design parameters such as the number of lanes and their width, curvature, overtaking sight distances and the influence of heavy vehicles depending on their proportion in the traffic stream, and the gradient determined the capacity value. So, in fact, the capacity value already included the influence of these parameters on travel speed. In the updated guideline RVS 03.01.11, the calculation of travel speed is still very similar even though factors such as lane width and overtaking sight distance are not used any more.

According to the parameters used, this means that, if travel speed is too low and the road design has to be changed, more or wider lanes and/or modifications of alignment and/or gradients are necessary. Since longer sections through urban areas reduce average operational speed, bypass roads are also a frequent measure to meet the speed target.

Defining an Appropriate Performance Target

Assessing road performance basically means evaluating if and how well a road fulfils its function. But what measure represents the function? Is travel speed the right measure?

Some experts, for instance in Germany (Brilon, et al., 1994), addressed problems related to road design in hilly or mountainous regions resulting in unreasonable alignment and expensive road construction. Problems in justifying large-scale road reconstruction with little speed differences were reported by practitioners. For this reason, it seems better to use a measure which allows more efficient road design.

Since many parameters of road design as well as traffic composition have significantly lower effects on capacity than on travel speed, the load factor would meet this requirement.

However, bringing traffic back to its roots, i.e. serving people to carry out activities at different places to fulfill their basic needs, quality measures have to be derived from user needs. A transport planning perspective based on the findings of constant travel budgets, as outlined by Schafer (1998, 2000) for instance, leads to questioning the importance of travel speed as well (Mailer, 2001). Thus, rather the user's perspective that transport infrastructure has to ensure that a destination can be reached at the expected time of arrival is emphasized. Surveys asking for driver ratings support the importance of reliability in the context of transport quality (Mailer et al., 2008). However, this also supports using the load factor as a primary measure. Reflecting the utilization of capacity, the load factor is directly related to reliability. Lower load factors mean that traffic flows are farther away from the unstable traffic conditions at capacity and therefore offer users a higher reliability of travel times.

These considerations are also in line with the updated procedures of the fourth edition of the HCM (TRB, 2001) and with the first edition of its German equivalent *Handbuch für die Bemessung von Straßen* HBS (FGSV, 2002).

The following academic example should illustrate the differences resulting from changing the performance target from travel speed to load factor. In order to make these differences particularly clear, the comparison deliberately refers to the method of the old Austrian standard which is based on parameters that explicitly and also implicitly referred to travel speed as the quality target. The setting is the assessment of a road segment with four sections that have different characteristics (see Figure 3).

Figure 3. Example 1: road segment with four sections

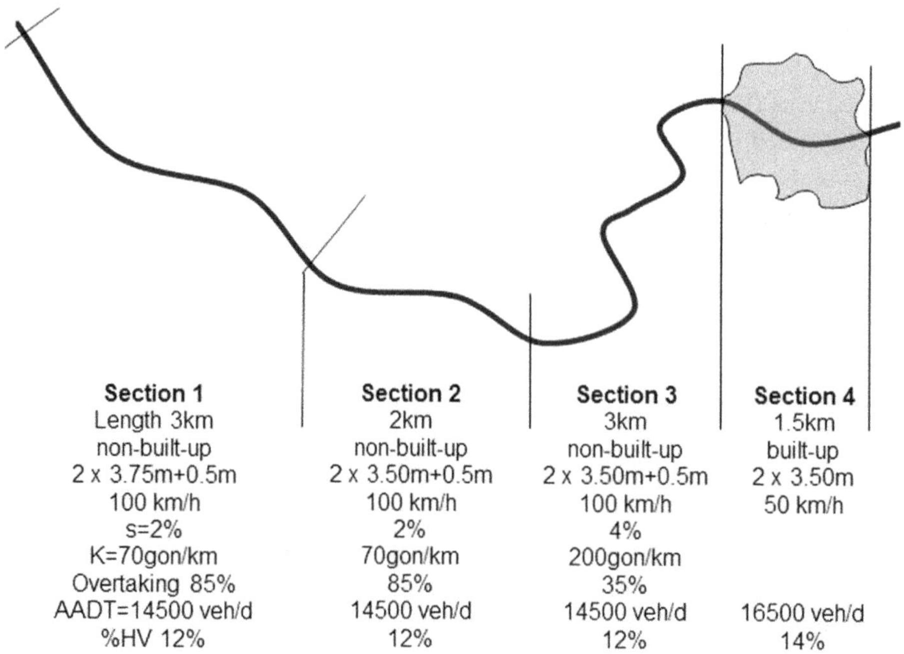

Table 1 shows the design parameters of the four sections and the factors derived from them which are needed for the calculation. The abbreviations are used according to the Austrian standard. The calculation of operational speed (V_V) is based on a value that represents capacity (BVS) in a volume-speed relationship. However, since it is calculated by multiplying a basic capacity (L_0) that is dependent on widths of lanes and hard shoulders with factors which also represent the influence on speed rather than on capacity (f_{SK} related to curvature and overtaking opportunities, f_V related to heavy vehicles and gradient), BVS already includes a quality requirement. When comparing the values of section 1 with those of section 2, it becomes obvious that a small change in lane width already causes a considerable reduction in BVS and operational speed. The latter even falls below the quality target of a category II road (60 km/h). In section 3, higher curvature and gradient cause a further reduction in BVS even falling below the design volume. The speed limit in section 4, which is located in a built-up area, is below the target speed. Thus, even if measures are taken to increase capacity in section 3, the average speed in this road segment will be below the target value resulting in the need for further measures, e.g. a bypass road in section 4.

The bottom of the table shows results calculated using the new approach developed for the revision of the guideline. Since the assessment is based on the load factor, the basic capacity (L_0) is not influenced by lane width, curvature, and overtaking opportunities and the influence of heavy vehicles and gradient

Table 1. Example 1: assessment of the performance of a 2-lane road using the approach of the former Austrian Guideline RVS 3.7 and an alternative concept (new approach)

	Section 1		Section 2		Section 3		Section 4	
	Non-Built-Up		Non-Built-Up		Non-Built-Up		built-up	
	Design Parameters	Performance	Design Parameters	Performance	Design Parameters	Performance	Design Parameters	Performance
Length	3 km		2 km		3 km		1.5 km	
Lanes	2 x 3.75m + 0.5m		2 x 3.50m + 0.5m		2 x 3.50m + 0.5m		2 x 3.50m	
L_0 [cars/h]		2500		2200		2200		
Speed limit [km/h]	100		100		100		50	
Curvature [gon/km]	70	f_{SK}=1.0	70	f_{SK}=1.0	200	f_{SK}=0.7		
Overtaking	85%		85%		35%			
Gradient	2%	f_V=0.67	2%	f_V=0.67	4%	f_V=0.63		
% HV	12%		12%		12%		14%	
DHV [veh/h]	1450		1450		1450		1650	
BVS [veh/h]		1667		1467		1100		
V_V [km/h]		66		54		over-loaded		50
New Approach								
L_0 [cars/h]		2800		2800		2800		2800
gradient + %HV		f_{SV}=0.88		f_{SV}=0.88		f_{SV}=0.88	various parameters	f_u=0.71
Load factor (v/c)		0.6		0.6		0.6		0.83

(factor f_{SV}) is considerably lower. Thus, the resulting loading factors are relatively low in all sections. Even if a value of 2500 cars per hour (as in the German HBS) was used for the basic capacity instead of 2800 cars per hour (as in HCM94), the values in sections 1, 2, and 3 would be well below 0.7 and in section 4 just above 0.9. So, the quality target can be met without improvement (expensive road construction).

Multi-Modal Assessment on the Road

Aiming at an approach that includes different modes of transport, it becomes obvious that vehicle units or personal car units are not suitable to describe traffic volumes, since they can neither be compared nor directly transformed to other modes. A multi-modal approach has to focus on transport demand, hence considering people and goods that are to be transported. In a multi-modal assessment, this transport demand can be related to all available modes. The modal split can be interpreted as an indicator of the extent to which the different modes meet user needs and preferences (availability, accessibility, travel times, costs, reliability, safety, comfort, etc.). A multi-modal view needs to consider origin-destination connections and the time of travel as well as actual options in mode choice.

With the change of the reference unit to persons and goods in addition to the capacity of the infrastructure itself, (free) capacity of the means of transport can also be considered. Low car occupancy rates and low loading of goods vehicles drastically reduce the achievable transport volume. So, the relation between transport demand (persons, goods) and its manifestation in traffic demand (vehicles) provides a first leverage point for traffic management measures aiming at increasing efficiency in road transport. Measures that increase car occupancy or loading factors of goods vehicles can raise transport capacity in terms of persons and goods even if road capacity in terms of vehicles remains unchanged. Thus, the increase of vehicle occupancy or utilization is an improvement in efficiency by better use of existing seat capacity or space in the vehicles. This goes hand in hand with reduced traffic load and consumption of resources (space, energy, environment, etc.) (Mailer, 2004). So, this first extension already results in alternative options to meet the quality benchmark. As shown in Figure 4, in addition to road improvements, measures can be taken to increase the efficiency of goods vehicles and of private car (PC) traffic as well as measures to improve public transport (PT) and increase its modal share as a response to failing the quality target.

Since roads usually have to accommodate both the transport of passengers and goods, the management scenarios are closely related. They all focus on transporting the same amount of passengers and goods with fewer vehicles. A reduction in the vehicle volume in order to meet the traffic quality target can be achieved with measures increasing car occupancy resulting in fewer cars as well as with measures for more efficient goods transport reducing the number of goods vehicles needed to transport the given quantity of goods. These options are connected with different measures such as promoting carpooling or ridesharing, logistics concepts, prioritization, and acceleration of public transport, aiming to accomplish the same transport demand with fewer vehicles. However, such measures require more detailed information on the circumstances and requirements of transport demand. The actual time constraints have to be considered as well as origin-destination relationships. A purely cross-sectional view and a determination of vehicle volumes have to be waived. Instead of an abstract hour-volume design used in many guidelines such as a score of the average annual daily traffic (AADT) or the thirtieth highest hourly traffic volume of the year, an identifiable dimensional hour such as a weekly recurring peak hour has to be used. Only with data on the main parameters determining transport demand can it be estimated to what extent this demand can be fulfilled jointly by different transport modes. However, of course there are differences in

Figure 4. Extended concept considering transport demand and capacity in passengers and goods allowing for different management scenarios for more efficient road transport
Mailer, 2004.

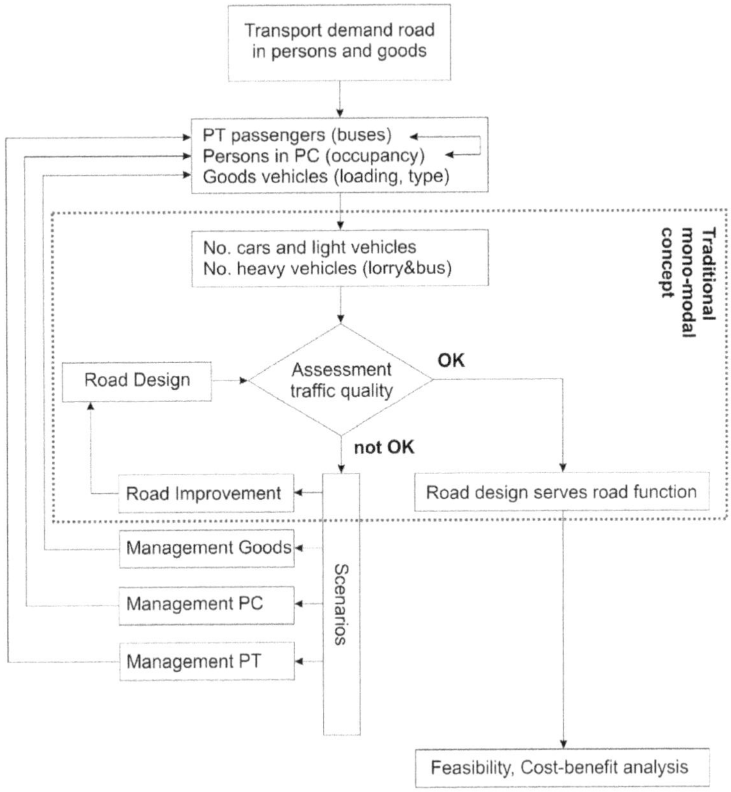

respect of the effectiveness, efficiency, and possibility of realizing the various measures. Thus, finally, feasibility and cost-benefit analyses have to be carried out to identify those measures that can be realized, are most effective, and most efficient in bringing traffic quality to the desired level (Mailer, 2004).

Multi-Modal Assessment in a Corridor

Multi-modality allows a further extension of system boundaries. In addition to the section of road that has to be assessed, other transport systems in the corridor can also be taken into account. The total demand for transport in the corridor has to be compared with the entire transport capacity. From the perspective of road transport, the amount of passengers and goods transported by alternative modes can be seen as a reduction in traffic load. The traffic demand to be accommodated on the road can be determined by subtracting the transport demand that is or should be covered by other modes from the total transport demand in the corridor. So, the road only has to take what is not (or cannot be) transported by other modes of transport. From this perspective, overloaded roads can be interpreted as insufficient capacity (or performance) of alternative transport systems. This approach results in a further scenario to deal with the inadequate traffic quality on the road. By increasing the availability of alternative transport systems or through appropriate management measures, the potential for a shift in the modal split (MS) can be used in the corridor (see Figure 5). Thus, for example, a shift in transport demand to rail can result in less traffic load on the road and thereby increase the quality of road transport. So, under certain circumstances, investment in the improvement of road capacity can be saved through better use of existing (rail) infrastructure.

Of course, this method illustrated from the perspective of the extension of existing approaches of road transport can also be derived from the point of view of rail traffic in a corridor. Figure 6 presents the complete picture. It clearly shows the reciprocal influences. The scenarios in both subsystems are extended by the options in the other system and the shift of the modal split in the corridor.

Due to the heterogeneity of measures related to the different scenarios, it has to be emphasized again that feasibility and cost-benefit analyses are needed to identify practicable, effective, and efficient measures.

Figure 5. Further extended concept including all means of transport in a corridor and different modal spilt scenarios
Mailer, 2004.

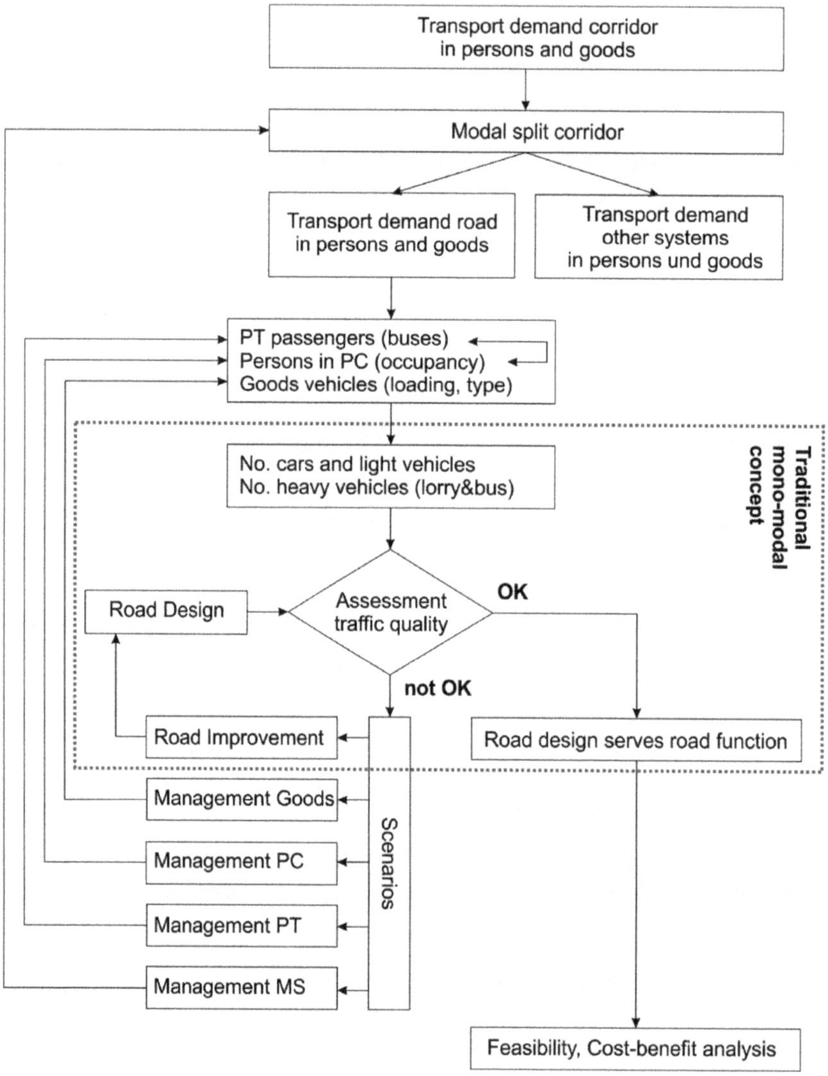

Figure 6. Concept for a multi-modal assessment of transport quality in a corridor
Mailer, 2002.

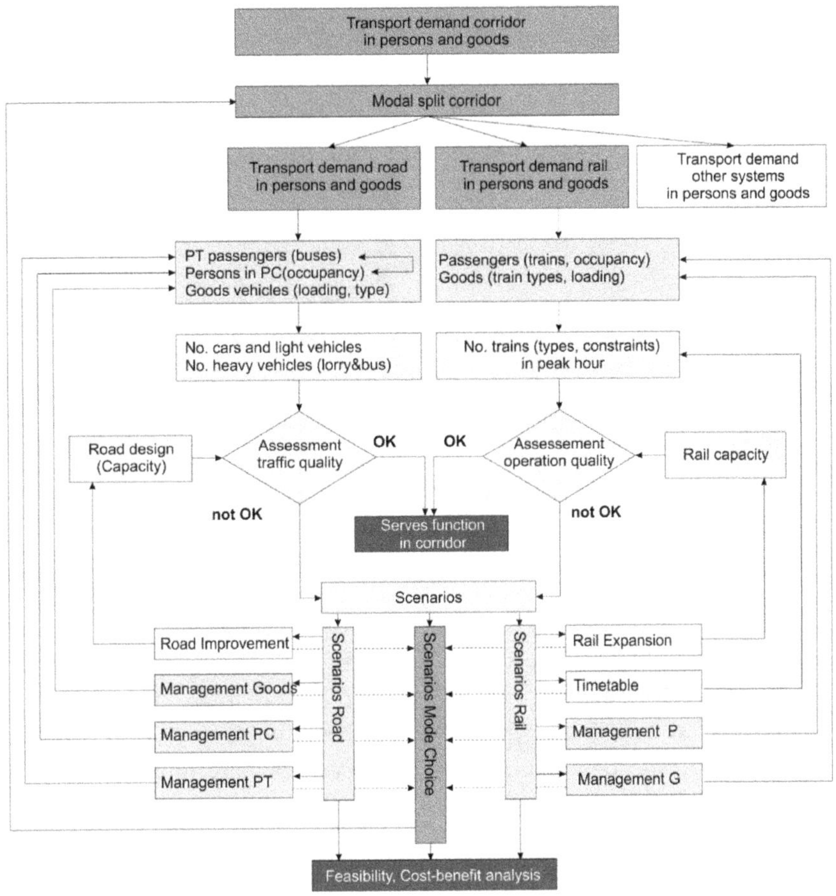

Internal Capacity

In a multi-modal assessment, transport capacity (people and goods units) has to be considered instead of traffic capacity (vehicles or trains which are only essential for the assessment of traffic flow). Even a rough comparison on the basis of a simple extension of the traffic flow relationships taking into account the number of seats in passenger transport makes clear how much the transport dimension lies above the simple vehicle dimension. For the comparison, each car is initially allocated 5 seats and each train is allocated 1000 seats (Figure 7).

This estimation, based on basic capacity, also shows the great potential for rail transport. While the hourly train numbers cannot even be displayed on this scale, the passenger performance of a railway track exceeds that of a road lane by more than 100%.

In the multimodal approach, the conversion of passenger and freight units to vehicle units is of central importance. In road traffic, the characteristics of the vehicles used and the utilization of their transport capacity are essential for accommodating the volumes of passenger and freight traffic. In passenger transport, the split of private car to public bus transport is crucial. Due to the comparatively high capacity of buses, the internal transport capacity can be increased. The transport performance of passenger

Figure 7. Extension of the traffic flow (possible vehicle performance) for a lane or a track with the number of seats (possible passenger capacity)

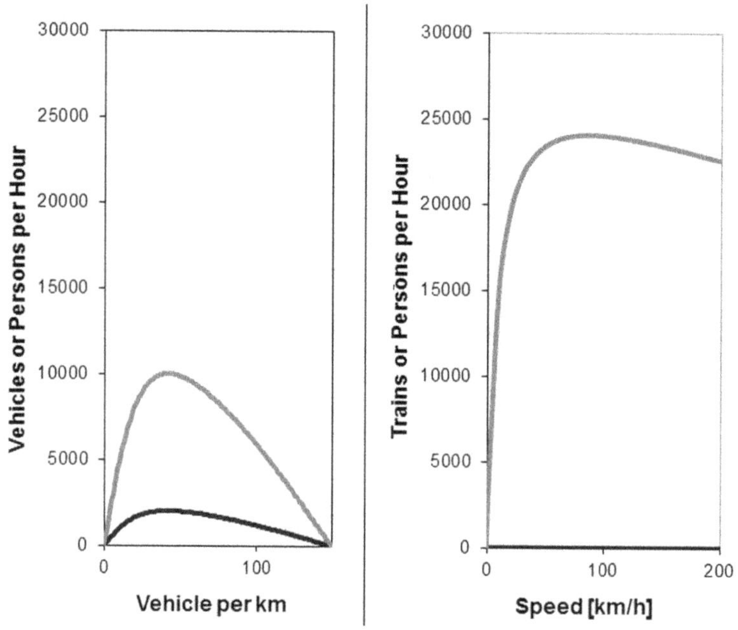

vehicles can be characterized by the average number of persons (average occupancy rate) transported in one vehicle. However, the speed characteristics of the vehicles are important for road capacity and traffic flow. The allocation of persons to car traffic (cars and car-like vehicles) and heavy traffic (buses) must be taken into account.

Goods are very heterogeneous with regard to their transport characteristics. The reference to the weight of the goods and the reference to transported metric tons seem inappropriate. Depending on the nature of the goods, the maximum transportable quantity in a vehicle can be limited by weight or volume. Often, however, the quantity transported is not limited by the vehicles but by the order. This differentiation is also reflected in the different types of vehicles used in freight transport (e.g. articulated trucks, truck trains, small trucks, delivery vehicles, station wagons, passenger cars). When transferring goods transport into vehicle units, account must be taken of the dependencies on different goods or groups of goods, vehicle types, and vehicle utilization (proportion of empty runs). Due to the heterogeneity of freight transport, the appropriate reference unit for freight transport is therefore goods vehicles. To assess the flow of traffic, the distinction between heavy vehicles and passenger-car-like vehicles is necessary. Accordingly, the handling of freight transport in the multimodal process requires data from detailed analyses.

In line with these arrangements, the transport performance of a road can be illustrated by the comparison of passenger transport and goods transport (Figure 8). The type of presentation is intended to illustrate the internal capacity reserves (degree of occupancy, utilization, public transport, etc.) and to demonstrate the dependencies of the multimodal influencing variables. The reference unit for passenger traffic is persons transported. The reference unit for freight transport is the number of goods vehicles used to transport the goods. This number is introduced as a share of all vehicles in the design hour. This proportion corresponds to a specific number of goods transport and thus describes goods transport as a

Figure 8. Transport capacity of persons and goods on a lane with emphasis on the internal transport capacity
Mailer, 2002.

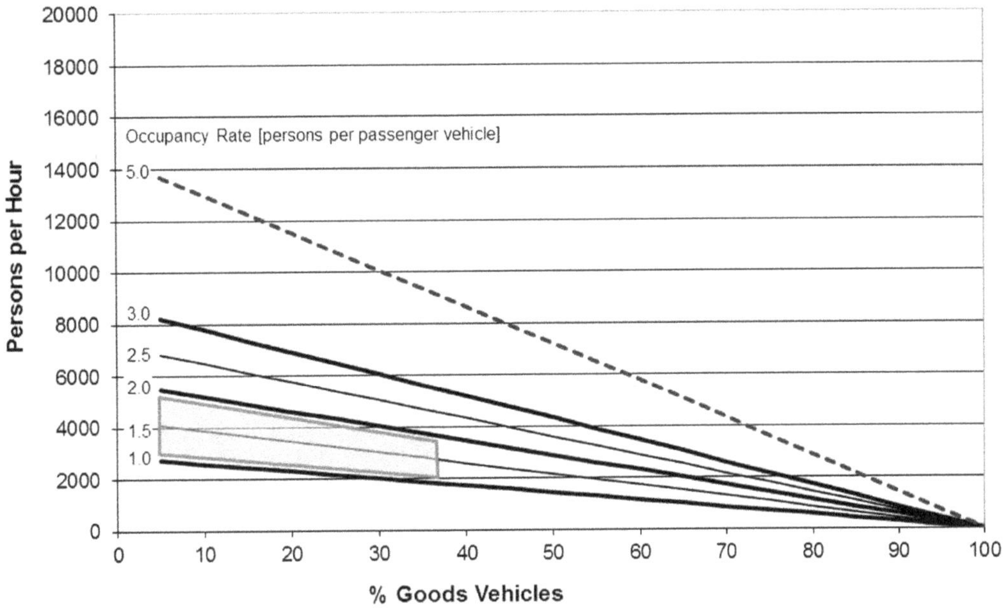

relative quantity to the traffic volume. The lines indicate the number of persons who can be transported at a certain vehicle traffic level depending on the occupancy rates. As a result, the decline in the number of persons transported with an increasing share of goods vehicles must not be seen as a decrease in transport performance. On the contrary, a high proportion of goods vehicles means a correspondingly high transport performance in freight transport. The utilization of vehicle capacities is indicated in passenger transport by the average occupancy rate. In freight transport, the average capacity utilization is already included in the number of goods vehicles (Mailer, 2002).

In Figure 8, a large range of parameters has been chosen to illustrate the relationships of the transport performance. The area of interest in practice is marked. The representation of relatively high average occupancy levels should show the possibilities of increasing transport performance by high occupancy levels in public transport.

On the basis of the functional relationships between occupancy rates and load factors, they can always be transformed into each other. Calculating occupancy rates has the advantage that changes in occupancy rates are usually associated with traffic management measures. They express direct changes in vehicle use, in the choice of means of transport or changes in vehicle composition. The advantages of displaying occupancy rates are therefore emphasis on multimodality and efficient vehicle utilization, and that they are directly observable.

The relevance of occupancy rates in the assessment can be clearly shown based on the previously introduced illustration of passenger transport and proportion of goods vehicles (Figure 9). The lines are based on the traffic capacity of the road and the maximum load factor. On the one hand, the maximum of passenger transport can be read as a function of the occupancy rate and percentage of goods vehicles. On

Figure 9. Example to illustrate the assessment based on the occupancy rate
Mailer, 2002.

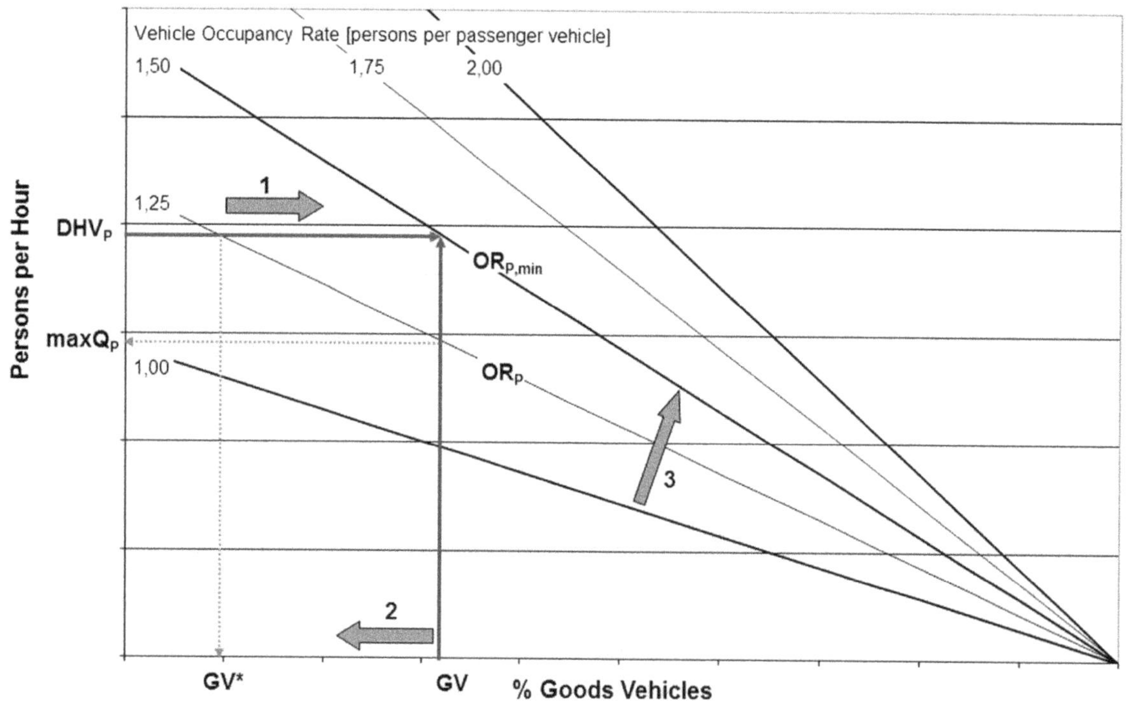

the other hand, a minimum average occupancy rate of passenger vehicles can be determined for specific requirements for passenger transport volume (DHV_P) and percentage of goods vehicles (GV). In this example, it is 1.5. If the actual occupancy rate is below this value (e.g. at 1.25), the transport demand cannot be met by the maximum possible vehicle traffic volume. For a given GV, only the maximum passenger volume $maxQ_P$ can be transported. On the other hand, with a given DHV_P only a smaller number of goods vehicles can be accommodated (GV *).

If the specified transport demand cannot be met, measures must be taken. This figure firstly illustrates the approaches of traffic management. In addition to an increase in the occupancy rate in passenger transport (1), e.g. through more attractive public transport, it is also possible to reduce the number of goods vehicles by improving their capacity utilization (for example, fewer empty runs) (2). The picture shows how the second measure would also reduce requirements to the minimum occupancy rate. In addition to these measures, there is also the possibility of road construction. This would result in new, steeper lines (3). For example, if road capacity increased by 50%, the current line for occupancy rate of 1.5 would then apply to the occupancy rate of 1.0. This means that this road construction would be equated with a 50% increase in the occupancy rate achieved by traffic management measures (Mailer, 2002).

Finally, another academic example should illustrate the different options in numbers. It is based on a 2-lane road again with a capacity of 2400 vehicles per hour. The given transport demand is 1720 persons per hour and 160 light and 250 heavy vehicles transporting goods. With a given occupancy rate of 1.1, this results in a Design Hour Volume (DHV) of 1950 vehicles and a load factor above the target value that is defined at 0.85 (Table 2). Table 3 shows how additional public transport services attracting 210

Table 2. Example 2a: traffic resulting from given transport demand (persons and goods) misses quality target

Type of Vehicle	Persons	Occupancy Rate [pers per veh]	Number of Vehicles	Share of Heavy Vehicles	Load Factor [V/C]
Car	1690	1.10	1538		
Light goods vehicle	0	0	160		
Lorry (HGV)	0	0	250		
PT bus	30	15	2		
	1720	0.88	1950	13%	0.92 > 0.85

Table 3. Example 2b: target is fulfilled due to modal shift resulting from additional public transport services

Type of Vehicle	Persons	Occupancy Rate [pers per veh]	Number of Vehicles	Share of Heavy Vehicles	Load Factor [V/C]
Car	1480	1.10	1347		
Light goods vehicle	0	0	160		
Lorry (HGV)	0	0	250		
PT bus	240	30	8		
	1720	0.97	1765	15%	0.84 < 0.85

people to use the bus can result in reduced DHV and a load factor fulfilling the target. Table 4 shows how even better values can be achieved by a mix of measures (higher car occupancy resulting from ride share promotion, better goods management with fewer empty runs and better vehicle utilization, and an additional bus service).

These examples demonstrate again that the multi-modal method results in various alternatives to infrastructure improvement to fulfil the target value of traffic quality on the road without road extension. For road transport, this is to improve efficiency in individual and freight transport as well as improvements in public road transport (see Figure 4). Looking at the corridor measures in the railway system can also be taken into account (see Figure 6).

Table 4. Example 2c: target is clearly fulfilled due to modal shift resulting from a mix of measures (PT, car occupancy, goods management)

Type of Vehicle	Persons	Occupancy Rate [pers per veh]	Number of Vehicles	Share of Heavy Vehicles	Load Factor [v/c]
Car	1480	1.13	1310		
Light goods vehicle	0	0	140		
Lorry (HGV)	0	0	220		
PT bus	240	30	8		
	1720	1.03	1678	14%	0.79 < 0.85

Various measures are associated with these variants, all of them aiming to accommodate the same transport demand (persons and goods) with fewer vehicles. However, such measures require more detailed information on the conditions of transport demand. In addition to the time frame, the origin-destination relationships have to be taken into account. This illustrates the fact that it is necessary to leave a merely cross-section-specific determination of traffic demand. As a consequence, the different scenarios are to be examined in any case regarding their feasibility. Not only feasibility but also the effects of the measures as well as the cost-benefit ratio must be evaluated.

FUTURE RESEARCH DIRECTIONS

The multi-modal concept extends the traditional assessment of highway performance. It allows multi-modal capacity utilization and capacity management to meet road traffic quality targets. Different scenarios to accommodate transport demand adequately can be worked out by extending the range of suitable measures to respond to unsatisfactory road traffic quality. In addition to road improvements, management measures can be taken to control a given demand rather than just accommodate it. However, the method is still purely demand oriented. Transport demand is not questioned. Measures to usefully reduce transport demand by changing travel patterns are not included. To do so, the relationship between transport and land use had to be considered. So, the model needed to assess the measures had to be extended to an integrated transport and land-use model to be able to evaluate the effects of land-use measures on transport demand. At the planning level, the function of a road or corridor can be reconsidered as well. The method pinpoints the need for these further developments for an overall assessment of roads and other transport infrastructure (Mailer, 2004).

CONCLUSION

The extended approach provides a systematic framework to assess highway performance taking into account the potentials of all transport means and systems. It still has the quality of road traffic at its center, but makes it clear that this quality is the result of complex interdependencies. Accordingly, in addition to road infrastructure improvements, it shows alternative scenarios to meet road quality benchmarks. Behind these scenarios, there is a wide range of measures. They include infrastructure improvements as well as organizational and legal measures. So, the multi-modal perspective forms a basis for the integration of traffic and mobility management in the assessment and planning of traffic facilities.

Since the possible measures now have very different potentials and effects, it is absolutely necessary to perform a scenario comparison. On the one hand, this must prove that the measures are feasible and have the desired effects. On the other hand, the measures must be compared using cost-benefit analyses evaluating their cost effectiveness and economic efficiency. Thus, the new approach gives not only a clear structure to the planning process and increases its transparency, but also forms a suitable basis for economic decisions.

Of course, the process becomes more complex due to the extension considering multi-modality. The distribution of transport demand to the different transport modes and traffic assignment requires more information (origin-destination) and a suitable transport model. This has to take into account the parameters that are decisive for mode choice comprehensively (for example spatial and temporal availability, accessibility of origin and destination, frequency of service, rated travel times including access time, costs, etc.) to reflect the effects of various measures.

Overall, the method has higher demands on its users. A multimodal approach not only needs more effort, but also requires more expertise. However, using this approach, the educated engineer regains the planning competence that is increasingly lost by using "recipe-like" guidelines.

REFERENCES

Beckmann, J. et al.. (2006). *Multimodale Verkehrsmittelnutzer im Alltagsverkehr* (Vol. 4). Internationales Verkehrswesen.

Brilon, W., Grossmann, M., & Blanke, H. (1994). Verfahren zur Berechnung der Leistungsfähigkeit und Qualität des Verkehrsablaufes auf Straßen. *Forschung Straßenbau und Straßenverkehrstechnik, 669*, 1994.

Chlond, B.; Lipps, O. (2000): *Multimodalität im Personenverkehr im intrapersonellen Längsschnitt.* Arbeitsbericht des Instituts für Verkehrswesen, Universität Karlsruhe no. 00-7.

FGSV. (2015). *Handbuch für die Bemessung von Straßenverkehrsanlagen.* Köln: Forschungsgesellschaft für Straßen- und Verkehrswesen.

FGSV. (2002). *Handbuch für die Bemessung von Straßenverkehrsanlagen.* Köln: Forschungsgesellschaft für Straßen- und Verkehrswesen.

FSV (2012). *RVS 03.01.11: Beurteilung des Verkehrsablaufs auf Straßen.* Forschungsgesellschaft Straße-Schiene-Verkehr, 2012.

FSV. (1994). *RVS 3.7: Überprüfung der Anlageverhältnisse von Straßen.* Wien: Forschungsgemeinschaft für Verkehrs- und Straßenwesen.

Franke, S. (2004). *Die „neuen Multimodalen* (Vol. 3). Internationales Verkehrswesen.

Hupkes, G. (1982, February). The Law of Constant Travel Time and Trip Rates. *Futures.*

Knoflacher, H. (1994). Do We Use the "Level of Service" Concept in the Right Way?. *2nd International Symposium on Highway Capacity.*

Knoflacher, H., Mailer, M., Schopf, J.M., Fischer, P., Emberger, G., & Pfaffenbichler, P. (2008). Multimodale Dimensionierung von Straßen aus der Beschreibung des Verkehrsablaufes, der funktionellen Bedeutung sowie der verkehrlichen Auswirkungen. *Beiträge zu einer ökologisch und sozial verträglichen Verkehrsplanung, 1/2008.*

Mailer, M. (2002). *Zur Beurteilung von Verkehrsanlagen mit einem multimodalen Ansatz.* Dissertation an der TU Wien, Institut für Verkehrsplanung und Verkehrstechnik.

Mailer, M. (2004). Considering Multi-Modal Capacity in the Assessment of Road Design. In C. Focas (Ed.), *Transport Issues and Problems in Southeastern Europe. Ashgate Publishing Limited.*

Mailer, M., Munk, T., Spangler, M., & Haberer, T. (2008). Verkehrsqualität auf Hauptverkehrsstraßen – Empfindung, Messung und Bewertung. *Internationales Verkehrswesen,* (7 + 8), 270.

Schafer, A. (1998). The Global Demand for Motorized Mobility. *Transportation Research Board Part A: Policy and Practice, 32*(6).

Schafer, A. (2000, December). Regularities in Travel Demand: An International Perspective. *Journal of Transportation and Statistics*.

Schnabel, W., & Lohse, D. (1997). *Verkehrsplanung* (Vol. 2). Berlin: Grundlagen der Straßenverkehrstechnik und Verkehrsplanung.

TRB. (2001). *Highway Capacity Manual* (4th ed.). Transport Research Board.

TRB. (1985, 1994). Highway Capacity Manual, Third Edition 1985, Updated 1994. Transport Research Board.

TRB. (2010). *Highway Capacity Manual* (5th ed.). Transportation Research Board.

RVS 03.01.11. (2012). *Richtlinien und Vorschriften für den Straßenbau: Beurteilung des Verkehrsablaufs auf Straßen*. Österreichische Forschungsgesellschaft Straße – Schiene – Verkehr (FSV), Wien.

VCÖ. (2014). *Multimodale Lebensstile prägen Mobilität der Zukunft*. Factsheet.

Chapter 14
A Specific Issue on Sustainability of Transportation Planning in an Urban Region:
Ambulance Location Problem

Aykan Akıncılar
Gazi University, Turkey & Celal Bayar University, Turkey

Ece Akıncılar
Turkish Aerospace Industries Inc., Turkey

ABSTRACT

In this study, ambulance location problem, which is a significant research field in transportation in healthcare, is focused on. It is assumed that travelling time of vehicles between two demand points is uncertain and this uncertainty is represented as a crisp and as well as convex set. In order to deal with this uncertainty, robust counterpart approach, which is capable of providing conservatism, is employed. According to results of the simulation process, which is designed for the sake of benchmarking robust and nominal solutions, it can be easily claimed that conservative robustness is able to be achieved with significantly sufferable costs, even with no costs in most of the scenarios, by the proposed model. Ultimately, since robustness is one of the most important keywords for sustainability, on which discussions can be found in several parts of the text, this chapter is concluded with the claim that, by the proposed model, sustainability of transportation in healthcare is enhanced based on the adopted problem.

INTRODUCTION

In our day, there can be large-scale as well as small-scale disasters, terrorist attacks, and –in a more frequent manner- accidents in an urban region. Since those incidents may urgently require professional aiding, transportation in healthcare becomes a very important subfield of transportation planning. In point of a more general view, transportation in healthcare is essentially indebted its importance to having undeniable influence on lifesaving capabilities, standard of living, trustworthiness of the city, tourism incomes, etc.

DOI: 10.4018/978-1-5225-2116-7.ch014

It is clear that transportation in healthcare must be sustainable for a more livable place. Robustness is one of the most important key words for sustainability, which will be discussed later in this chapter. And so, it can be said that in order to achieve sustainability in transportation planning, the current plan must be robust which is definitely valid for transportation in healthcare, also. In the literature, robustness has been drawing great interest in many areas. Robustness can be briefly defined as being invulnerable to the anticipated disruptive affects. Thus, a robust plan should be invulnerable to be disrupted caused by certain affects in some user-defined, specific limits.

Many aspects related to transportation in healthcare can be encountered in the literature. This study is focused on facility location problem for emergency vehicles, more specifically ambulance location problem, which is one of the most popular problems in this field. While facility location problem is the problem where facilities that are required for the system should be established, ambulance location problem can be defined as finding places in a region for locating center(s) that ambulances stay and be ready. For sure, such a plan must be carefully designed in order to serve well since lifesaving is the point in question. Although location plan has been very carefully architected, even it has been optimized, there still can be disruptions due to uncertainties. This situation has a great potential of creating serious damages on sustainability. The main purpose of this chapter is to develop a model which provides a robust plan for ambulance location problem, and so, to contribute to sustainability of transportation planning in an urban region.

This chapter is organized as follow: related literature has been surveyed in the next section; then, possible relationship between robustness and sustainable transportation planning has been discussed; the proposed model has been defined after robust counterpart (RC) approach has been briefly introduced; then, the proposed model has been implemented on a real life problem with presenting the results; and finally, the conclusion part has been appeared.

RELATED WORKS

In the literature, lots of works which proposed model(s) providing a plan for ambulance location problem can be encountered. According to considering an environment under certainty or uncertainty, previous works can be classified into two main classes: *i.* works considering environment under certainty; and *ii.* works considering environment under uncertainty. In this section, the works that consider environment under uncertainty are tried to be given due to their capability of dealing with uncertainty, which is almost always inevitable.

Since 1970s, some works attempting to develop a robust model for facility location problem for emergency vehicles have appeared. And then, some researchers proceeded the previous studies in 1980s and 1990s with new ideas. And finally, with the recent developments after 2000, the literature in this field has been shaped as in our day. Some of the most cornerstone works are tried to be given below.

One of the earliest works in this field is Swoveland et al. (1973)'s work in which probabilistic branch-and-bound (B&B) procedure was developed for the ambulance location problem.

Aly and White (1978) assumed the locations of incidents were random variables. The probability distribution for rectilinear travel time between a new facility location and the random location of the incident (P_i) was developed for the case of P_i being uniformly distributed over a rectangular region. The impact of the proposed model was evaluated by outputs of the employed simulation model.

Then, Daskin (1983) proposed a new model with a new idea. The idea in this work was to maximize the expected covering locations, which was an innovative idea. Thus, the robustness of the obtained plan was enhanced. This problem, which was basically named maximum expected covering location problem (MECLP) would have being studied by many researchers later, even in 2000s.

Similarly, in Gendreau et al. (1997)'s work, the authors developed a model which covered each demand point at least k times and was solved via tabu search method after Hogan and ReVelle (1986) had proposed a model providing backup coverage. The model which was proposed in Gendreau et al. (1997)'s work was tested by real data and it was claimed that the results were satisfactory. Laporte et al. (2009)'s work may be suggested for more details about that model, which was named as double standard model.

Another significantly innovative idea was to develop a model considering the probability of being busy of service suppliers when emergency call came. Revelle and Hogan (1989)'s work was one of the most innovative works considering availability/unavailability of the emergency vehicles after Volz (1971)'s and Chapman and White (1974)'s works, which were the very first ones that consider availability of the emergency vehicles. Sectoral specific estimates of availability of the vehicles were incorporated to the developed two models and it was claimed that capability of the obtained plan in dealing with the case of that vehicles being busy was boosted.

In Aytug and Saydam (2002)'s work, the authors attempted to develop an efficient genetic algorithm (GA) and to write its code. MECLP was adopted, which was inspired by Daskin (1982, 1983). Proposed GAs, Daskin's heuristic and the theoretically optimal value were compared with each other. In the end, it was reported that although Daskin's heuristic produced better quality solutions, the second GA, which was named as GALS, produced near optimal solutions much faster than others.

Another remarkable approach was described in Araz et al. (2007)'s work. In that study, the models which had been proposed in Revelle and Hogan (1989)'s and Pirkul and Schilling (1991)'s works were hybridized as a multi-objective model and, then, the proposed model was reformulated according to different solution approaches such as lexicographic multi-objective linear programming (LMOLP) and several fuzzy goal programming (FGP) approaches in which the goals were represented in a possibilistic manner. As a result, it was suggested that one of the FGP models overcome the others in terms of more realistic and flexible optimal solution for facility location problem for emergency vehicles.

Iannoni et al. (2009) studied on locating ambulances along the highway and districting of the response segments. Two algorithms, which hybridized hypercube algorithm and GA, were employed to solve the proposed two problems to near optimal. While one of those algorithms was only deciding on the location of ambulance centers along the highway, the other was also determining the areas that the centers were responsible. Computational results were analyzed by applying the developed algorithms to two case studies and it was reported that adopted objective functions could be improved by the proposed algorithms.

Beraldi and Bruni (2009) proposed a two-stage framework, which was a stochastic programming framework being embedded probabilistic constraints into. It was claimed that replacing stochastic constraints with their probabilistic counterparts allowed the system manager to evaluate different solutions by varying reliable level. A B&B and three heuristic algorithms were developed and their performances were compared. A similar approach can be found in Beraldi et al. (2004)'s work.

Canbolat and von Massow (2011) investigated the facility location problem for emergency vehicles on both one-dimensional and two-dimensional spaces. In that study, demand points were assumed random and a model which minimizes the expected maximum rectilinear distance from the facility to the demand points was developed. It was reported that a solution was provided at a lower risk.

Berman et al. (2013)'s work should also be mentioned. With travel time uncertainty, three approaches, namely expected covering, robust covering, and expected p-robust covering were developed for maximum covering problem. Langrangian and greedy heuristics were employed to solve the developed models and, as a result, it was reported that p-robust covering model gave better solution for locating 4 fire stations in Toronto, Canada.

One of the most recently works is Nickel et al. (2016)'s work, in which a sampling approach for the ambulance location problem including the number of ambulances in a particular base was proposed in order to cover some region under stochastic demand with adopting the objective that minimizing the total cost associated with the facilities being installed.

Lastly, a very interesting study, which appeared in recent years, must be mentioned here. According to the best of the authors' knowledge, Zhang and Jiang (2014)'s work was the very first one that employs RC approach in emergency facility location problem. A bi-objective deterministic model under certainty was developed and its RC under uncertainty in terms of demand, which was a conic quadratic mixed-integer programming, was solved via branch-and-cut (B&C) approach. Uncertainties in demands were represented as two ellipsoidal sets in the RC and, consequently, the gaps between robust and nominal solutions were compared.

For a more comprehensive review, one may refer to the following surveys: Brotcorne et al. (2003)'s, Snyder (2006)'s, Li et al. (2011)'s, and Basar et al. (2012)'s works.

IS ROBUSTNESS REALLY NECESSARY?

Necessity of robustness for a sustainable transportation planning is worth to discuss in a general manner. Sustainability can be defined as *i.* maintaining at length without interruption or weakening; and *ii.* continuity according to *Advanced English Dictionary.* The phrases "without interruption" and "continuity" are the initial points of this section.

Indeed, the word "disruption" should be replaced with "interruption" when a plan or a system is in subject. Any effort for fixing may not be needed to carry out the subject, e.g. talking, walking, etc, in the word *interruption* while source of the interruption must be fixed in order to make the subject, e.g. implementing a plan, running a system, etc, progress in the word *disruption*. Thus, the word *disruption* defines the case better than the word *interruption* when the subject is a plan or a system.

According to the adopted initial points of this section, in order to achieve sustainability, one of the questions that is of top priority should be how a plan or a system can be made disruption-free. As its nature, a disruption occurs out of control and mostly unexpectedly. If everything is certain and unchanged, one may control every disruption. Of course it is a utopia, which is very far in today. In real, even for a static system, scientific community is still too weak to define the entire system even under certainty. These limitations lead the scientific community to dealing with uncertainty, which indeed means seeking robustness for the plan or system. Previously, definition of robustness was made as being invulnerable to the disruptive affects in the introduction of this chapter, which strengthens the abovementioned idea. In order to prevent the audience from belaboring, the authors find that this discussion is sufficient for this chapter.

ROBUST COUNTERPART (RC) APPROACH

In this section, RC approach is defined based on Ben-Tal et al. (2009)'s book, which is a very comprehensive and of high quality source for RC approach. Note that all the formulations in this section are in matrix form.

For the sake of clear understandability, generalization of linear optimization model under certainty, which is denoted as LO, is initially employed. Equation 1 demonstrates the formulation of that model.

$$LO: \min_{x \in \mathbb{X}} \left\{ c^T x \middle| Ax \preceq b \right\} \tag{1}$$

where $x \in \mathbb{X} \subseteq \mathbb{R}^n$ denotes decision variables while $\mathbb{R}^n$ is the space of n-dimensional real numbers; $c \in \mathbb{R}^n$ stands for objective function parameters; $A \in \mathbb{R}^{m*n}$ denotes parameters of constraints; and $b \in \mathbb{R}^m$ is right hand side value of constraints.

The given model in Equation 1 turns into the model which is denoted by $LO_{\mathfrak{U}}$, when environment under uncertainty is the matter. A generalized $LO_{\mathfrak{U}}$ is a collection of LOs as demonstrated in Equation 2.

$$LO_{\mathfrak{U}}: \min_{x \in \mathbb{X}} \left\{ c^T x \middle| Ax \preceq b \right\}_{(c,A,b) \in \mathfrak{U}} \tag{2}$$

where $\mathfrak{U}$ denotes uncertainty set, which is a well-defined and crisp set. In that point, the question that how those optimization models in the collection which are defined in Equation 2 can be meaningfully gathered should come to mind. This question is tried to be answered by Equations 3-5.

$x \in \mathbb{X}$ is called a robust solution for $LO_{\mathfrak{U}}$ if and only if it satisfies the conditions given below:

$$Ax \preceq b, \forall (c, A, b) \in \mathfrak{U} \tag{3}$$

Let $\hat{c}(x)$ be robust value of a feasible solution $x \in \mathbb{X}$. Robust value of $x \in \mathbb{X}$ is:

$$\hat{c}(x) = \sup_{(c,A,b) \in \mathfrak{U}} \left\{ c^T x \right\} \tag{4}$$

Thus, robust counterpart of $LO_{\mathfrak{U}}$, which is denoted by ξ, can be formulated as:

$$\xi: \min_{x} \left\{ \sup_{(c,A,b) \in \mathfrak{U}} \left\{ c^T x \right\} \middle| Ax \preceq b, x \in \mathbb{X}, \forall (c, A, b) \in \mathfrak{U} \right\} \tag{5}$$

Essentially, fundamentals of RC approach –which has also been called *robust optimization* by some researchers– is based on Soyster (1973)'s work, although roots of RC approach reach to Wald's maximin model (Wald, 1939, 1945, 1950). After several decades from Soyster (1973)'s work, there was almost no development related to RC approach. It can be said that main contributions has begun in this field with few works that were published in late 1990s (e.g. Ben-Tal & Nemirovski, 1998, 1999, 2000; El Ghaoui & Lebret, 1997; El Ghaoui et al., 1998; and Vandenberghe & Boyd, 1996). Since 2000s, RC approach has become a very young and active research field. For a more comprehensive review and detailed discussions related to RC approach, Gorissen et al. (2015)'s, Gabrel et al. (2014)'s, and Ben-Tal and Nemirovski (2008)'s works can be eagerly suggested.

The most significant advantages of RC approach may be that *i.* providing a deterministic model, which is generally much more tractable than stochastic and/or probabilistic ones; *ii.* providing conservatism, which entirely ensures feasibility in the case of every possible scenario in an uncertain environment. Since the problem in question has a direct and intense influence on lifesaving of human-beings, RC approach, which is described in this section, is employed in this study.

PROBLEM DEFINITION AND THE PROPOSED MODEL

In this study, an emergency vehicle location problem under uncertainty is considered. The uncertainty pertains to travelling time of the emergency vehicles, namely ground ambulances. It is assumed that travelling time of the ambulances between two demand points might fluctuate in certain limits and it is possible for travelling time to get any value within those limits, but not any value out of. In addition, only ambulance location problem without relocating or allocating the vehicles is taken into account. For the case study, the capital city of Turkey, Ankara, is chosen. As a result, a robust plan for ambulance location problem is claimed. After setting the framework of the considered problem as above, the proposed model is described in the rest of this section.

As ordinary, like all models, the developed model in this study has few assumptions. It may be more suitable to begin with those assumptions. Below, one can find the assumptions have been made as listed:

- Only ground ambulances are considered.
- A specific region of the city is considered, which is Cankaya (the center of Ankara), in the case study. This region is partitioned into 23 districts. So, each demand point, and at the same time each alternative, is a district of this specific region.
- In each district, only one center for emergency vehicles is allowed.
- The number of ambulances is sufficient at each possible center, and, so, there is no need for ambulance allocation during the day/week, etc.
- Travelling times are calculated only based on the shortest path between two districts and this path is the only one alternative.
- The calculated shortest path is symmetric for the opposite direction.
- There can be at most $\pm 10\%$ changes on the travel time of each ambulance between particular districts.
- There is no extraordinary situation, event, etc on any road.

For the sake of understandability, a model under certainty, which is denoted by δ, is formulated below:

$$\delta : \min_{x_i \in \mathcal{X}} \sum_{i=1}^{n} x_i \tag{6a}$$

s.t.

$$\sum_{i=1}^{n} a_{ij} x_i \geq 1, \forall j \tag{6b}$$

$$x_i \in \{0,1\} \tag{6c}$$

where i and j denote districts in the specific region; n is the total number of the districts in the specific region; $\mathcal{X}$ is the set of all districts; x_i is 1 if a center is located in district i, 0 if otherwise; a_{ij} is 1 if the travelling time between i and j –which is calculated according to the shortest path between i and j – is less than prespecified threshold being denoted by $\mathfrak{r}$, 0 if otherwise.

For dealing with the uncertainty in travelling time of ambulances, RC approach, which is described in Section 4, is employed. Before RC formulation of δ, some definitions are required. In Equation 7-8, the function, $g(\zeta, t)$; and uncertain travelling time between districts i and j, which is indicated as t_{ij}' are defined, respectively.

$$g(\zeta_{ij}, t_{ij}) \begin{cases} 1, & if\ (1+\zeta_{ij}) t_{ij} \leq \mathfrak{r} \\ 0, & ow \end{cases} \tag{7}$$

$$t_{ij}' = (1+\zeta_{ij}) t_{ij} \tag{8}$$

Then, RC formulation of δ is:

$$\xi : \min_{x_i \in \mathcal{X}} \sum_{i=1}^{n} x_i \tag{9a}$$

s.t.

$$\sum_{i=1}^{n} g(\zeta_{ij}, t_{ij}) x_i \geq 1, \forall \zeta_{ij} \in \mathfrak{U}, \forall j \tag{9b}$$

$$x_i \in \{0,1\} \tag{9c}$$

$$\mathfrak{U} = \{\zeta_{ij} : -0.1 \leq \zeta_{ij} \leq 0.1, \forall i, j\} \tag{9d}$$

where t_{ij} is the nominal travelling time between districts i and j; ζ_{ij} is the uncertainty operator for the travelling time between districts i and j; $\mathfrak{U}$ is the uncertainty set in which each ζ_{ij} is defined; $g(\zeta, t)$ is a function of ζ and t, which is 1 if the travelling time between districts i and j under uncertainty is less than τ, 0 if otherwise.

Some attention may be paid on the uncertainty set, which is defined in Equation 9d. Although some other shapes such as ellipsoid, ball, etc are available for the uncertainty set, a box-shaped $\mathfrak{U}$ is adopted for ξ since a box shaped uncertainty set is the most conservative one. Since the problem at hand has a direct influence on lifesaving, conservatism should be at high degree as possible.

In that form, ξ is intractable since ξ contains infinite number of constraints which are caused by ζ. In order to solve ξ, its tractable form must be obtained. Since the worst case formulation of ξ is possibly tractable, it should be attempted to be obtained. The worst case formulation of Equation 9b in ξ can be obtained as below:

$$\min_{\zeta_{ij} \in \mathfrak{U}} \left\{ \sum_{i=1}^{n} g(\zeta_{ij}, t_{ij}) x_i \right\} \geq 1, \forall j \tag{10}$$

$$\sum_{i=1}^{n} \min_{\zeta_{ij} \in \mathfrak{U}} \{ g(\zeta_{ij}, t_{ij}) \} x_i \geq 1, \forall j \tag{11}$$

$$\sum_{i=1}^{n} g\left(\max_{\zeta_{ij} \in \mathfrak{U}} \{ t'_{ij} \} \right) x_i \geq 1, \forall j \tag{12}$$

Equation 12 represents the worst case formulation of Equation 9b. Thus, the worst case formulation of ξ, which is denoted by ξ', can be formulated as:

$$\xi' : \min_{x_i \in \mathcal{X}} \sum_{i=1}^{n} x_i \tag{13a}$$

s.t.

$$\sum_{i=1}^{n} g\left(\max_{\zeta_{ij} \in \mathfrak{U}} \{ t'_{ij} \} \right) x_i \geq 1, \forall j \tag{13b}$$

$$x_i \in \{0,1\} \tag{13c}$$

$$\mathfrak{U} = \{\zeta_{ij} : -0.1 \le \zeta_{ij} \le 0.1, \forall i, j\} \tag{13d}$$

Since one can claim that ξ' is equivalent to ξ, and, is tractable, ξ' is adopted as the main model. In the next section, computational results of the case study are demonstrated.

COMPUTATIONAL RESULTS

For the case study, central region of Ankara, which is called Cankaya, has been chosen. Figure 1 demonstrates the location of Cankaya on a map of Ankara. Cankaya can be segmented into 23 major districts, which are demonstrated in Table 1.

Based on the assumptions given in the previous section, the proposed model, namely ξ', is coded in MATLAB environment for several scenarios. Those scenarios are constructed according to different values of average velocity of emergency vehicles –which are ground ambulances in this chapter– (AVA) and different values of arrival threshold (τ) which is defined as the maximum time that the ambulance

Figure 1. Location of Cankaya in Ankara

Table 1. Main districts in Cankaya

# of District	Name of District	# of District	Name of District	# of District	Name of District
1	Ahlatlibel	9	Dikmen	17	Ovecler
2	Ayranci	10	Esat	18	Sihhiye
3	Bahcelievler	11	Gaziosmanpasa	19	Sogutozu
4	Balgat	12	Hurriyet	20	Tbmm
5	Bilkent	13	Kavaklidere	21	Yildiz
6	Birlik	14	Kirkkonaklar	22	Yuzuncuyil
7	Cebeci	15	Kizilay	23	Zafertepe
8	Cayyolu	16	Maltepe		

must have to reach the event scene. In Table 2, alternative values for both average velocity and arrival threshold of ambulances are given.

The nominal model, δ, and the proposed model, ξ', are run for each scenario. In Table 3, the optimal solutions for each scenario are presented, in which *OV* stands for optimal value while *OC* stands for optimal centers; $\xi'_{OV} - \delta_{OV}$ denotes difference between optimal values of δ and ξ' while $\xi'_{OC} \setminus \delta_{OC}$ denotes whether there is difference between optimal solutions of δ and ξ'.

As it can be seen that, mostly, there is no much difference between nominal and robust solutions. Although opening a center may be highly expensive, at most 1 more center is required to provide robustness for some scenarios, even no additional center is required for the others. The promised robustness is generally worth to its cost since, by this way, it can be assured that ambulances are able to reach to the event scene in the given threshold even the associated travelling time fluctuates $\pm\%10$. In other words, by ξ', whatever fluctuation of the travelling time within the limits of $\pm\%10$ is, the ambulances are able to reach to the event scene in the given threshold, which is extremely important in lifesaving.

CONCLUSION

In this study, a model which is capable of dealing with the uncertainty in travelling time of ambulances between two demand points is proposed. More explicitly, by the proposed model, one can assure the ambulances to reach the event scene in a given threshold in uncertain environment. As a distinct from the previous works, according to the best of the authors' knowledge, the proposed model is the very first one that is capable of *entirely* tolerating the whole prespecified uncertainty in travelling time of

Table 2. Alternative values for average velocity and arrival threshold of ambulances

Average Velocity of Ambulances (AVA)		Arrival Threshold (τ)	
Alternative 1	70 km/h	Alternative 1	6 mins
Alternative 2	75 km/h	Alternative 2	8 mins
Alternative 3	80 km/h	Alternative 3	10 mins

Table 3. Optimal solutions for both δ and ξ' and some comparisons

Scenarios	AVA	τ	δ_{OV}	δ_{OC}	ξ'_{OV}	ξ'_{OC}	$\xi'_{OV} - \delta_{OV}$	$\xi'_{OC} \setminus \delta_{OC}$
1	70km/h	6	5	4,5,8,12,23		4,5,8,12,23	0	Indifferent
2	70km/h	8	4	8,12,21,23	4	8,12,19,23	0	Different
3	70km/h	10	2	8,22	3	1,4,23	1	Different
4	75km/h	6	5	5,8,12,19,23	5	4,5,8,12,23	0	Different
5	75km/h	8	3	8,12,21	4	8,12,21,23	1	Different
6	75km/h	10	2	8,22	2	8,22	0	Indifferent
7	80km/h	6	4	8,12,19,23	5	4,5,8,12,23	1	Different
8	80km/h	8	3	1,4,23	4	8,12,21,23	1	Different
9	80km/h	10	2	5,23	2	8,22	0	Different

ambulances, which was previously mentioned as conservatism. In other words, the proposed model gives an optimal solution which *always* remains feasible whatever travelling times vary within prespecified limits. Since worth of lifesaving is indisputable, such a claim is vital and extremely precious which is explicitly achieved in this study. Thereby, proposing a model achieving this claim can be defined as the main contribution of this study to the literature.

One of the other most significant advantage of the proposed model, besides successively dealing with the uncertainty in travelling times by providing conservatism, may be that being deterministic, which means potentially more tractable model than its competitors. In order to achieve such advantages, RC approach is employed for ambulance location problem. By utilizing RC approach, one not only can obtain a solution which is capable of providing conservative robustness, but also is able to solve the problem easily due to its deterministic nature.

When results of the case study is analyzed, it can be seen that the robust solution does not cause dramatic additional costs when is compared to nominal solutions although it provides a conservative robustness. While in some scenarios at most one additional center is required, no additional center is required for the most. This result can be interpreted as –conservative– robustness is assured without extreme costs for all scenarios which are considered in this study, even without any additional costs for most of them.

Necessity of robustness for sustainability was discussed in Section 3, in which robustness was decided to be necessary. In the light of this brief discussion, since –even conservative– robustness is provided for ambulance location problem by the proposed model, ultimately, it can be properly claimed that the proposed model significantly contributes to the sustainability of transportation in healthcare with sufferable costs.

In addition to those benefits of the proposed model, of course there are several limitations, important assumptions, etc. First of all, it is assumed that each center has sufficient number of ambulances, which is hard to be sure in real life. Without this assumption, number of ambulances in each center must be decided. In that point, demand for ambulances becomes an important component and it may vary during the day and week, which may require to change the plans even completely. In addition, availability/business of the vehicles are not considered although this issue has been discussing frequently in the literature. One of the other important assumptions is that considering only the shortest path between two demand points. Alternative paths may contribute to the punctuality when traffic jam or another problem has occurred on the shortest path. Lastly, uncertainty in travelling time of ambulances is assumed to be always fallen into a set which has prespecified limits. Exceeding such limits is likely in real. So, those issues indicated above are strong candidates to be considered in future works.

REFERENCES

Advanced English Dictionary. (2015). Available from https://www.microsoft.com/en-us/store/apps/advanced-english-dictionary/9wzdncrfhmvb

Aly, A. A., & White, J. A. (1978). Probabilistic formulation of the emergency service location problem. *The Journal of the Operational Research Society*, *29*(12), 1167–1179. PMID:10239405

Araz, C., Selim, H., & Ozkarahan, I. (2007). A fuzzy multi-objective covering-based vehicle location model for emergency services. *Computers & Operations Research*, *34*(3), 705–726. doi:10.1016/j.cor.2005.03.021

Aytug, H., & Saydam, C. (2002). Solving large-scale maximum expected covering location problems by genetic algorithms: A Comparative study. *European Journal of Operational Research*, *141*(3), 480–494. doi:10.1016/S0377-2217(01)00260-0

Basar, A., Catay, B., & Unluyurt, T. (2012). A taxonomy for emergency service station location problem. *Optimization Letters*, *6*(6), 1147–1160. doi:10.1007/s11590-011-0376-1

Ben-Tal, A., El Ghaoui, L., & Nemirovski, A. (2009). *Robust optimization*. Princeton, NJ: Princeton University Press. doi:10.1515/9781400831050

Ben-Tal, A., & Nemirovski, A. (1998). Robust convex optimization. *Mathematics of Operations Research*, *23*(4), 769–805. doi:10.1287/moor.23.4.769

Ben-Tal, A., & Nemirovski, A. (1999). Robust solutions of uncertain linear programs. *Operations Research Letters*, *25*(1), 1–13. doi:10.1016/S0167-6377(99)00016-4

Ben-Tal, A., & Nemirovski, A. (2000). Robust solutions of linear programming problems contaminated with uncertain data. *Mathematical Programming*, *88*(3), 411–421. doi:10.1007/PL00011380

Ben-Tal, A., & Nemirovski, A. (2000). Selected topics in robust convex optimization. *Mathematical Programming Series B*, *112*(1), 125–158. doi:10.1007/s10107-006-0092-2

Beraldi, P., & Bruni, M. E. (2009). A probabilistic model applied to emergency service vehicle location. *European Journal of Operational Research*, *196*(1), 323–331. doi:10.1016/j.ejor.2008.02.027

Beraldi, P., Bruni, M. E., & Conforti, D. (2004). Designing robust emergency medical service via stochastic programming. *European Journal of Operational Research*, *158*(1), 183–193. doi:10.1016/S0377-2217(03)00351-5

Berman, O., Hajizadeh, I., & Krass, D. (2013). The maximum covering problem with travel time uncertainty. *IIE Transactions*, *45*(1), 81–96. doi:10.1080/0740817X.2012.689121

Brotcorne, L., Laporte, G., & Semet, F. (2003). Ambulance location and relocation models. *European Journal of Operational Research*, *147*(3), 451–463. doi:10.1016/S0377-2217(02)00364-8

Canbolat, M. S., & von Massow, M. (2011). Locating emergency facilities with random demand for risk minimization. *Expert Systems with Applications*, *38*(8), 10099–10106. doi:10.1016/j.eswa.2011.02.001

Chapman, S., & White, J. (1974). *Probabilistic formulations of emergency service facilities location problems*. Paper presented at ORSA/TIMS, San Juan, Puerto Rico.

Daskin, M. S. (1982). Application of an ecpected covering location model to emergency medical system design. *Decision Sciences*, *13*, 416–139. doi:10.1111/j.1540-5915.1982.tb00159.x

Daskin, M. S. (1983). A maximal expected covering location model: Formulation properties and heuristic solution. *Transportation Science*, *17*(1), 48–69. doi:10.1287/trsc.17.1.48

El Ghaoui, L., & Lebret, H. (1997). Robust solutions to least-squares problems with uncertain data. *SIAM Journal on Matrix Analysis and Applications*, *18*(4), 1035–1064. doi:10.1137/S0895479896298130

El Ghaoui, L., Oustry, F., & Lebret, H. (1998). Robust solutions to uncertain semidefinite programs. *SIAM Journal on Optimization*, *9*(1), 33–52. doi:10.1137/S1052623496305717

Gabrel, V., Murat, C., & Thiele, A. (2014). Recent advances in robust optimization: An overview. *European Journal of Operational Research*, *235*(3), 471–483. doi:10.1016/j.ejor.2013.09.036

Gendreau, M., Laporte, G., & Semet, F. (1997). Solving an ambulance location model by tabu search. *Location Science*, *5*(2), 75–88. doi:10.1016/S0966-8349(97)00015-6

Gorissen, B. L., Yanıkoglu, I., & den Hertog, D. (2015). A practical guide to robust optimization. *Omega*, *53*, 124–137. doi:10.1016/j.omega.2014.12.006

Hogan, K., & ReVelle, C. (1986). Concepts and applications of backup coverage. *Management Science*, *32*(11), 1434–1444. doi:10.1287/mnsc.32.11.1434

Iannoni, A. P., Morabito, R., & Saydam, C. (2009). An optimization approach for ambulance location and the districting of the response segments on highways. *European Journal of Operational Research*, *195*(2), 528–542. doi:10.1016/j.ejor.2008.02.003

Laporte, G., Louveaux, F. V., Semet, F., & Thirion, A. (2009). Applications of the double standard model for ambulance location. In Innovations in Distribution (pp. 235-249). Springer. doi:10.1007/978-3-540-92944-4_12

Li, X., Zhao, Z., Zhu, X., & Wyatt, T. (2011). Covering models and optimization techniques for emergency response facility location and planning: A review. *Mathematical Methods of Operations Research*, *74*(3), 281–310. doi:10.1007/s00186-011-0363-4

MobiSystems, Inc. (2016). *Advanced English Dictionary* (Version 8.7.1) [Software]. Available from https://itunes.apple.com/tr/app/advanced-english-dictionary/id291070079?mt=8

Nickel, S., Reuter-Oppermann, M., & Saldanha-da-Gama, F. (2016). Ambulance location under stochastic demand: a sampling approach. *Operations Research for Health care, 8,* 24-32. .10.1016/j.orhc.2015.06.006

Pirkul, H., & Schilling, D. A. (1991). The maximal covering location problem with capacities on total workload. *Management Science, 37*(2), 233–248. doi:10.1287/mnsc.37.2.233

ReVelle, C., & Hogan, K. (1989). The maximum availability location problem. *Transportation Science, 23*(3), 192–200. doi:10.1287/trsc.23.3.192

Snyder, L. V. (2006). Facility location under uncertainty: A review. *IIE Transactions, 38*(7), 537–554. doi:10.1080/07408170500216480

Swoveland, C., Uyeno, D., Vertinsky, I., & Vickson, R. (1973). Ambulance location: A probabilistic enumeration approach. *Management Science, 20*(44-II), 686–698. doi:10.1287/mnsc.20.4.686

Vandenberghe, L., & Boyd, S. (1996). Semidefinite programming. *SIAM Review, 38*(1), 49–95. doi:10.1137/1038003

Wald, A. (1939). Contributions to the theory of statistical estimation and testing hypotheses. *The Annals of Mathematics, 10*(4), 299–326. doi:10.1214/aoms/1177732144

Wald, A. (1945). Statistical decision functions which minimize the maximum risk. *The Annals of Mathematics, 46*(2), 265–280. doi:10.2307/1969022

Wald, A. (1950). *Statistical Decision Functions*. John Wiley.

Zhang, Z.-H., & Jiang, H. (2014). A robust counterpart approach to the bi-objective emergency medical service design problem. *Applied Mathematical Modelling, 38*(3), 1033–1040. doi:10.1016/j.apm.2013.07.028

Chapter 15
Logistics for City and Regional Planning:
Urban and Regional Planning Without Taking into Account the Effects of Transport Logistics

Hermann Knoflacher
Vienna University of Technology, Austria

ABSTRACT

Logistics in transport considers the physical structure of a city as given and tries to optimize the performance of companies or the transportation system within the given conditions. On the other side, companies choose their location under the given conditions and the expected changes and influence the structure of the city and its economy in a continuous way. City and transport planning methods have not considered these effects in their work so far and are therefore influenced by the driving forces of the economy of scale and demand oriented traffic growth. The introduction of principles of logistics into the early stages of land use and city planning would change the "given conditions" and open the path for a more sustainable development, with more pressure for innovation and fairness in the market.

INTRODUCTION

Different disciplines are planning, building, maintaining and operating the urban, economic and transport system. Each has its special view to the system and use assumptions how it should behave. Land use planners allocate pieces of land special functions, urban planners design buildings and spaces, transport planners try to optimize the infrastructure for transport and traffic with the traffic modes of the time. Societies had the benefits from the transport infrastructure for several thousands of years by controlling the flow of goods and people. The effects of the transport system were limited till the railway came into use. To operate the railway system the transport logistics had to be adapted to the new artificial transport mode on one side for internal operation and also for the user, the customers, the industry the cities. Land

DOI: 10.4018/978-1-5225-2116-7.ch015

use, economic and urban structures could be planned in agreement with the railway system in the 19th and the first half of the 20th century. Railways are a "closed system" under the control of the society and physically with controlled accessibility.

The modern road and motorway system planned and build for car and truck traffic in the second half of the 20th century is the infrastructure for an open system with no access control. This gave new opportunities for the development of new strategies and methods for transport logistics in our competitive "market economy" as never before. National borders were removed to give goods and people free flow across countries and regions. Beside this desired effects unexpected and unwanted effects occurred. Urban, settlement and economic structures have changed not only the form, but also the functions in the last century much more than in thousands of years of urban history before. Most of the changes were not expected, not planned and not wanted, like damage of local urban shops, workshops etc. Jane Jacobs war the first activist, which published these effects in her book "The Death and Life of Great American Cities" 1961. More than fifty years later the underlying causes of the effects have not been treated in traditional sciences. One of the reasons is the cross-disciplinary matter of the problem described. This chapter is a contribution to one branch of the causes.

LOGISTICS AS A TOOL AND ITS EFFECT ON STRUCTURES

Land use planning, city planning and transport planning are branches of applied logistics if we use the traditional definition for logistics as "the process of planning, implementing, and controlling the efficient, effective flow and storage of goods, services, and related information from point of origin to point of consumption for the purpose of conforming to customer requirements." Land use, urban and transport planners are doing exactly the same – at least in their mental or computer models. But finally they build rigid structures, which seem to be rather stabile – at least for some time.

Traditionally transport logistics does not take into account its interrelationship with these disciplines and vice versa. Within given patterns of activities in a static spatial distribution, transport logistics tries to optimise transport activities.

The basic assumption is: settlement structures are stable and static. Under short time system-view logistics has to:

- Optimize either the transport structure or
- Optimize locations for distribution, trade or production activities under given conditions of transport infrastructure.

Conditions of the transport system are dependent on settlement and urban structures.

In general transport logistics is seen as "software" in a "hardware" environment.

The effect of logistics is expressed with indicators like:

- Saving of travel time,
- Saving of distances,
- Saving of costs,
- Saving of resources,

- Efficiency, and
- Competitive advantage.

These indicators are also used in transport and settlement planning. It is therefore obvious that there are some correlations with transport logistics. Basically transport logistics is a short term strategy. City planning is dealing with long term effects and transport planning has influence on short term changes on the operational level and long term effects on the strategic and project level. If measures of these different disciplines are interconnected, feed-back effects can occur which can produce positive or negative effects. The common assumption or hypothesis is that logistics does not change urban or economic structures. It is a tool for a better service for customers, the industry and the society. But we know from Goethe´s Ballad "Der Zauberlehrling" (the sorcerer's apprentice) an artificial servant can produce unexpected and unwanted effects if he is not controlled by a wizard.

It will be interesting to have first an empirical look into the transport system from the macroscopic point of view if the expectations come true.

MACROSCOPIC ANALYSIS OF TRANSPORT AND ECONOMIC DEVELOPMENT

If logistics is effective in organising transport, this must be recognised in the data. The national industrial and economic system is based on the transport system, which can contribute to the changes of GDP in one or the other way. If our logistics works well, the physical effort for good transport should decline over the years. A simple method to evaluate this effect is the calculation of an indicator for the system-efficiency. The average travel distance for one ton of goods is a usable indicator.

A system becomes a better performance, if its entities are closer connected. This reduces the cost of transport and decrease dependency from oil imports – and has benefits for the environment and local working opportunities, if the system goes into this direction. If we make this analysis for Germany for the time period 1975 – 2015 we can see the results in Figure 1.

Figure 1. Average travel distance of 1 ton of goods on roads in Germany between 1975 and 2015

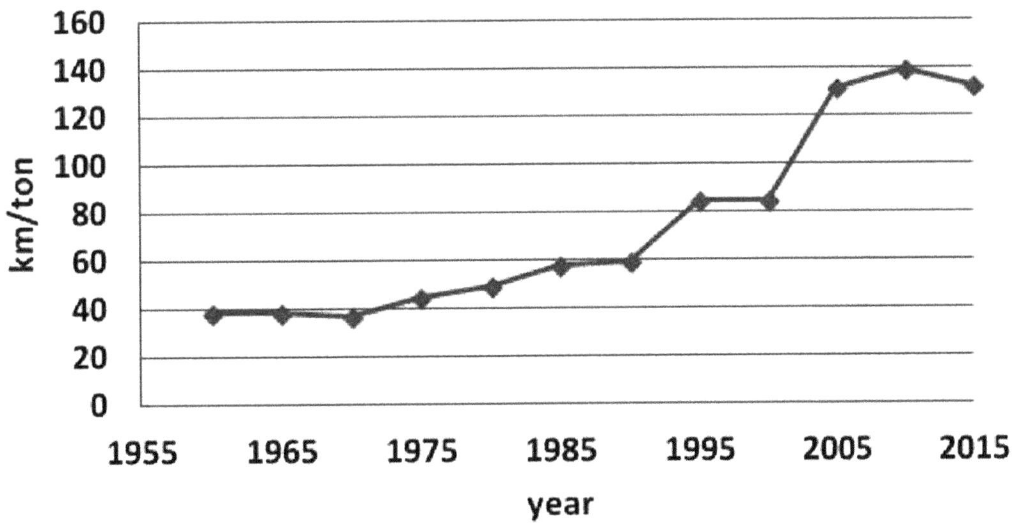

Between 1997 and 2014 the goods transported on German roads have increased about 10%, while the kilometres of road transport have increased by about 54%. (Verkehr in Zahlen, 2016) The efficiency of road transport has declined by about 40% compared to the situation in 1997. If logistics would have efficiency-supporting effects, the line of dots must go downwards. But this is not the case. The distances have increased continuously in the last 60 years.

On the macroscopic level there is no positive effect of any logistic measure visible. If we take the year 1960 as frame of reference with 100%, the German system efficiency in the good transport sector has decreased. Today the transport efficiency of good transport in Germany is less than one third of the efficiency in the 1960ies.

Private car traffic has nothing like logistics; it is more a kind of adaptation. In Germany the kilometre per person travelled in a car increased from 1970 to 2000 1.8-times. Transport distance of one ton of goods increased 2.3 times. Figure 2 shows that this adaptation seems to work better than logistics on the macro scale. Before we analyse this phenomenon, we will start with a simple theoretical approach.

On the other hand Germany is since many years leading the "Logistic Performance Index" globally specified by the World Bank LPI (2016). This shows the difference in the approach to the transport system. For the World Bank transport logistics is a tool for economic growth, without taking into account effects on urban and local economic structures, excluding effects on the social and ecological environment, while the approach in this chapter is dealing with these effects.

The World Bank follows the lines of traditional economic view on the transport system, not questioning the underlying fundamental theories of economy. This chapter analyses these theories under the aspects of space, competitiveness on different scale and structure-effects on human settlements and economy, first on the macroscopic level and further on the local level too.

Figure 2. Relative efficiency of road transport in %, base year 1960 (Germany)

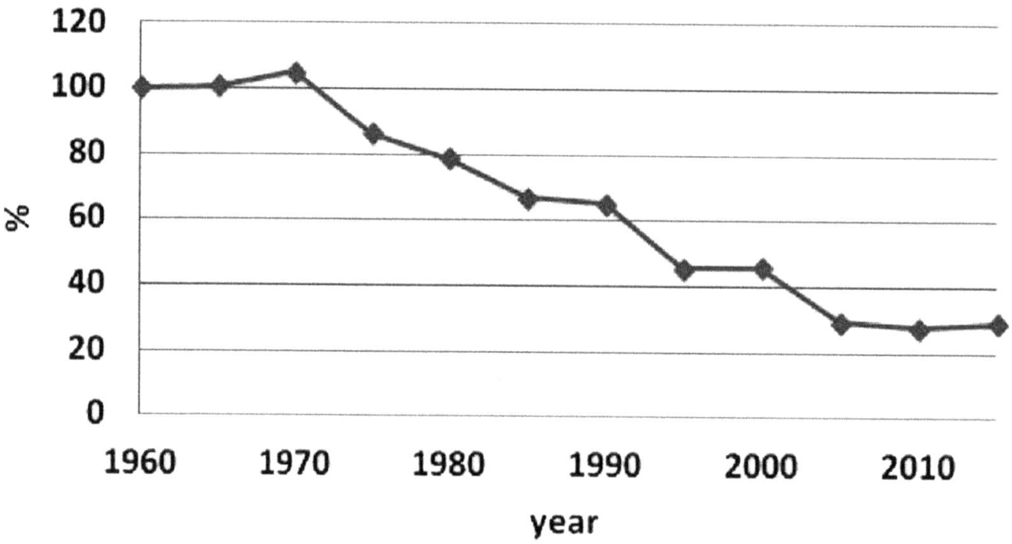

THEORETICAL APPROACH

Lill's Law

Transportation planners are familiar with Lill's law from 1889. By analysing trip frequency h and trip length l, Lill discovered the following relationship:

$$h * l = const \tag{1}$$

In (1) the frequency of trips multiplied by the length of the trips could be interpreted as the costs of transport or the budget for transport BT. Equation (1) can be reformulated to the following equation (2):

$$l_i = \frac{B_T}{h_i} \tag{2}$$

The instruments used in logistics are influencing the frequency of trips in many ways. Interesting is the effect of logistics on the trip length. Differentiating formula (2) by time we get formula (3).

$$\partial l_i = -\frac{\partial h_i}{h_i^2} * B_T + \frac{\partial B_T}{h_i} \tag{3}$$

Under the assumption that logistics have no effect on the transport system and the city structure $\partial l / \partial t$ must be zero. This brings us to formula (4):

$$\frac{\partial h_i}{h_i} = \frac{\partial B_T}{B_T} \tag{4}$$

Equation (4) shows that a stable transport and urban structure is only possible if cost reductions from logistics are accompanied by an equivalent reduction of number of trips. If less trips are necessary, less trucks are needed. But this is not the case. The number of trucks increased from 681.000 in 1960 to 2,7 million in 2015. There are more trips and more kilometres in the system instead of less. Logistics increase the length of trips by decreasing the costs of frequent trips. It trip length increases, origins and destinations must have changed. Does this also change the economic structures?

Gravity Model

The gravity model calculates the number of trips Fij between two cells i and j as follows:

$$F_{ij} = \frac{P_i * P_j}{W_{ij}} \tag{5}$$

P_i: Potential of cell i,
P_j: Potential of cell j,
W_{ij}: Impedance function between i and j.

The traditional interpretation of the gravity model is made under the assumption, that the potentials in zone i and j are independent from the number of trips Fij between i and j and the impedance Wij. In contradiction to this, practice Pi and Pj are depending on Fij and Wij. This could be easily seen in the fact that if Pi and Pj are not accessible they cannot produce or attract trips. Therefore the gravity model has to be written as follows:

$$F_{ij} = \frac{P_i\left(F_{ij}\right) * P_j\left(F_{ij}\right)}{W_{ij}} \tag{6}$$

Logistic instruments are influencing Wij. The duty of logistics is to reduce Wij. A differentiation of equation (6) gives equation (7):

$$\partial F_{ij} = -\frac{\partial W_{ij}}{W_{ij}^2} * P_i\left(F_{ij}\right) * P_j\left(F_{ij}\right) + \frac{1}{W_{ij}} * \left[\partial P_i\left(F_{ij}\right) * P_j\left(F_{ij}\right) + P_i\left(F_{ij}\right) * \partial P_j\left(F_{ij}\right)\right] \tag{7}$$

For simplification the potential Pj is kept constant:

$$\partial P_j\left(F_{ij}\right) = 0 \tag{8}$$

This gives the following equation:

$$\partial F_{ij} = -\frac{\partial W_{ij}}{W_{ij}^2} * P_i\left(F_{ij}\right) * P_j\left(F_{ij}\right) + \frac{\partial P_i\left(F_{ij}\right) * P_j\left(F_{ij}\right)}{W_{ij}} \tag{9}$$

Under the hypothesis that logistics have no effect on the city structure ∂F_{ij} has to be 0. This gives the equation (10):

$$\frac{\partial P_i\left(F_{ij}\right)}{P_i\left(F_{ij}\right)} = \frac{\partial W_{ij}}{W_{ij}} \tag{10}$$

Each change of W_{ij} changes the urban and/or economic structures. Again we get the result that if logistics would have no effect on the outside structures of the city and the transport system, all improvements in logistics, which reduce the costs of the transport system are compensated by a reduction of production which is obvious not the case. Empirical analysis results indicate that logistics is a driving force for structural changes in economic activities by increasing distances. The benefits of logistics are therefore compensated or even overcompensated by spatial and structural elasticities. From the theoretical analysis it is now clear, why no positive effects of logistics could be observed on the macro-level.

Transport is an important and crucial factor for the economy. A good transport infrastructure and low transport cost are therefore a continuous requirement from economists and representatives of industry and trade. The background of their expectations is one of the core pillars of economy: The Theory of Economy of Scale.

Economy of Scale and Spatial Structures

In microeconomics, economies of scale are the cost advantages that enterprises obtain due to size, output, or scale of operation, with cost per unit of output generally decreasing with increasing scale as fixed costs are spread out over more units of output. Often operational efficiency is also greater with increasing scale, leading to lower variable cost as well.

Economies of scale apply to a variety of organizational and business situations and at various levels, such as a business or manufacturing unit, plant or an entire enterprise. For example, a large manufacturing facility would be expected to have a lower cost per unit of output than a smaller facility, all other factors being equal, while a company with many facilities should have a cost advantage over a competitor with fewer. Some economies of scale, such as capital cost of manufacturing facilities and friction loss of transportation and industrial equipment, have a physical or engineering basis. The economic concept dates back to Adam Smith and the idea of obtaining larger production returns through the use of division of labor.

Economies of scale often have limits, such as passing the optimum design point where costs per additional unit begin to increase. Common limits include exceeding the nearby raw material supply, such as wood in the lumber, pulp and paper industry. A common limit for low cost per unit weight commodities is saturating the regional market, thus having to ship product uneconomical distances. Other limits include using energy less efficiently or having a higher defect rate (Economies of Scale, 2016).

THE TRANSPORT SYSTEM IN THE THEORY OF ECONOMY OF SCALE

Introduction

To optimize transport between supply and demand, sophisticated systems of logistics have to be applied. Traditionally logistics did not question structures and infrastructures and so far they are a part of the "hardware", the so-called built infrastructure, houses, factories, roads, railways etc. The result of the traditional approach in logistics can therefore only be sub-optimal, if the underlying structure is not developed in such a way that principles of optimal logistic strategies can be applied easily (Ihde, 1991).

The organization of the functions of cities, in space and time, has therefore been planned in such a way that, together with the dynamic of the logistic concepts, an optimum can occur.

This optimization process must not only take into account the traditional methods of short-term logistic strategies but must take care of long-term organization procedures. Short-term optimization as well as local optimized systems will not lead to an overall optimum.

The Problem of Different Disciplines

Land use management, city architecture, structural planning and the design of infrastructure for different kinds of land transport modes have their own goals, which are sometimes contradictory and in most cases not coordinated in reality (Bauordnung für Wien, 1997). Each discipline uses its own set of indicators to fulfill their duties. Some of the core disciplines (and faculties) in this system have no contact with logistics at all, although they are producing elements which influence logistics of today substantially.

Very often these disciplines are separated into different ministries or other administrative bodies. The outcome of this practice or policy is a kind of accumulation of "sectoral- optima" with more or less gaps between them. This deficit has then to be compensated with "Logistics". Normally the result can never reach an overall optimum - because it is not needed.

Basic Relationship

Human settlement structures exist in a great variety of shapes. These structures have always been dependent on the energy of the transport system. As long as human bodies energy was the main source of power, the distances and the size of settlements where limited to the travel-energy budget of the inhabitants. The density was only bearable if it was compensated with high quality of settlement structures and excellent logistic system, combining store rooms with a variety of workshops, specialization and a good organized external transport system (Mumford, 1984). The development of aqueducts, post stations, rules and privileges for trade are elements of this historical development of logistics.

Just in time was not the only a problem of the transport system, it was the farsighted planning and the forecast of demand and supply, the trade-off between store rooms, transportation possibilities and actual needs. The result of this optimization process: high quality cities which resulted from logistic optimization over a long time in a complex system. This is still reflected in the ratings scale of world capitals, as it was done by Kenworthyand Newman (1989) (Figure 3).

200 people per hectare and even more characterize the sustainable city. Many districts of Vienna, one of the cities with the best "Quality of Life" ranking in the world have such density figures. If people voluntarily relinquish to own a car, the city must have gained progress in urban logistics. More and more cities have the intention to reduce the modal spit for car traffic to 15% or less till 2030. This is only possible if the long term logistic of urban structures is organized in an intelligent manner.

It is obvious that the main differences in city structures are caused by differences in the transport system, mainly transport costs and land use regulations.

Inter-Relations

There exists close and inter-relationships between settlement-structures, transport system, industry and logistic, which have not been realized so far. The whole history of invention in transport shows traces of this inter-relationship. Rail and the invention of traffic lights before, the sophisticated computerized traffic management system, including GIS and GPS, are the result of these efforts, not knowing the underlying driving forces of the system, which have segregated well-coordinated functions in historical cities during the last 150 years.

Logistics helps to reduce costs in the system. However, the definition of cost depends on the defined system borders. What is outside these borders, economists call it "external costs". These costs are borne

Figure 3. Relationship between urban density and energy demand for transport

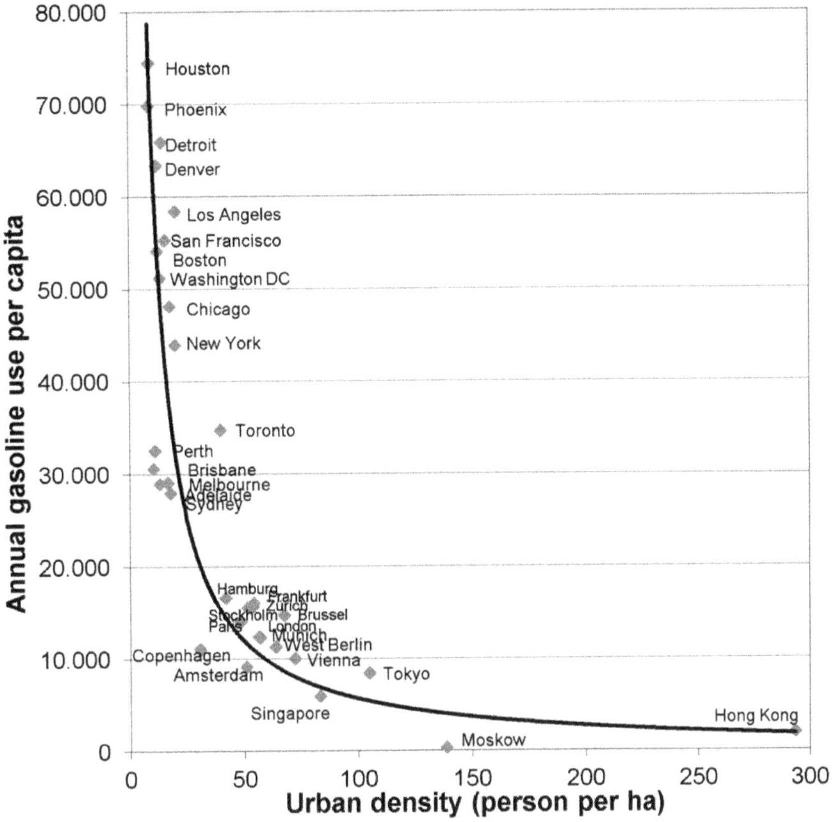

by the society, the environment and/or the future. Studies in Germany indicate that external costs for good transport on roads are 77- 89% of the real costs (Teufel et al., 1989). Traditional logistics in the transport sector operate with "internal cost" in general, since it is used on the company level and not on the level of municipalities or states. This is one of the reasons why industrial lobbies and representatives of economy believe on the magic of "low transport cost". But externalization of internal cost distort not only the market, but also urban and transport structures. If the market does not reflect the reality of the system then, current optimal solutions are far away from the real optimum. Economists are working with scales of quantity and not with distances or space. For transport engineers, distances and spatial dimensions are core variables and tools of their work. Transport logistics is connecting economy with the field of engineering, since it has one pillar of the bridge on engineering and the other one on economic activities. A closer look from the system point of view comprising both disciplines might be helpful to understand what is happening in the reality of today in the transport and economic field.

Structures and Economy of Scale

Economies of Scale: Definitions

Competitiveness of our economy depends very much upon the advantages relating to economy of scale. "The bigger the scale, the cheaper the products" (Samuelson & Nordhaus, 1987). There are many definitions of "Economy of scale" such as "Economies of scale is the cost advantage that arises with increased output of a product. Economies of scale arise because of the inverse relationship between the quantity produced and per-unit fixed costs; i.e. the greater the quantity of a good produced, the lower the per-unit fixed cost because these costs are spread out over a larger number of goods. Economies of scale may also reduce variable cost per unit because of operational efficiency and synergies. Economies of scale can be classified into two main types: Internal – arising from within the company; and External – arising from extraneous factors such as industry size" (Staff, 2016). Wikipedia has the following version: "In microeconomics, economies of scale are the cost advantages that enterprises obtain due to size, output, or scale of operation, with cost per unit of output generally decreasing with increasing scale as fixed costs are spread out over more units of output (Economies of Scale, 2016). Often operational efficiency is also greater with increasing scale, leading to lower variable cost as well. Economies of scale apply to a variety of organizational and business situations and at various levels, such as a business or manufacturing unit, plant or an entire enterprise. For example, a large manufacturing facility would be expected to have a lower cost per unit of output than a small facility, all other factors being equal, while a company with many facilities should have a cost advantage over a competitor with fewer.

Some economies of scale, such as capital cost of manufacturing facilities and friction loss of transportation and industrial equipment, have a physical or engineering basis. The economic concept dates back to Adam Smith and the idea of obtaining larger production returns through the use of division of labor. Economies of scale often have limits, such as passing the optimum design point where costs per additional unit begin to increase. Common limits include exceeding the nearby raw material supply, such as wood in the lumber, pulp and paper industry. A common limit for low cost per unit weight commodities is saturating the regional market, thus having to ship product uneconomical distances. Other limits include using energy less efficiently or having a higher defect rate." In most of common definitions of Economy of Scale, transport costs are mentioned not directly as key limiting factor.

Combining Transport System and Economy of Scale

With the new tool for transport engineering we know constrains in the transport system concerning mobility and travel time saving. Are the dramatic changes in economic structures from local, small retailers in municipalities and cities to shopping centers on motorway intersections and suburbs, just to mention one example, only the effect of greater efficiency or is there an influence from transport and logistics also a cause. It is interesting, that trade association continually asking for a reduction of transport cost as a kind of silver bullet for their problems. The background for this demand is the idea, that low transport cost will enhance the catchment area for everybody. The first analysis, taking into consideration economy of scale principles and transport was carried out 1996.[1]

By analyzing cost for production including infrastructure, material, manpower, production cost increase only logarithmic with the volume of production. The production price decrease therefore with the greater scale. The indicator is the number of units produced. To bring the output to the customers, transport is needed. Transport costs have to be added to the production cost. Using this simple logic, without going into details, transport cost will limit the scale (See Figure 4).

Transport cost are inducing pressure on companies for innovation and competition on local or regional level and are the secret for the variety and richness of local products and sustainable technologies in the past. To reduce the cost it is necessary for the company either to economize internal production processes and/or enhance the quality of the product in comparison to competitors. If the transport cost remain the

Figure 4. Transport cost limits the scale-effects of economic activities

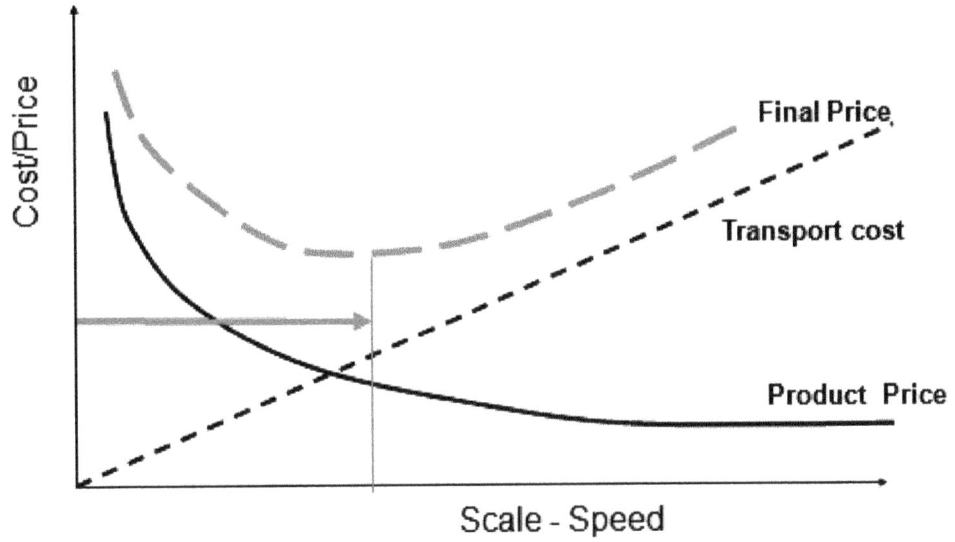

same, the company can gain profit by extending its market area till transport cost compensate the cost or quality advantages (See Figure 5).

When competitors catch up with their production or quality, a new equilibrium situation occur: The size of the catchment area is now smaller than before. There are more competitors in the system and competition is harder (See Figure 6).

Figure 5. Under stabile conditions and high transport cost competition between producers reduce either the price and succeed on a smaller scale or enhance the quality of its product to gain more customers

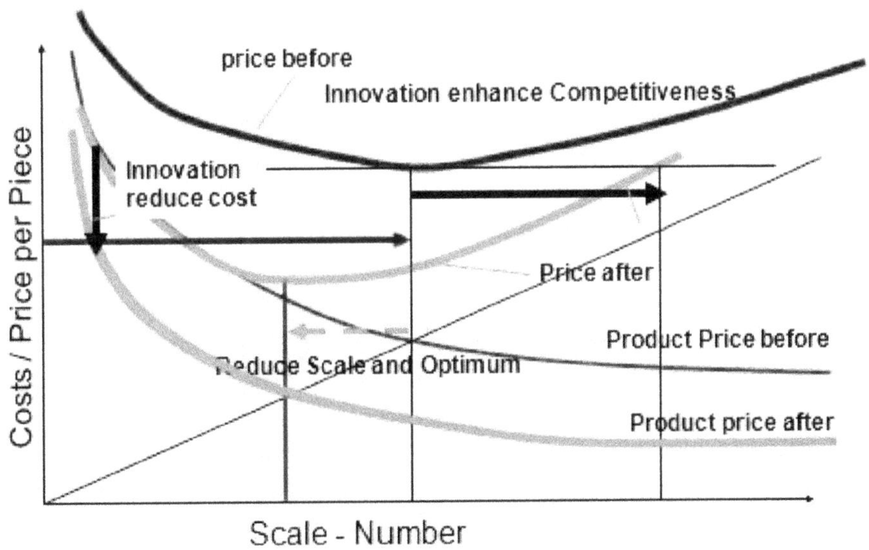

Figure 6. If innovation and competition lead to local optimization and specialization and finally to smaller scale

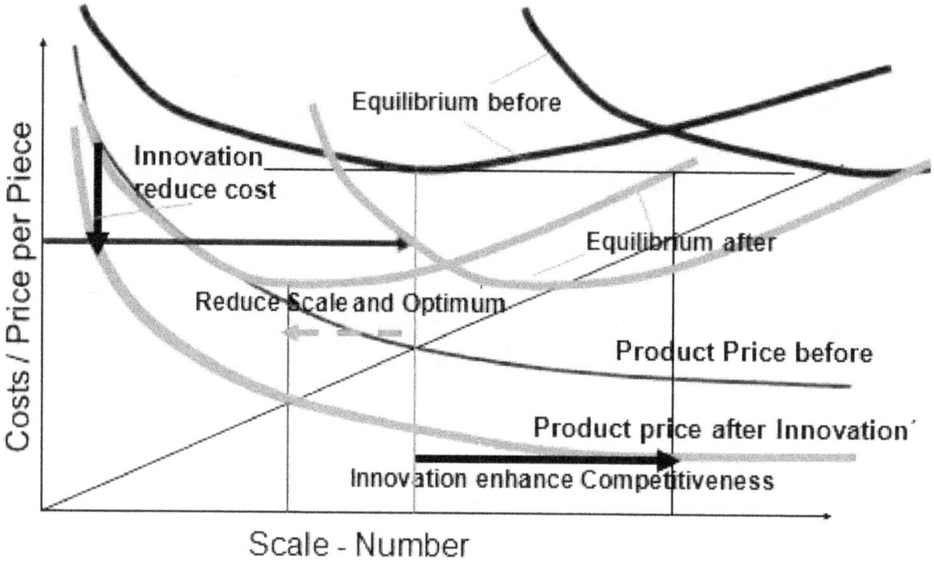

Human economic activities adapt to local conditions, available resources, human skills, conditions of landscape and climate. Distinctive cultures in the broadest sense are the outcome of this process. Transport costs have strong influence on the economic structures and therefore on urban and regional structures also. This was the case, before mechanical modes, driven by external energy came in use and the secret for the variety of products, each perfect developed for the local situation and region. European richness has its roots in slow and expensive transport systems for several hundreds of years. There was of course international long distance trade also, but only for special and rare goods. If the process continues, the scale of competing units is becoming smaller and smaller and the variety of solutions, as the result of continuous innovation-pressure, increases nearly infinite – like in the evolution of nature (See Figure 7).

It is obvious that there is an enormous pressure in the system to reduce transport cost. This can be done by innovation in logistics or use of external opportunities. Wind and water are well documented examples in human history. Feudal systems externalize transport cost to the society to gain local profits. If transport cost are reduced, the company which use this benefits, gain profit from this fact without changing its production or improving the quality of its product (See Figure 8).

The catchment area is extended and competitors are pushed out of the market. The company can grow and use its advantages of scale even better. Competitors can only gain ground, if they are able to reduce the cost substantially or enhance the quality of product fundamentally. Minor improvements like before have no chance under these conditions. (Emberger, 2005; Emberger & Pfaffenbichler 2016) The bigger unit will always be the winner, if transport costs are negligible. In the physical reality transport cost are related to transport energy and don´t change under comparable conditions. But if the companies can externalize transport cost to the society, the bigger survivors become “low-cost-addicts” and put pressure on politicians for a reliable and cheap transport system, because it is the guarantee for their profit. Low transport cost for the corporations by putting the financial burden on taxpayers are the main reason,

Figure 7. Under of fair transport cost, the evolutionary process of competition for limited resources reduce the scale and enhance the variety of solutions, products in agreement with the local ecological, social and economic conditions

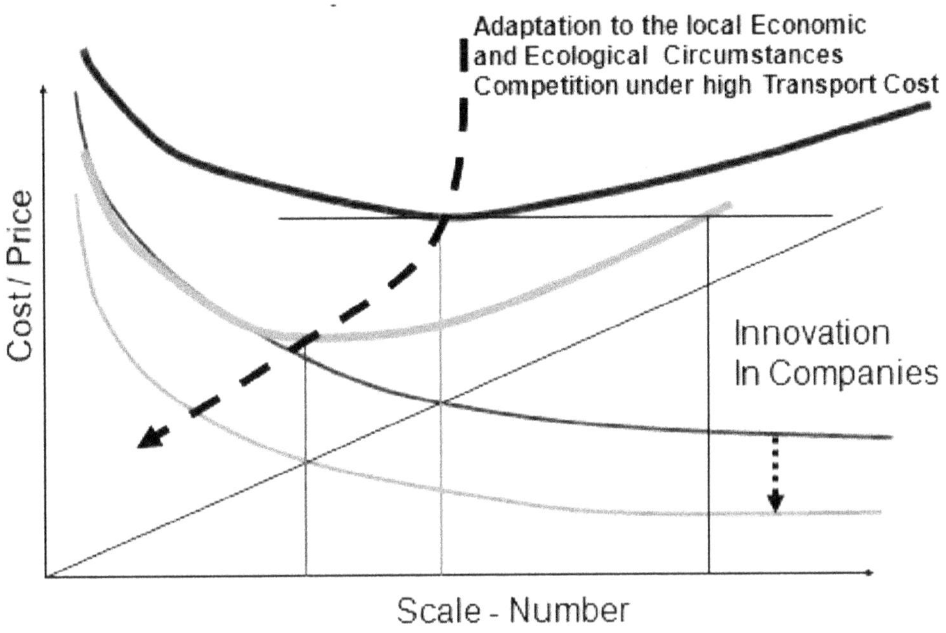

Figure 8. If transport cost are reduced, the competitor using this benefits first, can extend his catchment area without innovation, simply by using the benefits of economies of scale in combination with low transport cost

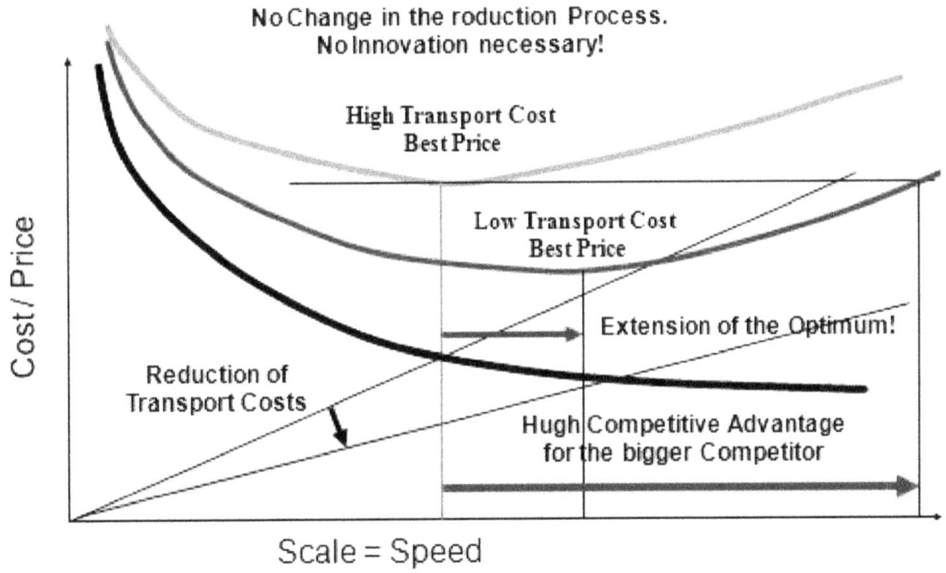

why local shops, workshops have disappeared in the last century and the variety of products, methods and techniques to solve problems on the local scale were lost. Logistics is contributing to this process of destructive competition and the predominance of international corporations over municipalities, cities and even states. See Figure 9.

If corporations can penetrate existing trade barriers, like national borders and customs, they can use differences of labor, environmental and social standards, between countries. They can now make profit be importing goods and products with a lower price, by exploiting the people and the nature. This is only possible, if somebody else pay the price for cheap transport. The economic system gets out of control of democratic structures. The transport system has therefore effects not only on built structures, but also social standards, regulations, employment and environment.

Long Term Logistics and the Effects on the Individual Level

"Logistics means having the right thing, at the right place, at the right time." But the long term effects have not been recognized in our modern societies and its short time horizon in the contrary to our ancestors. They have chosen their locations and urban structures under consideration of generations to have the right things, at the right place, at the right time for centuries. Long term logistics for transport and urban planning has to be applied, to prevent systematic disparities in space and time.

On the other side, low transport costs in a system with high speed increase the opportunities for choice for the individuals too. They react in the same way like corporations, but for other reasons. They search for cheap building land outside of cities accessible only by car. The result in the system is known as urban sprawl, an inorganic growth of settlements around cities, called also "commuter belt". Over the years these structures go through a process of demands since young families are the main cohort settling there. First there is a demand for kindergarten, followed by basic schools and finally nursing homes for

Figure 9. If the low-transport-cost-situation remains, companies grow to corporations and finally the economic system ends in monopoly

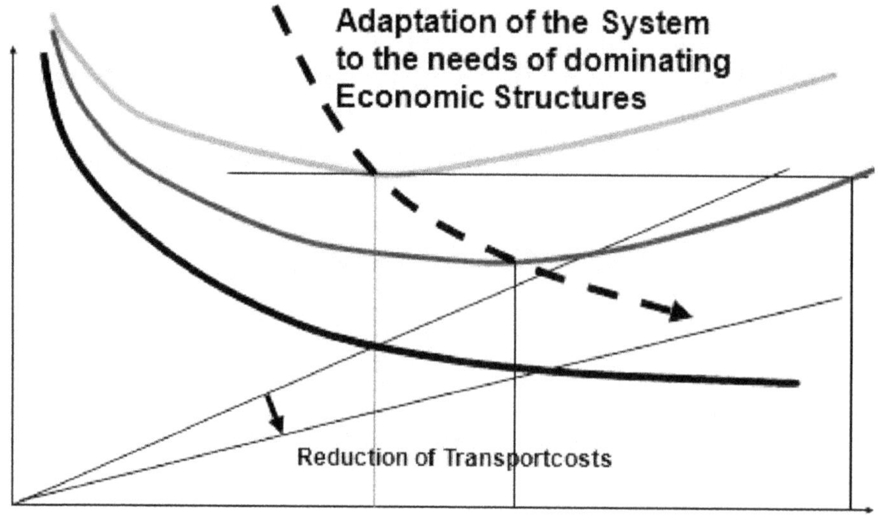

the elderly. When cost of individual traffic increase due to higher gasoline price, toll-roads, parking fee at the destination they recognize the lack of public transport. The people recognize that they don´t have the "right things, at the right place, at the right time with the right price". The former "freedom of choice" for housing becomes a trap and is suddenly converted in less freedom of choice. They are missing the infrastructure of a well-organized society, schools, pharmacies, doctors nearby, cultural opportunities etc. short term benefits are converted in long term disadvantages. This is exactly the situation, when urban long term logistics has failed. If land-use, urban planners and transport engineers would understand that logistics is not "just in time" in the scale of minutes, but also crucial in the long term for the society and stability of urban structures, these problems would not happen. On the urban scale, the process of planning, implementing, and controlling the efficient, effective flow and storage of goods, services, and related information for the purpose of conforming to customer requirements is also valid for generations.

Urban sprawl can only happen with parking provisions direct at homes and other human activities. If individuals have direct and convenient access to fast and cheap motorized individual modes, they can easily escape administrative borders and local regulation. Municipal and city administrations are losing their power to keep the municipality or city in shape. With the concentration of economic activities outside of cities they lose their tax revenue. In the system vehicle kilometers are enhanced however logistics methods are applied. Failing long term logistics in urban planning and transport engineering interfuse short term logistics.

It is easy to recognize that such a system can only occur if fundamental principles of logistics are not applied in land use and city planning. Costs are postponed or transferred to society, the future and to other regions of the world, since the system is too complex and long lasting that the effects on economy, ecology and society are not direct visible.

So far transport logistics does not care about city planning systems and city planning principles are in general free of the principles of logistics. If we apply better logistics in usual way, we can mitigate some problems, but we have to consider that negative side effects might occur from the changes in our built structures caused by the transport system. The more attention is paid to logistics in the early stages of city planning, the less effort is needed to compensate the deficit between land use, city and transport planning. This can be proved by analyzing the development of settlement structures, industry and transport indicators of technical transport systems during the time period of the last 200 years. On the other hand, highly developed systems in cities and regions can be found in structures preceding this period of mass motorization).

KEY FACTORS

An optimum in logistics can only be achieved, if hardware (built structures) elements and software (information system and modes) are taken into account in an early stage. This is the land-use and city-planning stage. However this is not yet the key element. This is the knowledge about human behavior in this man made artificial environment with a transport system, driven by externally energy, whatever source it has. There is a lack of understanding of this non-linear law of behavior in all related disciplines which creates an exponential growth of problems of different kinds (congestion, time losses and a decrease of efficiency of the whole system).

WHERE DOES THE MISTAKE HAPPEN?

Today the system is not driven by system principles but much more by the sum of principles from the single elements which try to optimize their positions without taking into account the effects on the whole system. The relationship between system users, speed and the amount of transport is described by the basic formula:

$$Q = D \cdot V \qquad (11)$$

Q: Amount of transport (Km per time),
D: Number of system users (car-units),
V: Speed of system users.

If we want to reduce the amount of traffic we have two possibilities: we can reduce D and or V (On the level of a transport company the goal is opposite, since Q is the indicator for its turnover, its success in the system.). However, where is the most effective leverage point?

If we take into account real human behavior, the organization of the local structures is the key factor. How parking is organized in relation to the places of human activities is shown in Figure 10.

Today companies as well as individual people try to optimize their logistics within given conditions. If the conditions are biased, which is the case for transport costs, the whole system will not be able to achieve the best overall performance. The decision for the best mode is crucial in this context. Modal split is strongly influenced by the organization of parking places, railway stations or public transport

Figure 10. Acceptance-function for different environments, based on empirical data. This function is based on human body energy consumption and its perception
Knoflacher, 1987.

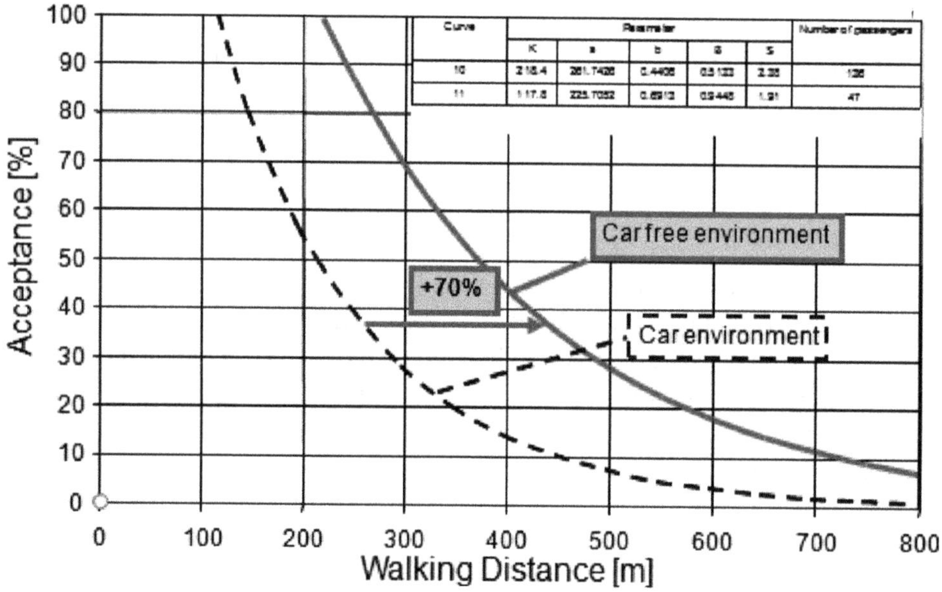

stops and the location of human and other economic activities. The influence of this '"micro-level-structures" on speed and modal split decisions is much more important than it was in the past, before the car came into mass-use.

The function in Figure 10 is the reciprocal of the "cost function" for an individual. It is based on the fundamental law from Weber & Fechner (1871) which is a useful tool to describe human behavior in this artificial environment (The Encyclopedia Americana, 1975).

This law is behind many well-known relations in transport. If the whole system is analyzed by using these inter-relationships, city logistics has to be applied in a much wider context, taking into account settlement structures and parking organization (accessibility of activities in great details).

This new approach is extremely helpful in analyzing existing systems and in developing optimal strategies for solutions, as it was done partly in the research projects of the European Union OPTIMA, FATIMA, EMRECU and PROSPECTS (2000). There are also data from several empirical analyses available such as from the· city of Vienna (WIZK, Vienna international Conference for the Future, Knoflacher, 1995).

RECOMMENDATIONS

If the relationships between logistics, city structures, industry and transport are known, it becomes obvious, that city planning must include in each step of its work at least the principles of logistics. Since the key elements of the system are humans, these principles must take care on their inherent behavioral structure in a rational and not an emotional way.

If these principles are applied, different pattern of city structures will result, industrial and productions structures will follow which are much more sustainable than the structures of today: In principle this is the application of logistic strategies in a long-term context of the dynamic evolution of cities. The empirical results show effects on a much greater magnitude, compared to the traditional field of applied logistics. The theoretical background of this approach will greatly assist practitioners to develop systems which are not only useful for the application of city logistics for city planning purposes, but also to move the whole development into a more sustainable direction. Very few researchers use this approach. If this issue is striped by an author, the existing spatial housing and transport structures are not questioned (Conveny & Dwyer 2009)

Since existing strategies in city logistic tend to reduce the resistance function, to optimize the economic efficiency of the companies there is less (or nearly no) need awareness for the performance of the whole urban system. This practice reduces efficiency of the whole, or could even lead to more instability in the future. City logistics, as it is applied today is influencing the core factors of the transport system- the resistance function. This must be taken into account to prevent this important instrument from producing adverse effects in the long term, while it produces strong positive effects in the short-term cycle.

What about ITS (Intelligent Transport Systems)?

There are thousands of new digital technologies and services that have come to market in recent years that are turning transportation from a brick-and-mortar business into an information-based and informatics- based activity[2]. Disruption is happening throughout the value chain. Some examples include:

- **Infrastructure Networks:** Adaptive traffic signaling, dynamic electronic road pricing
- **Vehicles:** Autonomous motor vehicles, autonomous aircraft, programmable vehicle performance characteristics
- **Business Models:** Car sharing, peer-to-peer taxi services
- **Interfaces:** Electronic taxi hailing, crowdsourced traffic reporting, integrated multi-modal services (route planning, booking, and payment)
- **iPhone©:** Navigation.

The result of all this investment and innovation is that city dwellers are increasingly dependent upon a growing array of digital services and technologies to make and manage their travel choices, and often to actually take the trip. These services are having not only real impacts on travel behavior – changing why, when, where and how people take journeys to work, home and other locations and activities, they restrain the perception of the real world and narrow. They are not only changing both the supply and the demand side of the mobility equation but also the recognition of the environment and accidental discoveries even in the transport system.

The Coming Crisis in Transportation Planning

The technological transformation of travel behavior is a rather new thing for transportation planners to grasp intuitively. Until now, transportation systems have largely been planned, designed and managed as purely physical infrastructure systems. Both supply and demand were treated as largely static, and demand forecasts were pretty crude and episodic. You basically designed the system, built it, and let it run, making infrequent adjustments.

Beginning in the 1980s and accelerating in the 1990s, transportation planning organizations began integrating digital technology into operations, developing a growing body of research, education and training, and prototypes they call "Intelligent Transportation Systems" (ITS for short). In the U.S. in Europe and Japan, investment in ITS is almost exclusively focused on roads and motor vehicles, and the development of large-scale improvements to existing publicly-owned transportation systems such as automated toll collection, signal controls, dynamic road pricing schemes, and anti-collision technologies. The driving force behind are corporations of the electronic industry.

From a broader perspective, ITS misses the forest for the trees, write A. M. Townsend. As an operations technology, it isn't focused on anything other than increasing throughput and promising to improve safety, in the process neglecting broader transportation planning goals such as sustainability, livability, and accessibility. Often, a successful ITS deployment actually exacerbates the problems planners and politicians seek to address. Ironically, where ITS failed to address the complex transportation-land use nexus, recent private-sector innovation in transportation technologies and services are actually starting to address key concerns of transportation planning. For instance, carpooling services, which are in use since decades, highlight potential business models for new mobility technologies. By providing tools to incentivize carpooling, large amounts of parking to be freed up for development – potentially creating a virtuous circle of densification that could support transit if planning decisions follow immediately. Similarly, the first autonomous features widely deployed in private vehicles may simply be for remote parking, or high-density garages with less area wasted on circulation. To an urban planner's eye, it seems then that the true killer app for such technologies is unlocking hidden land value. Giant corporations like Google, Facebook and others can become players in the real estate business.

Traditional transportation planning is only slowly waking up to these developments since it does not recognize the system effects. Internal debates in the field remain focused on infrastructure and urban form rather than the direct behavioral impacts of these new technologies. As a result, these high-value, high-impact innovations in transportation are coming from the private sector with little coordination or planning. And their collective impact and potential unintended consequences are not being adequately explored.

Take, for instance, the conflicts between Uber and local regulators. It is clear that the conflicts between innovators and regulators that we are seeing now are just an early indicator of much larger challenges to come. Modern urban transportation systems are being quite literally re-programmed by a new wave of information and communications systems developed and deployed by new technology firms directly to consumers. These are just the initial skirmishes in much bigger conflicts that will arise over how transportation systems work in 21st century cities, and the roles and relationships between public and private sector providers. One of the strategies of these players is to get free access to municipal controlled data.

Why are planners not paying attention? It's important to distinguish between the terms mobility and transportation. When we talk about mobility, we are talking about the potential to move. When we talk about transportation, we are talking about the system – and the use of that system – to actually move. Very little of the innovation we have seen in recent years is in transportation; most of it is in mobility via tools that allow us to navigate the options and opportunities in the transportation system. It's a not a subtle distinction as it looks like, but it highlights where planning is falling short: it mostly focuses on the transportation, not the mobility. Startups are building services, not infrastructure, so they focus on the mobility piece.

The following paragraph is written from the perspective of traditional transport and urban planning, citing A.M. Townsend:

Equipping the Field for Disruption

All of this raises some very serious existential questions for transportation management and planning. Why invest in formal transit when informal systems can achieve a similar impact? How can capital investment plans be made as revenue sources and forecasts change day-to- day due to disruptive innovations? How can exhaust data from private-sector operations be used to inform long-term planning?

Challenges to the status quo in transportation are emerging faster than we can even formulate these questions. Existing priorities are sure to shift, yet scholars and practitioners lack the capacity to assess and incorporate the implications of rapid advances in informatics. The transportation planning field's focus on ITS has been myopic at best. To expect that fiscally-strapped public sector organizations burdened by long planning cycles will lead is an act of hubris. The re-programming of mobility isn't something that is being engineered by planners; it will be something they respond to on a constant crisis basis. For years, transportation planners have resisted this change, but this is no longer a tenable position.

Yet as much as these new technological band- aids are filling gaps in older physical networks, making up for 50 years of transportation planning failures, they are inadequate for the future. Transportation planning is just as important, if not more so, but the kinds of plans and their areas of inquiry must adapt to this new reality. There have been promising advances, such as using mobile phone location records to improve demand modeling, and using GPS trace data to map informal jitney networks in the global south. But transportation planning remains largely siloed within specific modes such as road or rail, while the greatest potential positive impacts of re-programmed mobility systems may be in the power to shift people between modes and make multi-modalism more feasible. This threatens planning organizations and planners embedded within that institutional structure.

Finally, in the U.S., transportation policy continues to revolve around investments in different modes and infrastructures and continues to treat ITS as just another type of infrastructure investment. There is no scope for dealing with the transformational aspects of new technologies, leaving efforts to deal with the public consequences of largely private actors in the arena to ad-hoc state and local efforts such as the lawsuits against operators of electronic hailing systems or bans on texting while driving.

After Paradigm Change the View Change

ITS-tools are made with the promise to improve the transport system, by reducing mobility barriers for motorized system users. After Paradigm change, planners must know, they are responsible for the modal split, sustainability and social, as well as economic goals. Crucial is the built structure – how people are connected to different modes and what are the barriers for free modals choice. If the local urban or regional economy should flourish, origins and destinations in the city or region have to be maximized. The transport indicator to monitor the transformation into this direction is the travel time. The more travel time is captured within the system borders, the greater is the number of activities of the urban or regional body. This is totally in contradiction of traditional transport engineering efforts. Without knowledge and responsibility for the system transport engineers were proud if they could maximize the travel speed and reduced the travel time in the system. ITS is still doing the same only with the tools of the information society. From the system point of view, most of the ITS applications are counterproductive to sustainability and should be called as STS – Stupid Technology Systems. They have to be replaced by IPS – Intelligent Planning Systems. Intelligent planning systems, based on the tools of paradigm change take into account long term logistics, having knowledge about human behavior and human constrains and use these knowledge to design, build and organize sustainable structures, which are resistant against damage from temporary technological innovations, even they are fascinating for the individuals like the car, if we want to have a future for the human society. As transport engineers we have to respect the people as they are and not the supermen they seem to be, when they are using the artificial physical and mental limbs of today without recognizing the effects outside their limits of perception in space and time.

CONCLUSION

Transport logistics is reducing the costs, the time or the distance between economic activities at different locations. Due to time constancy in the system these short term effects produce medium and long them effects on the macro scale of urban patterns and economic activities. Time reduction is not possible in the system – therefore the system reacts on time savings by increasing distances. Time and distance are indicators for transport costs. The effect of logistics can be compensated only if the total amount of cost reduction in transport is put into innovation. If the money is used to enhance production or expansion, either the number of trips is increased and/or the distances. The benefits from better "external logistics" are therefore producing medium and long term effects in the city structure. This goes hand in hand with changes in economic structures and results in concentration and centralisation in order to take advantage of the effects of economy of scale.

The theoretical analysis of the effects of logistics on urban structures and spatial economic activities shows the compensating effects of the flexible spatial structures of the western society. This is backed by empirical analysis on the macro-level using data from Germany. Nobody working in logistics is interested to produce a more inefficient system by investing a lot of effort. But it is not recognised that logistics on the level of producers and distributors are pushing the urban structures into a direction which is compensating all its efforts.

Transport logistics can only work successfully in the long term, if transport and urban planning takes into account the opportunities and possibilities of advanced logistical systems in their models. The compensational effects of progress in logistics have to be considered. Additional measures to stabilise or recover healthy urban economic structures have to be introduced. This can be, for instance, taxes on transport intensive structures. Finally, it has to be recognised that not everything what is good for a single unit must be good for the system.

REFERENCES

Advanced Logistics and Road Freight Transport. (1992). *Report prepared by an OECD Expert Group*. Paris: Organisation for Economic Co-Operation and Development.

Baum, H. (1995). *Verringerung von Leerfahrten im Straßengüterverkehr. Buchreihe des Instituts für Verkehrswissenschaft an der Universität zu Köln, Nr. 58. Verkehrs-Verlag J.* Düsseldorf: Fischer.

Bauordnung für Wien (Landuse Regulation example). (1997). Textausgabe nach dem Stand 1.1.1997. Linde Verlag, Wien.

Brewer, A. M., Button, K. J., & Hensher, D. A. (2001). Handbook of Logistics and Supply-Chain Management. Pergamon.

Conveney, J., & O`Dwyer, L. A. (2009). *Effects of mobility and location on food access*. Healthplace.

Economies of Scale. (2016). Retrieved from https://en.wikipedia.org/wiki/Economies_of_scale

Emberger, G. (2005). *Freiht Transport – A Holistic Approach. "Beiträge zu einer ökologisch und sozial verträglichen Verkehrsplanung*. Institut für Verkehrsplanung und Verkehrstechnik, Technische Universität Wien.

Emberger, G., & Pfaffenbichler, P. (2016). *Wechselwirkungen und Rückkopplungen im Inividualverkehr und im Güterverkehr*. Wien: Paradigmenwechsel im Verkehrswesen, Österreichische Verkehrswissenschaftliche Gesellschaft - ÖVG.

EMRECU. (1998). *EU- DG XVll Project*. Author.

Hesse, M. (1997). *Wirtschaftsverkehr stadtverträglich. Der Strukturwandel in der Logistik und seine Bedeutung für die Stadtentwicklung. Stadtforschung aktuell Nr. 61*. Basel: Birkhäuser Verlag.

Hidber, C. (1992). Optimierung in Verkehrsplanung, Transporttechnik und Logistik. Schriftenreihe des Instituts für Verkehrsplanung, Transporttechnik, Straßen- und Eisenbahnbau (IVT) – Nr. 91, Zürich.

Ihde, G. B. (1991). *Transport, Verkehr, Logistik: gesamtwirtschaftliche Aspekte und einzelwirtschaftliche Handhabung. 2., völlig überarbeitete und erweiterte Auflage*. München: Vahlen Verlag.

Knoflacher, H. (1987). *Verkehrsplanung für den Menschen* (Vol. 1). Wien: Grundstrukturen. Verlag Orac.

Knoflacher, H. (1993). Does the Development of Mobility in Traffic Follow a Pattern? *History and Technology*, *15*, 125–140.

Knoflacher, H. (1995). Economy of Scale - Die Transportkosten und das Ökosystem. *GAlA*, *4*(2), 100–108.

Knoflacher, H. (1995). *Sustainable Development- Öko-City. Projektgruppe 1: "Mobilität in der Stadt" (Stadt und Verkehr). Durchgeführt im Auftrag der Wiener Internationalen Zukunftskonferenz*. WIZK.

Mumford, L. (1984). *Die Stadt*. DTV Wissenschaft Nr. 4326.

New Trends in Logistics in Europe. (1997). Report of the Hundred and Fourth Round Table on Transport Economics, held in Paris on 3rd and 4th October, 1996. European Conference of Ministers of Transport (ECMT), Paris, France.

Newman, P. W. G., & Kenworthy, J. R. (1989). *Cities and Automobile Dependence: A Sourcebook*. Essex, UK: Gower Publishing Company Limited.

Samuelson, P.A., & Nordhaus, W.D. (1987). Volkswirtschaftslehre. Grundlagen der Makro- und Mikroökonomie. Achte, grundlegend überarbeitete deutsche Neuauflage. Bund-Verlag, Köln.

Schmitz, S. (2001). Revolution der Erreichbarkeit. Gesellschaft, Raum und Verkehr im Wandel. Stadtforschung aktuell Nr. 83. Leske + Budrich, Opladen. doi:10.1007/978-3-663-11306-5

H. M. Soekkha (Ed.). (1990). Telematics – Transportation and Spatial Development.*Proceeding of an International Symposium organized under the auspices of the Centre for Transportation Engineering*. DelftUniversity of Technology.

Sonntag, H. (1999). Städtischer Wirtschaftsverkehr und logistische Knoten. Wirkungsanalyse von Verknüpfungen der Güterverkehrsnetze auf den städtischen Wirtschafts- und Güterverkehr. Berichte der Bundesanstalt für Straßenwesen. *Schriftenreihe Verkehrstechnik*, *V*, 68.

Städtebau und Verkehr. (1996). *Dokumentation der Modellvorhaben. Stand: Juli 1996. ExWost-Informationen Nr. 06.10*. Bonn: Bundesforschungsanstalt für Landeskunde und Raumordnung.

Staff, I. (2016). *Economies of Scale*. Retrieved from http://www.investopedia.com/terms/e/economiesofscale.asp?ad=dirN&%3Bqo=investopediaSiteSearch&%3Bqsrc=0&%3Bo=40186

Taniguchi, E. (2001). City Logistics. Network Modelling and Intelligent Transport Systems. Pergamon. doi:10.1108/9780585473840

Taniguchi, E., & Thomson, R. G. (Eds.). (2001). *City Logistic II*. The Second International Conference on City Logistics, 27th – 29th 2001, Okinawa, Japan.

Teufel, D., Bauer, P, Beker, G., & Wagner, T. (1989). *Gesellschaftliche Kosten des Straßengüterverkehrs. Kasten-Deckungsgrad im Jahr 1987 und Vorschläge zur Realisierung des Verursacherprinzips*. UPI Umwelt- und Prognose-Institut Heidelberg, UPI-Bericht Nr. 14.

Townsend, A. M. (2014). Re-Programming Mobility: how the Tech-Industry Is Driving Us Towards A Crisis In Transportation Planning. In Cities on the Move. New Cities Foundation Geneva.

Uber. (2016). Retrieved from https://www.uber.com/

Weber-Fechner Law. (1975). In The Encyclopedia Americana. Americana Corporation.

World Bank. (2016). Retrieved from http://lpi.worldbank.org/international/global

ENDNOTES

1 Knoflacher

2 Townsend A.M. Re-Programming Mobility: how the Tech-Industry Is Driving Us Towards A Crisis In Transportation Planning. In Cities on the Move, New Cities Foundation Geneva 2014

Compilation of References

Abley, S. (2005). *Walkability scoping paper*. Academic Press.

Advanced English Dictionary. (2015). Available from https://www.microsoft.com/en-us/store/apps/advanced-english-dictionary/9wzdncrfhmvb

Advanced Logistics and Road Freight Transport. (1992). *Report prepared by an OECD Expert Group*. Paris: Organisation for Economic Co-Operation and Development.

Aggarwal, C. C. (2003). Towards systematic design of distance functions for data mining applications. In *Proceedings of the ninth ACM SIGKDD international conference on Knowledge discovery and data mining*. ACM. doi:10.1145/956750.956756

Aggarwal, C. C., & Reddy, C. K. (Eds.). (2013). *Data Clustering: Algorithms and Applications*. CRC Press.

Ahmad, S., & de Oliveira, J. A. P. (2016). Determinants of urban mobility in India: Lessons for promoting sustainable and inclusive urban transportation in developing countries. *Transport Policy*, *50*, 106–114. doi:10.1016/j.tranpol.2016.04.014

Akin, D. (2014). Is Istanbul's growth monocentric or multicentric? Analyses of intraurban travel data between 1996 and 2006. In *OPT-i, 1st International Conference on Engineering and Applied Sciences Optimization* (pp. 2773-2799). Institute of Structural Analysis and Antiseismic Research, School of Civil Engineering National Technical University of Athens (NTUA), Greece.

Akın, D. (2011). *Kentsel Yapının Değişiminin Kentsel Yolculukların Hiyerarşik Kümeleme Analizi ile İncelenmesiyle Ortaya Konması: İstanbul Metropoliten Alanı 1996-2006 Yılları* [Studying the Change of Urban Structure by Hierarchical Cluster Analysis: Istanbul Metropolitan Region between 1996 and 2006]. *2nd Symposium of Urban and Regional Studies*, Ankara, Turkey.

Akin, D., & Alasalvar, S. (2016). Estimate Urban Growth and Expansion by Modeling Urban Spatial Structure Using Hierarchical Cluster Analyses of Interzonal Travel Data. *International Journal of System Dynamics Applications*, *5*(4), 16–40. doi:10.4018/IJSDA.2016100102

Alan, A. W. (1986). *Urban transit systems: guidelines for examining options*. Word Bank Technical Paper.

Albert, G., & Mahalel, D. (2006). Congestion tolls and parking fees: A comparison of the potential effect on travel behavior. *Transport Policy*, *13*(6), 496–502. doi:10.1016/j.tranpol.2006.05.007

Alkan, L. (2014). 1980 sonrası konut politikalarının mekansal yansımaları: Ankara örneği. *Journal of Urban Studies*, *12*, 103–131.

Alkan, L. (2015). Gated communities and social sustainability: A case study of Ankara. *2nd International Multidisciplinary Scientific Conference on Social Sciences & Arts Sgem 2015*, 423-430. Doi:10.5593/SGEMSOCIAL2015/B41/S15.052

Alpkokin, P., Cheung, C., Black, J., & Hayashi, Y. (2008). Dynamics of clustered employment growth and its impacts on commuting patterns in rapidly developing cities. *Transportation Research Part A, Policy and Practice*, *42*(3), 427–444. doi:10.1016/j.tra.2007.11.002

Alpkokin, P., Hayashi, Y., Black, J., & Gercek, H. (2005). Polycentric employment growth and impacts on urban commuting patterns: Case study of Istanbul. *Journal of the Eastern Asia Society for Transportation Studies*, *6*(1), 3835–3851.

Alsnih, R., & Hensher, D. A. (2003). The mobility and accessibility expectations of seniors in an aging population. *Transportation Research Part A, Policy and Practice*, *37*(10), 903–916. doi:10.1016/S0965-8564(03)00073-9

Altan, M., & Engin, O. (2003). The measurement of user satisfaction in a travel company. *Journal of Social and Economical Researches of Selcuk University*, *6*, 84–98.

Altman, D. G. (1990). *Practical statistics for medical research*. CRC Press.

Aly, A. A., & White, J. A. (1978). Probabilistic formulation of the emergency service location problem. *The Journal of the Operational Research Society*, *29*(12), 1167–1179. PMID:10239405

Andreassen, T. W. (1995). (Dis)satisfaction with public services: The case of public transportation. *Journal of Services Marketing*, *9*(5), 30–41. doi:10.1108/08876049510100290

Ankara Kalkınma Ajansı. (2013). *İstatistiklerle Ankara 2013, 36*. Retrieved from www.ankaraka.org.tr/tr/files/yayinlar/istatistiklerle-ankara-2013.pdf)

Araz, C., Selim, H., & Ozkarahan, I. (2007). A fuzzy multi-objective covering-based vehicle location model for emergency services. *Computers & Operations Research*, *34*(3), 705–726. doi:10.1016/j.cor.2005.03.021

Argyris, C. (1985). *Strategy, Change and Defensive Routines*. Boston, MA: Pitnam.

Asadi-Shekari, Z., Moeinaddini, M., & Shah, M. Z. (2014). A Pedestrian Level of Service Method for Evaluating and Promoting Walking Facilities on Campus Streets. *Land Use Policy*, *38*, 175–193. doi:10.1016/j.landusepol.2013.11.007

Asian Development Bank. (2008). *The Guidelines and Toolkits for Urban Transport Development*. Retrieved from http://sti-india-uttoolkit.adb.org/mod4/se5/006_2.html

Atkins, S. (1977). Transportation planning: Is there a road ahead? *Traffic Engineering & Control*, *18*(2).

Attoney General's Chambers. (2010). *Singapiore Statutes Onlie - Legislation History*. Retrieved from http://statutes.agc.gov.sg/non_version/cgi-bin/cgi_getdata.pl?actno=2004-REVED-276&doctitle=ROAD%20TRAFFIC%20ACT%0a&date=latest&method=part&sl=1&segid=1104745820-005150

AUAP. (2014). Ankara Metropoliten Alanı ve Yakın Çevresi Ulaşım Ana Planı (2013-2038). *Gazi Üniversitesi Ulaşım Ana Planı Ofisi*, 1-27, 34-41, 70.

Ausserer, K., & Risser, R. (2001). *Cycling in Vienna - Differences in attitudes between cyclists and politicians*. Paper presented at the Velo-city 2001, Edinburgh, UK.

Automatische Straßenverkehrszählung – Jahresauswertung 2000. (2001). Bundesministerium für Verkehr, Innovation und Technologie, Wien.

Aytug, H., & Saydam, C. (2002). Solving large-scale maximum expected covering location problems by genetic algorithms: A Comparative study. *European Journal of Operational Research*, *141*(3), 480–494. doi:10.1016/S0377-2217(01)00260-0

Azmi, D. I., Karim, H. A., & Ahmad, P. (2013). Comparative study of neighbourhood walkability to community facilities between two precincts in Putrajaya. *Procedia: Social and Behavioral Sciences*, *105*, 513–524. doi:10.1016/j.sbspro.2013.11.055

Azmi, D. I., Karim, H. A., & Amin, M. Z. M. (2012). Comparing the walking behaviour between urban and rural residents. *Procedia: Social and Behavioral Sciences*, *68*, 406–416. doi:10.1016/j.sbspro.2012.12.237

Babalik-Sutcliffe, E. (2013). Urban form and sustainable transport: Lessons from the Ankara case. *International Journal of Sustainable Transportation*, *7*(5), 416–430. doi:10.1080/15568318.2012.676152

Banister, D. (1996). *Barriers to implementation of urban sustainability.* Paper presented at the European Regional Science Association - 36th European Congress, ETH Zürich, Switzerland.

Basar, A., Catay, B., & Unluyurt, T. (2012). A taxonomy for emergency service station location problem. *Optimization Letters*, *6*(6), 1147–1160. doi:10.1007/s11590-011-0376-1

Baum, H. (1995). *Verringerung von Leerfahrten im Straßengüterverkehr. Buchreihe des Instituts für Verkehrswissenschaft an der Universität zu Köln, Nr. 58. Verkehrs-Verlag J.* Düsseldorf: Fischer.

Bauordnung für Wien (Landuse Regulation example). (1997). Textausgabe nach dem Stand 1.1.1997. Linde Verlag, Wien.

Beaudoin, J., Farzin, Y. H., & Lawell, C. Y. C. L. (2015). Public transit investment and sustainable transportation: A review of studies of transits impact on traffic congestion and air quality. *Research in Transportation Economics*, *52*, 15–22. doi:10.1016/j.retrec.2015.10.004

Beaumont, H. (2002). *SDS Maximum Parking Standards: Derivation of PTAL-based parking restraint.* SDS Technical Report Twenty.

Beckmann, J. et al.. (2006). *Multimodale Verkehrsmittelnutzer im Alltagsverkehr* (Vol. 4). Internationales Verkehrswesen.

Belwal, R. (2013). Peoples perception of public transport services in Oman. *Jurnal Teknologi*, *65*(3), 9–16. doi:10.11113/jt.v65.2140

Belwal, R., & Belwal, S. (2010). Public transportation services in Oman: A study of public perceptions. *Journal of Public Transportation*, *13*(4), 1–21. doi:10.5038/2375-0901.13.4.1

Ben-Tal, A., El Ghaoui, L., & Nemirovski, A. (2009). *Robust optimization.* Princeton, NJ: Princeton University Press. doi:10.1515/9781400831050

Ben-Tal, A., & Nemirovski, A. (1998). Robust convex optimization. *Mathematics of Operations Research*, *23*(4), 769–805. doi:10.1287/moor.23.4.769

Ben-Tal, A., & Nemirovski, A. (1999). Robust solutions of uncertain linear programs. *Operations Research Letters*, *25*(1), 1–13. doi:10.1016/S0167-6377(99)00016-4

Ben-Tal, A., & Nemirovski, A. (2000). Robust solutions of linear programming problems contaminated with uncertain data. *Mathematical Programming*, *88*(3), 411–421. doi:10.1007/PL00011380

Ben-Tal, A., & Nemirovski, A. (2000). Selected topics in robust convex optimization. *Mathematical Programming Series B*, *112*(1), 125–158. doi:10.1007/s10107-006-0092-2

Beraldi, P., & Bruni, M. E. (2009). A probabilistic model applied to emergency service vehicle location. *European Journal of Operational Research, 196*(1), 323–331. doi:10.1016/j.ejor.2008.02.027

Beraldi, P., Bruni, M. E., & Conforti, D. (2004). Designing robust emergency medical service via stochastic programming. *European Journal of Operational Research, 158*(1), 183–193. doi:10.1016/S0377-2217(03)00351-5

Berköz, L. (2000). Location of financial, insurance, and real estate firms in Istanbul. *Journal of Urban Planning and Development, 126*(2), 75–88. doi:10.1061/(ASCE)0733-9488(2000)126:2(75)

Berkoz, L., & Turk, S. S. (2010). Intrametropolitan determinants of foreign investment firms in Istanbul. *Journal of Urban Planning and Development, 136*(3), 234–242. doi:10.1061/(ASCE)UP.1943-5444.0000025

Berlin. (2012). *Stellplatzregelung bei Neubauten.* Retrieved April 04, 2016, from http://soup.fin.io/post/276091733/Berlin-hat-die-Pkw-Stellplatzregelung-bei-Neubauten

Berman, O., Hajizadeh, I., & Krass, D. (2013). The maximum covering problem with travel time uncertainty. *IIE Transactions, 45*(1), 81–96. doi:10.1080/0740817X.2012.689121

Bern. (1997). *Baugesetz (BauG), Fassung vom 18.6.1997, Änderungen bis inklusive 25. März 2002.* Bern, Schweiz: Großer Rat Kanton Bern.

Besussi, E., Chin, N., Batty, M., & Longley, P. (2010). The structure and form of urban settlements. In *Remote Sensing of Urban and Suburban Areas* (pp. 13–31). Springer Netherlands. doi:10.1007/978-1-4020-4385-7_2

Bhandari, K., Alpkokin, P., Kachi, N., Jia, P., & Kato, H. (2007). *Dynamics of employment distribution and accessibility index: Case of Delhi. Proc., Infrastructure Planning.* JSCE.

Bhandari, K., Jia, P., Alpkokin, P., Mukhopadhyay, D., & Gangopadhyay, S. (2007). Policies, commuting patterns and accessibility in a non-monocentric city: Case study of Delhi. In *Proceedings of the Eastern Asia Society for Transportation Studies* (pp. 192-192). Eastern Asia Society for Transportation Studies.

Bhat, C., Handy, S., Kockelman, K., Mahmassani, H., Chen, Q., Srour, I., & Weston, L. (2001). *Assessment of accessibility measure* (Research Report No. 7-4938). Texas Department of Transportation.

Bhat, C. R., & Guo, J. Y. (2007). A comprehensive analysis of built environment characteristics on household residential choice and auto ownership levels. *Transportation Research Part B: Methodological, 41*(5), 506–526. doi:10.1016/j.trb.2005.12.005

Bias, T. K., Leyden, K. M., Abildso, C. G., Reger-Nash, B., & Bauman, A. (2010). The importance of being parsimonious: Reliability of a brief community walkability assessment instrument. *Health & Place, 16*(4), 755–758. doi:10.1016/j.healthplace.2010.01.008 PMID:20303317

Black, W. R. (2010). *Sustainable transportation: Issues and solutions.* Academic Press.

BMLFUW. (2006). *Anpassung der Klimastrategie Österreichs zur Erreichung des Kyoto-Ziels 2008 -2012 - Entwurf 17.06.2006.* Wien, Österreich: Bundesministerium für Land- und Forstwirtschaft, Umwelt und Wasserwirtschaft.

BMLFUW. (2010). Maßnahmenvorschläge. Wien, Österreich: Bundesministerium für Land- und Forstwirtschaft, Umwelt und Wasserwirtschaft; Bundesministerium für Wirtschaft, Familie und Jugend. Author.

BMVIT. (2002). *Generalverkehrsplan Österreich 2002 (GVP-Ö 2002) verkehrspolitische Grundsätze und Infrastrukturprogramm; Bericht Januar 2002.* Bmvit.

BMVIT. (2013). *Gesamtverkehrsplan für Österreich*. BMVIT.

Bocarejo, S. J. P., & Oviedo, H. D. R. (2012). Transport accessibility and social inequities: A tool for identification of mobility needs and evaluation of transport investments. *Journal of Transport Geography*, *24*, 142–154. doi:10.1016/j.jtrangeo.2011.12.004

Brewer, A. M., Button, K. J., & Hensher, D. A. (2001). Handbook of Logistics and Supply-Chain Management. Pergamon.

Brilon, W., Grossmann, M., & Blanke, H. (1994). Verfahren zur Berechnung der Leistungsfähigkeit und Qualität des Verkehrsablaufes auf Straßen. *Forschung Straßenbau und Straßenverkehrstechnik*, *669*, 1994.

Brotcorne, L., Laporte, G., & Semet, F. (2003). Ambulance location and relocation models. *European Journal of Operational Research*, *147*(3), 451–463. doi:10.1016/S0377-2217(02)00364-8

Brown, B. B., Yamada, I., Smith, K. R., Zick, C. D., Kowaleski-Jones, L., & Fan, J. X. (2009). Mixed land use and walkability: Variations in land use measures and relationships with BMI, overweight, and obesity. *Health & Place*, *15*(4), 1130–1141. doi:10.1016/j.healthplace.2009.06.008 PMID:19632875

Buchanan, C. (1964). *Verkehr in Städten*. Essen, Deutschland: Vulkan-Verlag.

Bukhari, S. M. A., Hine, J., Gunay, B., & Blair, N. (2010, October). *Transport disadvantage and public transport network change: a case study of Belfast City*. In *12th World Conference on Transport Research*, Lisbon, Portugal.

Bundesinstitut für Bau-, Stadt- und Raumforschung. (2011). Ohne Auto einkaufen - Nahversorgung und Nahmobilität in der Praxis. Berlin, Deutschland: Bundesministerium für Verkehr, Bau und Stadtentwicklung, Werkstatt. *Praxis*, *76*, 29.

Bundeskanzleramt. (2015). *Landesrecht im Rechtsinformationssystem*. Wien, Österreich: Bundeskanzleramt.

Burchardt, T. (1999). Participation not consumption: The need for new participatory practices to address transport and social exclusion. *Transport Policy*, *10*, 272.

Burns, L. D. (1979). *Transportation, temporal and spatial components of accessibility*. Toronto: Lexington Books.

Burns, P. C. (1999). Navigation and the mobility of older drivers. *The Journals of Gerontology. Series B, Psychological Sciences and Social Sciences*, *54*(1), S49–S55. doi:10.1093/geronb/54B.1.S49 PMID:9934402

Buwana, E., Hasibuan, H. S., & Abdini, C. (2016). Alternatives selection for sustainable transportation system in Kasongan City. *Procedia: Social and Behavioral Sciences*, *227*, 11–18. doi:10.1016/j.sbspro.2016.06.037

Canbolat, M. S., & von Massow, M. (2011). Locating emergency facilities with random demand for risk minimization. *Expert Systems with Applications*, *38*(8), 10099–10106. doi:10.1016/j.eswa.2011.02.001

Car Loft. (2016). *Was ist Carloft?* Retrieved April 1, 2016, from http://www.carloft.de/carloft-gmbh/

Cass, N., Shove, E., & Urry, J. (2003). *Changing infrastructures, measuring socio-spatial inclusion/exclusion*. Final Report to The Department for Transport.

Cats, O., Wang, Q., & Zhao, Y. (2015). Identification and classification of public transport activity centres in Stockholm using passenger flows data. *Journal of Transport Geography*, *48*, 10–22. doi:10.1016/j.jtrangeo.2015.08.005

Chakwizira, J., Bikam, P., Dayomi, M. A., & Adeboyejo, T. A. (2011). Some missing dimensions of urban public transport in Africa: Insights and perspectives from South Africa. *The Built & Human Environment Review*, *4*(2), 56–84.

Chapman, S., & White, J. (1974). *Probabilistic formulations of emergency service facilities location problems*. Paper presented at ORSA/TIMS, San Juan, Puerto Rico.

Chlond, B.; Lipps, O. (2000): *Multimodalität im Personenverkehr im intrapersonellen Längsschnitt*. Arbeitsbericht des Instituts für Verkehrswesen, Universität Karlsruhe no. 00-7.

Chorherr, C. (2014). *Flexibilisierung der Stellplatzverpflichtung*. Retrieved April 04, 2016, from http://chorherr.twoday.net/stories/453140592/

Christopoulou, P., & Pitsiava-Latinopoulou, M. (2012). Development of a model for the estimation of pedestrian level of service in Greek urban areas. *Procedia: Social and Behavioral Sciences*, *48*, 1691–1701. doi:10.1016/j.sbspro.2012.06.1144

Church, A., Frost, M., & Sullivan, K. (2000). Transport and social exclusion in London. *Transport Policy*, *7*(3), 195–205. doi:10.1016/S0967-070X(00)00024-X

City of Vienna. (2008). Vienna Garage Ordinance 2008. *Wiener Garagengesetz 2008 - WGarG 2008*. Retrieved from http://www.wien.gv.at/recht/landesrecht-wien/rechtsvorschriften/html/b1000000.htm

Cole, R., Burke, M., Leslie, E., Donald, M., & Owen, N. (2010). Perceptions of representatives of public, private, and community sector institutions of the barriers and enablers for physically active transport. *Transport Policy*, *17*(6), 496–504. doi:10.1016/j.tranpol.2010.05.003

Cologne Website. (2012, December). Retrieved from http://www.koeln.de/koeln/ein_jahr_koeln_3__ parkplatznot_in_der_zweiten_reihe_ist_doch_ immer_was_frei_670143.htm

Commission of the European Communities. (2001). *White Paper - European Transport Policy for 2010: Time to decide*. Retrieved from http://ec.europa.eu/transport/white_paper/documents/doc/lb_texte_complet_de.pdf

Conveney, J., & O`Dwyer, L. A. (2009). *Effects of mobility and location on food access*. Healthplace.

COST Action Report No. 342. (2005). Parking policies and the effects on economy and mobility (Report on COST Action 342). European Cooperation in the Field of Scientific and Technical Research. The Netherlands: Technical Committee on Transport.

Cubukcu, E., Hepguzel, B., Onder, Z., & Tumer, B. (2015). Active Living for Sustainable Future: A Model to Measure Walk Scores via Geographic Information Systems. *Procedia: Social and Behavioral Sciences*, *168*, 229–237. doi:10.1016/j.sbspro.2014.10.228

Currie, G. (2004). Gap analysis of public transport needs: measuring spatial distribution of public transport needs and identifying gaps in the quality of public transport provision. *Transportation Research Record: Journal of the Transportation Research Board*, (1895), 137-146.

Daley, M., & Rissel, C. (2011). Perspectives and images of cycling as a barrier or facilitator of cycling. *Transport Policy*, *18*(1), 211–216. doi:10.1016/j.tranpol.2010.08.004

Dalkmann, H., & Brannigan, C. (2007). *Transport and climate change. In A Sourcebook for Policy-Makers in Developing Cities: Module (5th ed.)*. Gesellschaft für Technische Zusammenarbeit–GTZ Eschborn.

Dalvi, M. Q., & Martin, K. M. (1976). The measurement of accessibility: Some preliminary results. *Transportation*, *5*(1), 17–42. doi:10.1007/BF00165245

Daniel, B. D., Nor, S. N. M., Rohani, M. M., Prasetijo, J., Aman, M. Y., & Ambak, K. (2016, January). Pedestrian Footpath Level of Service (FOOT-LOS) Model for Johor Bahru. In *MATEC Web of Conferences* (Vol. 47). EDP Sciences.

Daniels, R., & Mulley, C. (2013). Explaining walking distance to public transport: The dominance of public transport supply. *Journal of Transport and Land Use*, *6*(2), 5–20. doi:10.5198/jtlu.v6i2.308

Daskin, M. S. (1982). Application of an ecpected covering location model to emergency medical system design. *Decision Sciences*, *13*, 416–139. doi:10.1111/j.1540-5915.1982.tb00159.x

Daskin, M. S. (1983). A maximal expected covering location model: Formulation properties and heuristic solution. *Transportation Science*, *17*(1), 48–69. doi:10.1287/trsc.17.1.48

Das, S., & Pandit, D. (2013). Importance of user perception in evaluating level of service for bus transit for a developing country like India: A review. *Transport Reviews*, *33*(4), 402–420. doi:10.1080/01441647.2013.789571

Davidge, W. R. (1924). *The London Traffic Act, 1924*. Liverpool University Press.

Davies, D., Brook-Carter, N., & Gardner, G. (2000). *Institutional and organisational attitudes to cycling (No. TRL Report 468)*. Crowthorne, UK: Transport Research Laboratory - TRL.

Davis, A. Y., Pijanowski, B. C., Robinson, K., & Engel, B. (2010). The environmental and economic costs of sprawling parking lots in the United States. *Land Use Policy*, *27*(2), 255–261. doi:10.1016/j.landusepol.2009.03.002

De Ona, J., de Ona, R., & Lopez, G. (2015). Transit service quality analysis using cluster analysis and decision trees: A step forward to personalized marketing in public transportation. *Transportation*, 1–23. doi:10.1007/s11116-015-9615-0

DellOlio, L., Ibeas, A., & Cecın, P. (2010). Modelling user perception of bus transit quality. *Transport Policy*, *17*(6), 388–397. doi:10.1016/j.tranpol.2010.04.006

DellOlio, L., Ibeas, A., & Cecın, P. (2011). The quality of service desired by public transport users. *Transport Policy*, *18*(1), 217–227. doi:10.1016/j.tranpol.2010.08.005

Department for Transport (DfT). (2000). *Transport 2010 - The 10 year Plan*. Author.

derstandard. (2013). *Der "Qualitätswahnsinn" und seine neuen Feinde*. Retrieved April 04, 2016, from http://derstandard.at/1363710562791/Der-Qualitaetswahnsinn-und-seine-neuen-Feinde?seite=2#forumstart

Development & Building Control Division Roads Group. L. T. A., Singapore. (2005). *Handbook on VEHICLE PARKING PROVISION in development proposals*. Retrieved from http://www.distillate.ac.uk/about/about.php

Dixon, L. (1996). Bicycle and pedestrian level-of-service performance measures and standards for congestion management systems. *Transportation Research Record: Journal of the Transportation Research Board*, (1538), 1-9.

Dodson, J., Gleeson, B., & Sipe, N. G. (2004). *Transport disavantage and social status: a review of literature and methods*. Brisbane: Urban Policy Program, Griffith University.

Dökmeci, V., & Berköz, L. (1994). Transformation of Istanbul from a monocentric to a polycentric city. *European Planning Studies*, *2*(2), 193–205. doi:10.1080/09654319408720259

Dorner, A., Herry, M., & Schuster, M. (1997). *Werkstattbericht Parkraumbewirtschaftung in Wien*. Wien, Österreich: Magistratsabteilung 18.

Dortmund. (2006). *Stellplatzkosten in Dortmund - Stand: 2006*. Dortmund, Deutschland: Stadtplanungs- und Bauordnungsamt Dortmund.

Düsseldorf. (2016). *ÖPNV-Lagegunst der Wohnstandorte nach Stadtteilen und ÖPNV-Einzugsbereiche ausgewählter Haltestellen in Düsseldorf*. Retrieved April 4, 2016, from https://www.duesseldorf.de/bauaufsichtsamt/baulexikon/stellplaetze.shtml

Eboli, L., & Mazzulla, G. (2012). Performance indicators for an objective measure of public transport service quality. *European Transport*, *51*, 1–21.

Economies of Scale. (2016). Retrieved from https://en.wikipedia.org/wiki/Economies_of_scale

EIFER. (2010-2013). *Software development and use Agreement Software regarding MARS - Application for Strasbourg and Mulhouse (France)*. EIFER.

El Ghaoui, L., & Lebret, H. (1997). Robust solutions to least-squares problems with uncertain data. *SIAM Journal on Matrix Analysis and Applications*, *18*(4), 1035–1064. doi:10.1137/S0895479896298130

El Ghaoui, L., Oustry, F., & Lebret, H. (1998). Robust solutions to uncertain semidefinite programs. *SIAM Journal on Optimization*, *9*(1), 33–52. doi:10.1137/S1052623496305717

Ellis, G., Hunter, R., Tully, M. A., Donnelly, M., Kelleher, L., & Kee, F. (2015). Connectivity and physical activity: Using footpath networks to measure the walkability of built environments. *Environment and Planning. B, Planning & Design*.

Eltis Website. (2014, September 29). Retrieved from http://www.eltis.org/discover/case-studies/ispark-parking-turkey

Emberger, G., & Ibesich, N. (2006). MARS in Asia - How a model can help and influence decision makers. *Proceedings of CORP 2005, 11th International Conference on Urban Planning & Regional Development in the Information Society*, 193-200.

Emberger, G. (2005). *Freiht Transport – A Holistic Approach. "Beiträge zu einer ökologisch und sozial verträglichen Verkehrsplanung*. Institut für Verkehrsplanung und Verkehrstechnik, Technische Universität Wien.

Emberger, G., & Pfaffenbichler, P. (2016). *Wechselwirkungen und Rückkopplungen im Inividualverkehr und im Güterverkehr*. Wien: Paradigmenwechsel im Verkehrswesen, Österreichische Verkehrswissenschaftliche Gesellschaft - ÖVG.

EMRECU. (1998). *EU- DG XVll Project*. Author.

Envall, P. (2009). *What is hindering good cycle planning: how do city planners think?* Paper presented at the Velo-city 2009 - Re-cycling cities, Bruxelles, Belgium.

Envall, P. (2012). *What is the role of Bicycle Parking for Increased Cycling in Large Cites? A Literature Review*. Stockholm: CyCity.

EPA Website. (n.d.). Retrieved from https://www.epa.gov/smartgrowth/smart-growth-and-transportation

European Commission Directorate-General Energy and Transport. (2011). *White paper on Transport - Roadmap to a single European transport area — Towards a competitive and Resource efficient transport system*. doi:10.2832/30955

Eurostat. Statistics explained. (2016). *Road safety statistics on regional level*. Retrieved from http://ec.europa.eu/eurostat/statistics-explained/index.php/Road_safety_statistics_at_regional_level

Everitt, B. (2011). *Cluster analysis*. Chichester, UK: Wiley. doi:10.1002/9780470977811

Fabian, T. (2009, November). *Quartier Vauban, Freiburg - Autoreduziertes Verkehrskonzept*. Paper presented at Innovatives Parkraummanagement – Chance für Wirtschaft & Umwelt (Lebensministerium), Bregenz, Österreich.

Fairnie, G. A., Wilby, D. J., & Saunders, L. E. (2016). Active travel in London: The role of travel survey data in describing population physical activity. *Journal of Transport & Health*, *3*(2), 161–172. doi:10.1016/j.jth.2016.02.003

Fancher, R.E. (1996). Pioneers of Psychology. New York: W.W. Norton & Company.

FATIMA. (1997-1998). *Financial Assistance for Transport Integration in Metropolitan Areas*. Retrieved from http://www.ivv.tuwien.ac.at/forschung/projekte/international-projects/fatima-1997.html

Fellesson, M., & Friman, M. (2008). Perceived satisfaction with public transport service in nine European cities. *Journal of the Transportation Research Forum*, *47*(3), 93–103.

Ferstl, K. (2013). *Verhandlungsschrift*. Baden: Bezirkshauptmannschaft Baden.

FGSV. (1991). *Empfehlungen für Anlagen des ruhenden Verkehrs (EAR 91)*. Köln, Deutschland: Forschungsgesellschaft für Straßen- und Verkehrswesen.

FGSV. (2015). *Handbuch für die Bemessung von Straßenverkehrsanlagen*. Köln: Forschungsgesellschaft für Straßen- und Verkehrswesen.

Fleiss, J. L. (1981). *Statistical Methods for Rates and Proportions* (2nd ed.). New York: Wiley.

Florez, J., Muniz, J., & Portugal, L. (2014). Pedestrian quality of service: Lessons from Maracanã Stadium. *Procedia: Social and Behavioral Sciences*, *160*, 130–139. doi:10.1016/j.sbspro.2014.12.124

Franke, S. (2004). *Die „neuen Multimodalen* (Vol. 3). Internationales Verkehrswesen.

Front Seat Management, L. L. C. Walk Score® Methodology. (2011). Retrieved from https://www.walkscore.com/methodology, at July 30, 2016.

FSV (2012). *RVS 03.01.11: Beurteilung des Verkehrsablaufs auf Straßen*. Forschungsgesellschaft Straße-Schiene-Verkehr, 2012.

FSV. (1994). *RVS 3.7: Überprüfung der Anlageverhältnisse von Straßen*. Wien: Forschungsgemeinschaft für Verkehrs- und Straßenwesen.

FSV. (2001). Grundlagen und Motive bezüglich der Organisation und Anzahl der Stellplätze für Fahrzeuge im Individualverkehr. Wien, Österreich: Österreichische Forschungsgesellschaft Straße-Schiene-Verkehr, Arbeitspapier Nr. 1.

FSV. (2001a). *RVS 03.02.13: Nicht motorisierter Verkehr - Radverkehr*. Wien: FSV - Österreichische Forschungsgesellschaft Straße, Schiene, Verkehr.

FSV. (2001b). *RVS 03.04.12: Stadtstraßenquerschnitte; Querschnittsgestaltung von Innerortsstraßen*. Wien: FSV - Österreichische Forschungsgesellschaft Straße, Schiene, Verkehr.

FSV. (2008). *Organisation und Anzahl der Stellplätze für den Individualverkehr – RVS 03.07.11*. Wien, Österreich: Forschungsgesellschaft Straße-Schiene-Verkehr.

FSV. (2010). *Entwurfsgrundlagen für Garagen – RVS 03.07.32*. Wien, Österreich: Forschungsgesellschaft Straße-Schiene-Verkehr.

Gabrel, V., Murat, C., & Thiele, A. (2014). Recent advances in robust optimization: An overview. *European Journal of Operational Research*, *235*(3), 471–483. doi:10.1016/j.ejor.2013.09.036

Gallimore, J. M., Brown, B. B., & Werner, C. M. (2011). Walking routes to school in new urban and suburban neighborhoods: An environmental walkability analysis of blocks and routes. *Journal of Environmental Psychology*, *31*(2), 184–191. doi:10.1016/j.jenvp.2011.01.001

Gallin, N. (2001). Quantifying pedestrian friendliness--guidelines for assessing pedestrian level of service. *Road & Transport Research*, *10*(1), 47.

Gan, G., Ma, C., & Wu, J. (2007). *Data clustering: theory, algorithms, and applications* (Vol. 20). Siam. doi:10.1137/1.9780898718348

Garcia Blasco, M. (2007). *Eine Innovative Verkehrsorganisaion für den 9. Bezirk (Master).* Wien: TU Wien.

Garrido, C., de Ona, R., & de Ona, J. (2014). Neural networks for analyzing service quality in public transportation. *Expert Systems with Applications*, *41*(15), 6830–6838. doi:10.1016/j.eswa.2014.04.045

Gehl, J. (2013). *Cities for people*. London: Island Press.

Gendreau, M., Laporte, G., & Semet, F. (1997). Solving an ambulance location model by tabu search. *Location Science*, *5*(2), 75–88. doi:10.1016/S0966-8349(97)00015-6

Germany. (2014). *Unfallentwicklung auf den Bundesautobahnen der Region im Fünfjahresvergleich.* Retrieved from www.pd-bs.polizei-nds.de/.../Verkehrsunfallstatistik_BAB_im_5-Jahresvergleich_bis_

Geurs, K. T., & Van Wee, B. (2004). Accessibility evaluation of land-use and transport strategies: Review and research directions. *Journal of Transport Geography*, *12*(2), 127–140. doi:10.1016/j.jtrangeo.2003.10.005

GHG Transpord. (2009-2011). *EU-Project, 7th FP.* Retrieved from https://tiss.tuwien.ac.at/research/project.xhtml?projectId=8402

Goodwin, P. (1997). Solving congestion: When we must not build roads, increase spending, lose votes, damage the economy or harm the environment, and will never find an equilibrium. Inaugural lecture for the Professorship of Transport Policy, University College London.

Gori, S., Nigro, M., & Petrelli, M. (2014). Walkability Indicators for Pedestrian-Friendly Design. *Transportation Research Record: Journal of the Transportation Research Board*, (2464), 38-45.

Gorissen, B. L., Yanıkoglu, I., & den Hertog, D. (2015). A practical guide to robust optimization. *Omega*, *53*, 124–137. doi:10.1016/j.omega.2014.12.006

Grant, J. (2005). Planning Responses to Gated Communities in Canada. *Housing Studies*, *20*(2), 273–285. doi:10.1080/026730303042000331772

Greater London Authority. (2008). *The London Plan, Spatial Development Strategy for Greater London Consolidated with Alterations since 2004.* Retrieved from www.london.gov.uk/thelondonplan

Gruber, J. (2009). Verkehrskonzept 15: zeitlich gestaffelte Umsetzung des Äquidistanzmodells am Beispiel Wien 15. *Verkehrskonzept, 15.*

Guo, Z. (2009). Does the pedestrian environment affect the utility of walking? A case of path choice in downtown Boston. *Transportation Research Part D, Transport and Environment*, *14*(5), 343–352. doi:10.1016/j.trd.2009.03.007

Guo, Z. (2013). Does residential parking supply affect household car ownership? The case of New York City. *Journal of Transport Geography*, *26*, 18–28. doi:10.1016/j.jtrangeo.2012.08.006

Haider, A. (2010). *Urbane europäische Ballungsräume – Potenzial für eine stadtverträgliche Mobilität (Unpublished master's theses).* Wien, Österreich: Technische Universität Wien.

Hajna, S., Dasgupta, K., Halparin, M., & Ross, N. A. (2013). Neighborhood walkability: Field validation of geographic information system measures. *American Journal of Preventive Medicine*, *44*(6), e55–e59. doi:10.1016/j.amepre.2013.01.033 PMID:23683990

Haken, H. (1983). *Advanced Synergetics - Instability Hierarchies of Self-Organizing Systems and Devices* (Vol. 20). Springer-Verlag.

Halkidi, M., Batistakis, Y., & Vazirgiannis, M. (2002). Cluster validity methods: Part I. *SIGMOD Record*, *31*(2), 40–45. doi:10.1145/565117.565124

Hamburg. (2013). *Senat beschließt weitere Erleichterungen für den Wohnungsbau*. Retrieved April 04, 2016, from http://www.hamburg.de/pressearchiv-fhh/4128858/2013-10-29-bsu-wohnungsbau/

Handy, S. L. (2002). Accessibility-vs. mobility-enhancing strategies for addressing automobile dependence in the US. Institute of Transportation Studies.

Handy, S. (1994). Highway blues: Nothing a little accessibility can't cure. *ACCESS Magazine*, *1*(5), 3–7.

Hansen, W. G. (1959). How accessiblity shapes land use. *Journal of the American Planning Institute*, *25*(2), 73–76. doi:10.1080/01944365908978307

Härdle, W., & Simar, L. (2007). *Applied multivariate statistical analysis*. Berlin: Springer.

Hasan, T., Siddique, A., Hadiuzzaman, M. D., & Musabbir, S. R. (2016). Determining the Most Suitable PLOS Methodology for Dhaka City through Synthesis of Subjective and Objective Measurements. *TRB 2015 Annual Meeting*. Transportation Research Board of the National Academies.

Haurin, D. R., & Brasington, D. (1996). School Quality and Real House Prices: Inter- and Intrametropolitan Effects. *Journal of Housing Economics*, *5*(4), 351–368. doi:10.1006/jhec.1996.0018

Heinen, E., van Wee, B., & Maat, K. (2010). Commuting by Bicycle: An Overview of the Literature. *Transport Reviews*, *30*(1), 59–96. doi:10.1080/01441640903187001

Hemedoglu, H. (2012). Measuring service quality in metro services: a comparison of perceived service quality and expected service quality. *Journal of Business Economics Institute of Business School of Istanbul University – Management*, *23*(72), 25-47.

Hernandez, S., & Monzon, A. (2016). Key factors for defining an efficient urban transport interchange: Users perceptions. *Cities (London, England)*, *50*, 158–167. doi:10.1016/j.cities.2015.09.009

Herry, M., Rosinak, W., & Schuster, M. (1999). Parkraumbewirtschaftung in Wien: Vorgehensweise und Erfahrungen. *Internationales Verkehrswesen*, *51*(6), 244–249.

Hesse, M. (1997). *Wirtschaftsverkehr stadtverträglich. Der Strukturwandel in der Logistik und seine Bedeutung für die Stadtentwicklung. Stadtforschung aktuell Nr. 61*. Basel: Birkhäuser Verlag.

Heusch/Boesefeldt. (1997). *Überprüfung der Bewertungsmodelle für die Bundesfernstraßen beim Bundesverkehrswegeplan 1992 anhand beobachteter Verhältnisse*. Aachen: Author.

Hidber, C. (1992). Optimierung in Verkehrsplanung, Transporttechnik und Logistik. Schriftenreihe des Instituts für Verkehrsplanung, Transporttechnik, Straßen- und Eisenbahnbau (IVT) – Nr. 91, Zürich.

Highway Capacity Manual (HCM). (1950). *Practical Application of Research*. Washington, DC: By the Committee on Highway Capacity Department of Traffic and Operations Research Board.

Highway Capacity Manual. (1965). *National Research Council (U.S.) Highway Research Board*. Committee on Highway Capacity.

Hine, J., & Mitchell, F. (2001). The role of transport in social exclusion in urban Scotland. *Scottish Executive*, 11-42.

Hine, J., & Grieco, M. (2003). Scatters and clusters in time and space: Implications for delivering integrated and inclusive transport. *Transport Policy*, *10*(4), 299–306. doi:10.1016/S0967-070X(03)00055-6

Hofstetter, F. (2015). *A Public Parking Management System for Zurich* (Master Thesis). University of Zurich, Zurich, Switzerland.

Hogan, K., & ReVelle, C. (1986). Concepts and applications of backup coverage. *Management Science*, *32*(11), 1434–1444. doi:10.1287/mnsc.32.11.1434

Hong Kong Transport Department. (2002). *The Second Parking Demand Study*. Hong Kong: Author.

Hoogendoorn, S. P., & Bovy, P. H. (2004). Pedestrian route-choice and activity scheduling theory and models. *Transportation Research Part B: Methodological*, *38*(2), 169–190. doi:10.1016/S0191-2615(03)00007-9

Horton, D. (2007). Fear of Cycling. In D. Horton, P. Rosen, & P. Cox (Eds.), *Cycling and Society* (pp. 133–152). Aldershot, UK: Ashgate Publishing.

Housing Asked Prices in Ankara. (2016). Available at: http://www.hurriyetemlak.com/

Hsiao, S., Lu, J., Sterling, J., & Weatherford, M. (1997). Use of geographic information system for analysis of transit pedestrian access. *Transportation Research Record: Journal of the Transportation Research Board*, (1604), 50-59.

Huber, J., & Yilmaz-Huber, N. (2013). *Leistbarer Wohnraum in Tirol*. Innsbruck, Österreich: Arbeiterkammer Tirol.

Humpel, N., Owen, N., Iverson, D., Leslie, E., & Bauman, A. (2004). Perceived environment attributes, residential location, and walking for particular purposes. *American Journal of Preventive Medicine*, *26*(2), 119–125. doi:10.1016/j.amepre.2003.10.005 PMID:14751322

Hupkes, G. (1982). The Law of Constant Travel Time and Trip-Rates. *Futures*, *14*(1), 38–46. doi:10.1016/0016-3287(82)90070-2

Hupkes, G. (1982, February). The Law of Constant Travel Time and Trip Rates. *Futures*.

Hurni, A. (2006). Transport and social disadvantage in Western Sydney: a partnership research project. *Western Sydney Community Forum*, 11-93.

Iannoni, A. P., Morabito, R., & Saydam, C. (2009). An optimization approach for ambulance location and the districting of the response segments on highways. *European Journal of Operational Research*, *195*(2), 528–542. doi:10.1016/j.ejor.2008.02.003

Ihde, G. B. (1991). *Transport, Verkehr, Logistik: gesamtwirtschaftliche Aspekte und einzelwirtschaftliche Handhabung. 2., völlig überarbeitete und erweiterte Auflage*. München: Vahlen Verlag.

Ispark Website. (n.d.). Retrieved from http://ispark.istanbul/projeler/otopark-master-plani/

Istanbul Population 2016. (n.d.). Retrieved December 23, 2016, from http://worldpopulationreview.com/world-cities/istanbul-population/

ITMP. (2011). *2023 Istanbul Metropolitan Area Transport Master Plan*. Istanbul: Istanbul Metropolitan Municipality.

Jacobsen, P. L. (2003). Safety in numbers: More walkers and bicyclists, safer walking and bicycling. *Injury Prevention*, *9*(3), 205–209. doi:10.1136/ip.9.3.205 PMID:12966006

Jaensirisak, S., Emberger, G., Pfaffenbichler, P. C., & Ibesich, N. (2006). *Application of the Metropolitan Activity Relocation Simulator (MARS). Model in Decision Making Process for Sustainable Development.* Paper presented at the Technology and Innovation for Sustainable Development Conference - TISD, Khon Kean, Thailand.

Jain, A. K., & Dubes, R. C. (1988). *Algorithms for clustering data.* Prentice-Hall, Inc.

Jain, A. K., Murty, M. N., & Flynn, P. J. (1999). Data clustering: A review. *ACM Computing Surveys, 31*(3), 264–323. doi:10.1145/331499.331504

Janik, W. (2002). Parkmöglichkeiten für Pkw - Mikrozensus Juni 2001. *Statistische Nachrichten, 4*, 294–300.

Jaskiewicz, F. (1999). *Pedestrian Level of Service Based on Trip Quality, TRB Circular.* TRB.

Jensen, S. (2007). Pedestrian and bicyclist level of service on roadway segments. *Transportation Research Record: Journal of the Transportation Research Board*, (2031), 43-51.

Jittrapirom, P., & Emberger, G. (2011). Off-street residential parking organisation - a review of current practices. *Proceedings of the Eastern Asia Society for Transportation Studies, 8*, 16.

Jones, M. (2001). Promoting cycling in the U.K. – Problems experienced by the practitioners. *World Transport Policy and Practice, 7*(3), 7–12.

Kadali, B. R., & Vedagiri, P. (2016). Review of Pedestrian Level of Service: Perspective in Developing Countries. *Transportation Research Record: Journal of the Transportation Research Board*, (2581), 37-47.

Käfer. (2011). *Ordnungspolitik und energieeffiziente Raumstrukturen: Evaluierung von Instrumenten und Least-Cost.* Wien, Österreich: Klima- und Energiefonds, Blue Globe Report, 1/2011

Kaiser, E. J., Lawrence, D. F., Peter, O. E., & Thomas, L. S. (1995). *Health and Community Design.* Urbana, IL: University of Illinois Press.

Kane, L., & Behrens, R. (2002). *Transport Planning Models - An Historical and Critical Review.* Paper presented at the 21st Annual South African Transport Conference "Towards Building Capacity and Accelerating Delivery", South Africa.

Karacasu, M., & Bilgic, S. (1999). Survey study sample in Eskisehir/Turkey and Solution offers for transportation problems for Eskisehir city.*Proceedings of 2nd National Urban Infrastructure Symposium in Turkey*, 87-99.

Karadeniz, M., & Unver, N. M. (2014). Customer's perception of service and satisfaction in maritime transport in the city: An application. *Journal of Offering of Marmara University, 11*(42), 43–68.

Karajan, J. (2001). Wirkung und Festlegung von Stellplatzvorschreibungen für Wohnnutzungen. Innsbruck, Österreich: Institut für Straßenbau und Verkehrsplanung, Universität Innsbruck, Schriftenreihe, Heft 34.

Karatas, P. (2015). *Determination of Pedestrian Level of Service for Walkways: METU Campus Example* (MS Thesis). Middle East Technical University, Ankara, Turkey.

Karatas, P., & Tuydes-Yaman, H. (2016). A pairwise Comparison of Different Pedestrian Level of Service (PLOS) Ratings. *Transportation Research Board (TRB)95th Annual Meeting.*

Karim, H. A., & Azmi, D. I. (2013). Convenience and Safety of Walking Experience in Putrajaya Neighbourhood Area. *Procedia: Social and Behavioral Sciences, 101*, 318–327. doi:10.1016/j.sbspro.2013.07.206

Kelly, J. A., & Clinch, J. P. (2009). Temporal variance of revealed preference on-street parking price elasticity. *Transport Policy, 16*(4), 193–199. doi:10.1016/j.tranpol.2009.06.001

Kenyon, S., Lyons, G., & Rafferty, J. (2002). Transport and social exclusion: Investigating the possibility of promoting inclusion through virtual mobility. *Journal of Transport Geography*, *10*(3), 207–219. doi:10.1016/S0966-6923(02)00012-1

Khalid, U. A., Bachok, S., Osman, M. M., & Mansor, I. (2014). User perceptions of rail public transport services in Kuala Lumpur, Malaysia: Ktm komuter. Amer International Conference on Quality of Life. *Procedia: Social and Behavioral Sciences*, *153*, 566–573. doi:10.1016/j.sbspro.2014.10.089

Kirby, A. (1982). *The Politics of Location*. London: Methuen.

Kleijnen, M., Lee, N., & Wetzels, M. (2009). An exploration of consumer resistance to innovation and its antecendents. *Journal of Economic Psychology*, *30*(3), 344–357. doi:10.1016/j.joep.2009.02.004

Knoflacher H. (1981). *Ungenutzte Möglichkeiten zur Erhöhung der Verkehrssicherheit*. ARBÖ-Verkehrsjurist, Nr 55/56, 1981

Knoflacher H. (2015). Road safety in urban areas, transport planning and traffic safety: Making cities, road and vehicles safer. *CRC 2016*.

Knoflacher H. Kloss H. (1980). *Verkehrskonzeption Wien. Konsulentengutachten – Radverkehr*. Magistratsabteilung 18 Wien.

Knoflacher, H. (1970). *Methoden der Unfallauswertung als Beitrag zur Erhöhung der Verkehrssicherheit auf Stadtstraßen*. Kleine Fachbuchreihe des Kuratoriums für Verkehrssicherheit (KfV), Band 9/1970.

Knoflacher, H. (1980). Zusammenhang zwischen Parkraumausstattung bzw. -organisation und Einwohner- sowie Arbeitsplatzentwicklung bzw. -verschiebung. *Straßenverkehrstechnik*, (1), 1-7.

Knoflacher, H. (1981). *Human Energy Expenditure in different modes, Implications for town planning*. Paper presented at the International Conference on Surface Transportation System Performance, Washington, DC.

Knoflacher, H. (1981). *Human Energy Expenditure in Different Modes: Implications for Town Planning*. Paper presented at the International Symposium on Surface Transportation System Performance, Washington, DC.

Knoflacher, H. (1981). *Human Energy Expenditure in Different Modes: Implications for Town Planning*. Paper presented at the International Symposium on Surface Transportation System Performance.

Knoflacher, H. (1982). Zum Problem der subjektiven Widerstände in Transportsystemen. *Int Verkehrswes*, *34*(6).

Knoflacher, H. (1985). Ungenutzte Möglichkeiten zur Erhöhung der Verkehrssicherheit. Österreichische Gemeindezeitung, Wien 1985/2

Knoflacher, H. (1989). A 23 Südost-Tangente. Strukturelle Auswirkungen 1978 – 1988. Durchgeführt im Auftrag des Magistrats der Stadt Wien, MA 18 – Stadtstrukturplanung. Wien.

Knoflacher, H. (1990): Wie sicher sind Autobahnen? *Verkehr & Umwelt*, *4*(2), 22-26.

Knoflacher, H. (2001). *Verkehrssicherheit Wien – Analyse des Bestands und Vorschläge für die Zukunft*. Durchgeführt im Auftrag des Magistrats der Stadt Wien – MA 46.

Knoflacher, H., & Schopf, J. M. (2001). Modellvorhaben Sanfte Mobilitäts-Partnerschaft. *Internationales Verkehrswesen*, *53*(3), 86-89.

Knoflacher, H., Mailer, M., Schopf, J.M., Fischer, P., Emberger, G., & Pfaffenbichler, P. (2008). Multimodale Dimensionierung von Straßen aus der Beschreibung des Verkehrsablaufes, der funktionellen Bedeutung sowie der verkehrlichen Auswirkungen. *Beiträge zu einer ökologisch und sozial verträglichen Verkehrsplanung, 1/2008*.

Knoflacher. (1987). *Verkehrsplanung für den Menschen Band 1 Grundstrukturen.* Wien: Verlag Orac.

Knoflacher, H. (1985). *Katalysatoren für Nichtmotorisierte.* Eigenverlag.

Knoflacher, H. (1986). Kann man Straßenbauten mit Zeiteinsparungen begründen. *Internationales Verkehrswesen, 38*(6), 454–457.

Knoflacher, H. (1986). Verkehrsplanung für den Menschen: Vol. 1. *Grundstrukturen.* Orac Verlag.

Knoflacher, H. (1987). *Verkehrsplanung für den Menschen* (Vol. 1). Wien: Grundstrukturen. Verlag Orac.

Knoflacher, H. (1993). Does the Development of Mobility in Traffic Follow a Pattern? *History and Technology, 15*, 125–140.

Knoflacher, H. (1994). Do We Use the "Level of Service" Concept in the Right Way?. *2nd International Symposium on Highway Capacity.*

Knoflacher, H. (1995). Economy of Scale - Die Transportkosten und das Ökosystem. *GAIA, 4*(2), 100–108.

Knoflacher, H. (1995). *Fußgeher- und Fahrradverkehr.* Planungsprinzipien. Böhlau Verlag Wien – Köln – Weimar.

Knoflacher, H. (1995). *Sustainable Development- Öko-City. Projektgruppe 1: "Mobilität in der Stadt" (Stadt und Verkehr). Durchgeführt im Auftrag der Wiener Internationalen Zukunftskonferenz.* WIZK.

Knoflacher, H. (1996). *Zur Harmonie von Stadt und Verkehr. Freiheit vom Zwang zum Autofahren. 2., verbesserte und erweiterte Auflage.* Böhlau Verlag Wien – Köln - Weimar.

Knoflacher, H. (1997). *Landschaft ohne Autobahnen. Für eine zukunftsorientierte Verkehrsplanung.* Wien: Böhlau Verlag.

Knoflacher, H. (2001). *Stehzeuge. Der Stau ist kein Verkehrsproblem.* Wien: Böhlau Verlag.

Knoflacher, H. (2006). A new way to organize parking: The key to a successful sustainable transport system for the future. *Environment and Urbanization, 18*(2), 387–400. doi:10.1177/0956247806069621

Knoflacher, H. (2007). *Grundlagen der Verkehrs- und Siedlungsplanung.* Wien, Köln, Weimar: Böhlau Verlag.

Knoflacher, H. (2013). *Virus Auto: die Geschichte einer Zerstörung.* Wien: Böhlau.

Knoflacher, H., & Macoun, T. (1989). *Ökologie und Strassenverkehr, Reports UBA-89-035.* Wien: Umweltbundesamt.

Knoflacher, H., & Ocalir, E. V. (2011). Eco-methodology for urban and transport planning for the future eco-technology. In O. Yalciner Ercoskun (Ed.), *Green and Ecological Technologies for Urban Planning: Creating Smart Cities* (pp. 154–173). IGI Global. doi:10.4018/978-1-61350-453-6.ch009

Knoflacher, H., Schopf, J. M., & Brezina, T. (2007). *Zusammenstellung und Analyse der Vorschriften und Richtlinien zu Anzahl und Organisation der Stellplätze im MIV für Österreich in Hinblick auf eine Verbesserung der Umweltverhältnisse.* Wien, Österreich: Bundesministerium für Land- und Forstwirtschaft, Umwelt und Wasserwirtschaft.

Kovacic, W. (1999). *Der österreichische Bundesverkehrswegeplan - Abgeschlossene und laufende Arbeitspakete.* Academic Press.

KTH, DTU, VTI, TOI, Austria Tech, & TSU. (2011). Deliverable 5: How to manage barriers to formation and implementation of policy packages in transport (OPTIC - Optimal Policies for Transport in Combination). Oslo: TOI.

Kulekci, O., Ayozen, Y. A., & Inac, H. (2015). *IstanbulOtopark Master Plani Calismalari.* ROTRASA Road and Traffic Safety Congress, Ankara, Turkey.

Kurtulus, H. (2011). Gated communities as a representation of new upper and middle classes in İstanbul. *İ.Ü. Siyasal Bilgiler Fakultesi Dergisi*, *44*, 49–65.

Lance, D., & McNeil, I. (1996). *Biographical Dictionary of the history of Technology*. Taylor & Francis.

Landis, B., Vattikuti, V., Ottenberg, R., McLeod, D., & Guttenplan, M. (2001). Modeling the roadside walking environment: pedestrian level of service. *Transportation Research Record: Journal of the Transportation Research Board*, (1773), 82-88.

Landman, K. (2000a). Gated communities and urban sustainability: taking a closer look at the future.*2nd Southern African Conference on Sustainable Development in the Built Environment*.

Landman, K. (2000b). *An overview of enclosed neighbourhoods in South Africa*. Pretoria: CSIR Publication.

Laporte, G., Louveaux, F. V., Semet, F., & Thirion, A. (2009). Applications of the double standard model for ambulance location. In Innovations in Distribution (pp. 235-249). Springer. doi:10.1007/978-3-540-92944-4_12

Leslie, E., Coffee, N., Frank, L., Owen, N., Bauman, A., & Hugo, G. (2007). Walkability of local communities: Using geographic information systems to objectively assess relevant environmental attributes. *Health & Place*, *13*(1), 111–122. doi:10.1016/j.healthplace.2005.11.001 PMID:16387522

Lewin, K. (1951). *Field theory in social science: Selected theoretical papers* (D. Cartwright, Ed.). Harper and Row.

Li, H. R. (2016). Study on Green Transportation System of International Metropolises. *Procedia Engineering*, *137*, 762–771. doi:10.1016/j.proeng.2016.01.314

Litman, T. (2003). Economic value of walkability. *Transportation Research Record: Journal of the Transportation Research Board*, (1828), 3-11.

Litman, T. (2008). *Parking Management: Strategies, Evaluation and Planning*. Victoria Transport Policy Institute. Retrieved from www.vtpi.org/park_man.pdf

Litman, T. A. (2013). *Transport Elasticities: Impacts on Travel Behaviour*. Retrieved from Bonn, Germany: http://www.sutp.org/files/contents/documents/resources/B_Technical-Documents/GIZ_SUTP_TD11_Transport-Elasticities_EN.pdf

Litman, T. (2011). *Evaluating accessibility for transportation planning: measuring people's ability to reach desired goods and activities*. London: Victoria Transport Policy Institute.

Liu, B., Wang, Z. W., & Zhou, Y. B. (2014). Investigating Pedestrian Behavior Differences of Walking Streets and their Effect Factors with Three Wuhu Cases. *Mechanics of Materials*, *587*, 1879–1883.

Li, X., Zhao, Z., Zhu, X., & Wyatt, T. (2011). Covering models and optimization techniques for emergency response facility location and planning: A review. *Mathematical Methods of Operations Research*, *74*(3), 281–310. doi:10.1007/s00186-011-0363-4

Low, N., Gleeson, B., & Rush, E. (2003). Making Believe: Institutional and Discursive Barriers to Sustainable Transport in Two Australian Cites. *International Planning Studies*, *8*(2), 93–114. doi:10.1080/13563470305158

Lucas, K. (2012). Transport and social exclusion: Where are we now? *Transport Policy*, *20*, 105–113. doi:10.1016/j.tranpol.2012.01.013

Luke, S., & MacDonald, M. (2006). Public transport mode selection: a review of international practice.*European Transport Conference*.

Maddock, J. E., Ramirez, V., Heinrich, K. M., Zhang, M., & Brunner, M. A. (2012) Statewide Observational Assessment of the Pedestrian and Bicycling Environment in Hawaii, 2010. *CDC- Preventing Chronic Disease Control and Prevention, 9*(1), 1–9.

Magistrat der Stadt Wien. (2003). Masterplan Verkehr Wien 2003. Wien: Im Auftrag des Magistrats der stadt Wien Magistratsabteilung 18 - Stadtentwicklung und Stadtplanung, Referat Verkehrsplanung und Regionalentwicklung.

Maheshwari, P., Kachroo, P., Paz, A., & Khaddar, R. (2015). Development of control models for the planning of sustainable transportation systems. *Transportation Research Part C, Emerging Technologies, 55*, 474–485. doi:10.1016/j.trc.2015.03.024

Mailer, M. (2002). *Zur Beurteilung von Verkehrsanlagen mit einem multimodalen Ansatz.* Dissertation an der TU Wien, Institut für Verkehrsplanung und Verkehrstechnik.

Mailer, M., Munk, T., Spangler, M., & Haberer, T. (2008). Verkehrsqualität auf Hauptverkehrsstraßen – Empfindung, Messung und Bewertung. *Internationales Verkehrswesen,* (7 + 8), 270.

Mailer, M. (2004). Considering Multi-Modal Capacity in the Assessment of Road Design. In C. Focas (Ed.), *Transport Issues and Problems in Southeastern Europe. Ashgate Publishing Limited.*

Manual, H. C. (2000). *HCM* (Vol. 1). Washington, DC: National Research Council, Transportation Research Board.

Manual, H. C. (2000). *Transportation research board.* Washington, DC: National Research Council.

Maraglino, V., dellOlio, L., Borri, D., & Portilla, A. I. (2014). dell Olio, L., Borri, D. & Portilla, A. I. (2014). Methodology for a study of the perceived quality of public transport in Santander. XI Congreso de Ingenieria del Transporte. *Procedia: Social and Behavioral Sciences, 160*, 499–508. doi:10.1016/j.sbspro.2014.12.163

Marchetti, C. (1994). *Anthropological Invariants in Travel Behavior.* Technological Forecasting and Social Change. IIASA Laxenburg.

Marsden, G. (2006). The evidence base for parking policies—a review. *Transport Policy, 13*(6), 447–457. doi:10.1016/j.tranpol.2006.05.009

Martens, M. J., & Griethuysen, S. v. (1999). *The ABC Location Policy in the Netherlands - 'the Right Business at the Right Place'.* Retrieved April 04. 2016, from https://fenix.tecnico.ulisboa.pt/downloadFile/3779572236303/abc.pdf

Mayerthaler, A., Haller, R., & Emberger, G. (2009). *Modelling land-use and transport at a national scale - the MARS Austria model.* Paper presented at the 49th European Congress of the Regional Science Assosiation International - Territorial Cohesion of Europe and Integrative Planning, Lodz, Poland.

McMillen, D. P., & Redfearn, C. L. (2010). Estimation and hypothesis testing for nonparametric hedonic house price functions. *Journal of Regional Science, 50*(3), 712–733. doi:10.1111/j.1467-9787.2010.00664.x

McNally, M. G. (2008). The Four Step Model. In D. A. H. K. J. Button (Ed.), Handbook of Transport Modelling (2nd ed.). Amsterdam: Elsevier.

Metz, D. H. (2000). Mobility of older people and their quality of life. *Transport Policy, 7*(2), 149–152. doi:10.1016/S0967-070X(00)00004-4

Miller, H. J. (2003). Travel chances and social exclusion. *10th International Conference on Travel Behavior Research*, 1-38.

Milligan, G. W. (1980). An examination of the effect of six types of error perturbation on fifteen clustering algorithms. *Psychometrika*, *45*(3), 325–342. doi:10.1007/BF02293907

Millington, C., Thompson, C. W., Rowe, D., Aspinall, P., Fitzsimons, C., & Nelson, N. et al.SPARColl—the Scottish Physical Activity Research Collaboration. (2009). Development of the Scottish walkability assessment tool (SWAT). *Health & Place*, *15*(2), 474–481. doi:10.1016/j.healthplace.2008.09.007 PMID:19041275

Milosavljević, N., & Simićević, J. (2016). User response to parking policy change: A comparison of stated and revealed preference data. *Transport Policy*, *46*, 40–45. doi:10.1016/j.tranpol.2015.11.007

MobiSystems, Inc. (2016). *Advanced English Dictionary* (Version 8.7.1) [Software]. Available from https://itunes.apple.com/tr/app/advanced-english-dictionary/id291070079?mt=8

Morfoulaki, M., Tyrinopoulos, Y., & Aifadopoulou, G. (2007). Estimation of satisfied customers in public transport systems: A new methodological approach. *Journal of the Transportation Research Forum*, *46*(1), 63–72.

Morikawa, S., Asao, K., Igo, T., & Kato, H. (2010). *Institutional system and current problems of car parking in Tokyo*. Japan: Academic Press.

Mōri, M., & Tsukaguchi, H. (1987). A new method for evaluation of level of service in pedestrian facilities. *Transportation Research Part A, General*, *21*(3), 223–234. doi:10.1016/0191-2607(87)90016-1

MTC Website. (2007, June). *Reforming Parking Policies to Support Smart Growth*. Retrieved from http://mtc.ca.gov/sites/default/files/Toolbox-Handbook.pdf,14/04/2016

Mumford, L. (1984). *Die Stadt*. DTV Wissenschaft Nr. 4326.

Muraleetharan, T., Adachi, T., Uchida, K. E., Hagiwara, T., & Kagaya, S. (2003). *A study on evaluation of pedestrian level of service along sidewalks and at intersections using conjoint analysis*. In Annual Meeting of Japanese Society of Civil Engineers (JSCE) Infrastructure Planning, Toyohashi, Japan.

Nakamura, H., Schäfer, P., & Trieb, A. (2005). Parkraumbewirtschaftung - Was macht Japan anders? *Internationales Verkehrswesen*, *57*(10), 434-437.

Nash, A. (2004). Traffic calming in three European cities: Recent experience.*Proceedings of the 84th meeting of the Transportation Research Board*.

New Trends in Logistics in Europe. (1997). Report of the Hundred and Fourth Round Table on Transport Economics, held in Paris on 3rd and 4th October, 1996. European Conference of Ministers of Transport (ECMT), Paris, France.

Newman, P. W. G., & Kenworthy, J. R. (1989). *Cities and Automobile Dependence: A Sourcebook*. Essex, UK: Gower Publishing Company Limited.

Nickel, S., Reuter-Oppermann, M., & Saldanha-da-Gama, F. (2016). Ambulance location under stochastic demand: a sampling approach. *Operations Research for Health care*. 8, 24-32. .10.1016/j.orhc.2015.06.006

Niederländisches Ministerium für Wohnungswesen (1991). *The Right Business in the Right Place*. Den Haag, Niederlande: Niederländisches Ministerium für Wohnungswesen.

Niemeier, D. A. (1997). Accessibility: An evaluation using consumer welfare. *Transportation*, *24*(4), 377–396. doi:10.1023/A:1004914803019

OECD. (1981). *Evaluation of urban parking systems*. Paris, France: OECD.

OECD. (2009). *Transport, Energy and CO2: Moving Towards Sustainability*. Paris: OECD International Energy Agency.

Ojeda-Cabral, M., & Chorus, C. G. (2016). Value of travel time changes: Theory and simulation to understand the connection between Random Valuation and Random Utility methods. *Transport Policy*, *48*, 139–145. doi:10.1016/j.tranpol.2016.03.006

Olsen, J. R. (2016). *Effects of new urban motorway infrastructure on road traffic accidents on the local ares: a retrospective longitudinal study in Scotland. Journal on Epidemiology and Community Health*. doi:10.1136/jech-2016-207378

OPTIMA. (1996-1997). *Optimisation of Policies for Transport Integration in Metropolitan Areas*. Retrieved from http://www.ivv.tuwien.ac.at/forschung/projekte/international-projects/optima-1995.html

ÖVG. (2005). *Merkblatt Chancengleichheit der Verkehrsträger*. Wien, Österreich: Österreichische Verkehrswissenschaftliche Gesellschaft.

Owen, N., Cerin, E., Leslie, E., duToit, L., Coffee, N., Frank, L. D., & Sallis, J. F. et al. (2007). Neighborhood Walkability and the Walking Behavior of Australian Adults. *American Journal of Preventive Medicine*, *33*(5), 387–395. doi:10.1016/j.amepre.2007.07.025 PMID:17950404

Ozus, E., Akın, D., & Çiftçi, M. (2012). Hierarchical cluster analysis of multicenter development and travel patterns in Istanbul. *Journal of Urban Planning and Development*, *138*(4), 303–318. doi:10.1061/(ASCE)UP.1943-5444.0000121

Panel 3. Sustainable transport and land use. (n.d.). Retrieved from http://www.eceee.org/library/conference_proceedings/eceee_Summer_Studies/2005c/Panel_3

Park, S. (2008). *Defining, Measuring, and Evaluating Path Walkability, and Testing Its Impacts on Transit Users' Mode Choice and Walking Distance to the Station* (PhD. Dissertation). University of California, Berkeley, CA.

Park, S., Deakin, E., & Lee, J. (2014). Perception-Based Walkability Index to Test Impact of Microlevel Walkability on Sustainable Mode Choice Decisions. *Transportation Research Record: Journal of the Transportation Research Board*, (2464), 126-134.

Parlament. (2015). *Preistreiber Stellplatzverordnung*. Retrieved April 04. 2016, from http://www.parlament.gv.at/PAKT/VHG/XXV/J/J_02362/fnameorig_362885.html

Parma, K. D., Koprowski, Y., & Zhang, W. (2013). Origin-Destination and Travel Time Measurement Using Bluetooth Technology: Results and Lessons Learned. ITE Technical Conference and Exhibit.

Peiravian, F., Derrible, S., & Ijaz, F. (2014). Development and application of the Pedestrian Environment Index (PEI). *Journal of Transport Geography*, *39*, 73–84. doi:10.1016/j.jtrangeo.2014.06.020

Peperna, O. (1982). *Master Thesis:Die Einzugsbereiche von Haltestellen öffentlicher Nahverkehrsmittel im Straßenbahn und Busverkehr*. Wien.

Peperna, O. (1982). *Die Einzugsbereiche von Haltestellen öffentlicher Nahverkehrsmittel im Straßenbahn- und Busverkehr (Master thesis)*. Technische Universität Wien.

Peperna, O. (1982). *Die Einzugsbereiche von Haltestellen öffentlicher Nahverkehrsmittel im Straßenbahn- und Busverkehr (Unpublished master's theses)*. Wien, Österreich: Technische Universität Wien.

Petritsch, T., Landis, B., McLeod, P., Huang, H., Challa, S., & Guttenplan, M. (2005). Part 2: Pedestrians: Level-of-service model for pedestrians at signalized intersections. *Transportation Research Record: Journal of the Transportation Research Board*, (1939), 53-62.

Pfaffenbichler, P. (2003). The strategic, dynamic and integrated urban land use and transport model MARS (Metropolitan Activity Relocation Simulator) - Development, testing and application (Doctoral thesis). Vienna: Vienna University of Technology. Retrieved from http://www.ivv.tuwien.ac.at/publications/online/MARS_smallest_size.pdf

Pfaffenbichler, P. C., & Emberger, G. (2005). A reform of car parking organisation as an instrument for creating sustainable urban regions. *ECEEE 2005 Summer Study.*

Pfaffenbichler, P. C., Emberger, G., & Shepherd, S. P. (2008). The Integrated Dynamic Land Use and Transport Model MARS. *Networks and Spatial Economics*, 3.

Pfaffenbichler, P. C., Emberger, G., & Shepherd, S. P. (2010). A system dynamics approach to land use transport interaction modelling: The strategic model MARS and its application. *System Dynamics Review*, *26*(3), 262–282. doi:10.1002/sdr.451

Pfaffenbichler, P., & Schopf, J. M. (2011). *Einfluss der Parkraumorganisation und der Anzahl der Stellplätze auf die Nutzung des motorisierten Individualverkehrs und die Erreichung verkehrs-, umwelt- und siedlungspolitischer Ziele (PAN).* Wien, Österreich: Österreichische Energieagentur – Austrian Energy Agency.

Pfister, C. (1994). *Das 1950-er Syndrom, GAIA.* Trnsdiziplinäre Zeitschrift, Oekom Verlalg München.

Pirkul, H., & Schilling, D. A. (1991). The maximal covering location problem with capacities on total workload. *Management Science*, *37*(2), 233–248. doi:10.1287/mnsc.37.2.233

Planning Department of Hong Kong. (2006, June). *Parking Standard.* Retrieved from http://www.pland.gov.hk/pland_en/tech_doc/hkpsg/full/ch8/ch8_tbl_11.htm

Polus, A., Schofer, J. L., & Ushpiz, A. (1983). Pedestrian flow and level of service. *Journal of Transportation Engineering*, *109*(1), 46–56. doi:10.1061/(ASCE)0733-947X(1983)109:1(46)

Preston, J. (2009). Epilogue: Transport policy and social exclusion—Some reflections. *Transport Policy*, *16*(3), 140–142. doi:10.1016/j.tranpol.2009.04.003

PROSPECTS. (2000-2003). *PROSPECTS - Procedures for Recommending Optimal Sustainable Planning of European City Transport Systems.* Retrieved from http://www.ivv.tuwien.ac.at/forschung/projekte/international-projects/prospects-2000.html

Rajé, F. (2003). The impact of transport on social exclusion processes with specific emphasis on road user charging. *Transport Policy*, *10*(4), 321–338. doi:10.1016/S0967-070X(03)00038-6

Rauh, W., & Maierbrugger, G. (2003). *Grenzen überwinden im Verkehr - Maßnahmen und Beispiele.* Wien: VCÖ.

Reichsarbeitsministerium. (1939). *Verordnung über Garagen und Einstellplätze (Reichsgaragenordnung).* Berlin: Author.

ReVelle, C., & Hogan, K. (1989). The maximum availability location problem. *Transportation Science*, *23*(3), 192–200. doi:10.1287/trsc.23.3.192

Riedel, R. (2014). Grundsätze der Stadtplanung im Ruhenden Verkehr. *Zeitschrift für Verkehrsrecht, 12a*, 453-461.

Riedl, R. (1984a). *Biology of Knowledge: The Evolutionary Basis of Reason.* Chichester, UK: John Wiley & Sons.

Riedl, R. (1984b). Evolution and Evolutionary Knowlege. In F. M. Wuketits (Ed.), *Concepts and Approaches in Evolutionary Epistemology* (pp. 35–50). Dordrecht: Reidel. doi:10.1007/978-94-009-7127-1_2

Rios, R. A., Vicentini, V. L., & Acevedo-Daunos, R. (2013). *Practical Guidebook: Parking and Travel Demand Management Policies in Latin America, Despacio and Institute for Transportation and Development Policy (ITDP).* Interamerican Development Bank.

Rodrigue, J. P., Comtois, C., & Slack, B. (2006). *The Geography of Transport Systems*. Hofstra University, Department of Economics & Geography.

Rodrigue, J. P., Comtois, C., & Slack, B. (2013). *The geography of transport systems*. Routledge.

Rodríguez, D. A., & Joo, J. (2004). The relationship between non-motorized mode choice and the local physical environment. *Transportation Research Part D, Transport and Environment*, *9*(2), 151–173. doi:10.1016/j.trd.2003.11.001

Roitman, S. (2005). Who Segregates Whom? The Analysis of a Gated Community in Mendoza, Argentina. *Housing Studies*, *20*(2), 303–321. doi:10.1080/0267303042000331790

Rojo, M., dellOlio, L., Gonzalo-Orden, H., & Ibeas, A. (2013). Interurban bus service quality from the users viewpoint. *Transportation Planning and Technology*, *36*(7), 599–616. doi:10.1080/03081060.2013.845432

Rojo, M., dellOlio, L., Gonzalo-Orden, H., & Ibeas, Á. (2015). Inclusion of quality criteria in public bus service contracts in metropolitan areas. *Transport Policy*, *42*, 52–63. doi:10.1016/j.tranpol.2015.05.002

Román, C., Martín, J. C., Espino, R., Cherchi, E., Ortúzar, J. D., Rizzi, L. I., & Amador, F. J. et al. (2014). Valuation of travel time savings for intercity travel: The Madrid-Barcelona corridor. *Transport Policy*, *36*, 105–117. doi:10.1016/j.tranpol.2014.07.007

Rosenberg, D., Ding, D., Sallis, J. F., Kerr, J., Norman, G. J., Durant, N., & Saelens, B. E. et al. (2009). Neighborhood Environment Walkability Scale for Youth (NEWS-Y): Reliability and relationship with physical activity. *Preventive Medicine*, *49*(2), 213–218. doi:10.1016/j.ypmed.2009.07.011 PMID:19632263

Rosen, E., Stigson, H., & Sander, U. (2011). Literature review of pedestrian fatality risk as a function of car impact speed. *Accident; Analysis and Prevention*, *43*(1), 25–33. doi:10.1016/j.aap.2010.04.003 PMID:21094293

Rosenkranz, B. (2009). *Resolution betreffend NÖ Bauordnung 1996 und NÖ Bautechnikverordnung 1997 bzgl. weiterer Erhöhung des Radverkehrsanteiles am Gesamtverkehrsaufkommen* (P. des Landtages von Niederösterreich, Ed.). St. Pölten.

Roy, M. (2009). Planning for sustainable urbanisation in fast growing cities: Mitigation and adaptation issues addressed in Dhaka, Bangladesh. *Habitat International*, *33*(3), 276–286. doi:10.1016/j.habitatint.2008.10.022

RVS 03.01.11. (2012). *Richtlinien und Vorschriften für den Straßenbau: Beurteilung des Verkehrsablaufs auf Straßen*. Österreichische Forschungsgesellschaft Straße – Schiene – Verkehr (FSV), Wien.

Sammer, G. (2005). *Instrumente zur Steuerung des Stellplatzangebotes für den Zielverkehr – IN-STELLA, Teil 1: Analyse nationaler und internationaler Umsetzungsbeispiele*. Wien, Österreich: Universität für Bodenkultur, Department für Raum, Landschaft und Infrastruktur.

Samuelson, P.A., & Nordhaus, W.D. (1987). Volkswirtschaftslehre. Grundlagen der Makro- und Mikroökonomie. Achte, grundlegend überarbeitete deutsche Neuauflage. Bund-Verlag, Köln.

Schafer, A. (1998). The Global Demand for Motorized Mobility. *Transportation Research Board Part A: Policy and Practice*, *32*(6).

Schafer, A. (2000). Regularities in travel demand: An international perspective. *Journal of Transportation and Statistics*, *3*(3), 1–31.

Schafer, A. (2000, December). Regularities in Travel Demand: An International Perspective. *Journal of Transportation and Statistics*.

Schlossberg, M. (2006). From TIGER to audit instruments: Measuring neighborhood walkability with street data based on geographic information systems. Transportation Research Record: Journal of the Transportation Research Board, (1982), 48-56.

Schmitt, H., & Landgraf, S. (2009). *Beobachtungsstudie: Einkaufsverhalten in Graz und Umgebung*. Graz: ARGUS Steiermark.

Schmitz, S. (2001). Revolution der Erreichbarkeit. Gesellschaft, Raum und Verkehr im Wandel. Stadtforschung aktuell Nr. 83. Leske + Budrich, Opladen. doi:10.1007/978-3-663-11306-5

Schnabel, W., & Lohse, D. (1997). Article.*Grundlagen der Strassenverkehrstechnik und der Verkehrsplanung*, *1*, 2.

Schnabel, W., & Lohse, D. (1997). *Verkehrsplanung* (Vol. 2). Berlin: Grundlagen der Straßenverkehrstechnik und Verkehrsplanung.

Scholas, M. (1988). *Auswirkungen von Ortsumgehungen im Zuge von Bundes- und Landesstraßen in Nordrhein-Westfalen – unter besonderer Berücksichtigung des Unfallgeschehens*. Dissertation Universität Dortmund 1988.

Schopf, J. M., et al. (2015). *Wohnbau, Wohnumfeld und Mobilität*. Wien, Österreich: Verkehrsclub Österreich, Mobilität mit Zukunft 3/2015.

Schopf, J. M. (2016). Der ruhende Verkehr in der RVS. In H. Knoflacher & H. Frey (Eds.), *Paradigmenwechsel im Verkehrswesen* (pp. 145–166). Wien: Österreichische Verkehrswissenschaftliche Gesellschaft.

Schopf, J. M., & Brezina, T. (2015). *Umweltfreundliches Parkraummanagement – Leitfaden für Länder, Städte, Gemeinden, Betriebe und Bauträger*. Wien, Österreich: Bundesministerium für Land- und Forstwirtschaft, Umwelt und Wasserwirtschaft.

Schuster, A., Sattler, J., & Hoffmann, St. (2012). Benötigen wir ein neues Pkw-Bemessungsfahrzeug für den Entwurf von Anlagen des ruhenden Verkehrs? *Straßenverkehrstechnik*, *56*(1), 5-10.

SCS Extension. (1988). Retrieved from https://de.wikipedia.org/wiki/Shopping_City_S%C3%BCd

Sebhatu, S. P., & Enquist, B. (2016). Sustainable Public Transit Service Value Network for Building Living Cities in Emerging Economies: Multiple Case Studies from Public Transit Services. *Procedia: Social and Behavioral Sciences*, *224*, 263–268. doi:10.1016/j.sbspro.2016.05.458

SEU. (2002). *Making the connections: final report on transport and social exclusion*. London: Social Exclusion Unit.

SEU. (2003). *The future of social exclusion: drivers, patterns and policy challenges*. London: Social Exclusion Unit.

Shoup, D. C. (1997). The High Cost of Free Parking. *Journal of Planning Education and Research*, *17*(1), 3–20. doi:10.1177/0739456X9701700102

Shoup, D. C. (1999). The trouble with minimum parking requirements. *Transportation Research Part A, Policy and Practice*, *33*(7-8), 549–574. doi:10.1016/S0965-8564(99)00007-5

Simićević, J., Vukanović, S., & Milosavljević, N. (2013). The effect of parking charges and time limit to car usage and parking behaviour. *Transport Policy*, *30*, 125–131. doi:10.1016/j.tranpol.2013.09.007

Singh, K., & Jain, P. K. (2011). Methods of Assessing Pedestrian Level of Service, Journal of Engineering. *Research Studies (Pullman, Wash.)*, *2*(1), 116–124.

Sisiopiku, V. P., Byrd, J., & Chittoor, A. (2007) Application of Level of Service Methods for the Evaluation of Operations at Pedestrian Facilities, In *TRB 2007 Annual Meeting*. Transportation Research Board of the National Academies. doi:10.3141/2002-15

Snyder, L. V. (2006). Facility location under uncertainty: A review. *IIE Transactions*, *38*(7), 537–554. doi:10.1080/07408170500216480

H. M. Soekkha (Ed.). (1990). Telematics – Transportation and Spatial Development.*Proceeding of an International Symposium organized under the auspices of the Centre for Transportation Engineering*. DelftUniversity of Technology.

Sohn, J. (2005). Are commuting patterns a good indicator of urban spatial structure? *Journal of Transport Geography*, *13*(4), 306–317. doi:10.1016/j.jtrangeo.2004.07.005

Sonntag, H. (1999). Städtischer Wirtschaftsverkehr und logistische Knoten. Wirkungsanalyse von Verknüpfungen der Güterverkehrsnetze auf den städtischen Wirtschafts- und Güterverkehr. Berichte der Bundesanstalt für Straßenwesen. *Schriftenreihe Verkehrstechnik*, *V*, 68.

SPARKLE. (2004-2006). *Sustainability Planning for Asian Cities making use of Research, Know-How and Lessons from Europe*. Retrieved from http://www.ivv.tuwien.ac.at/forschung/projekte/international-projects/sparkle-2004.html

Stadt Wien - Stadtentwicklung. (2002). *Masterplan Verkehr Wien Positionspapier* (Vol. Nr. 43). Wien: Stadtentwicklung Wien, Magistratsabteilung 18.

Städtebau und Verkehr. (1996). *Dokumentation der Modellvorhaben. Stand: Juli 1996. ExWost-Informationen Nr. 06.10.* Bonn: Bundesforschungsanstalt für Landeskunde und Raumordnung.

Stadtplanung Wien. (1994a). *step 1994 - Stadtentwicklungsplan für Wien*. Author.

Stadtplanung Wien. (1994b). *Verkehrskonzept Wien - Generelles Maßnahmenprogramm*. Author.

Stadtwerke-Generaldirektion. (1984). *Energie für Wien, Energiekonzept der Stadt Wien 1, Fortschreibung*. Wien: Stadtwerke-Generaldirektion.

Staff, I. (2016). *Economies of Scale*. Retrieved from http://www.investopedia.com/terms/e/economiesofscale.asp?ad=dirN&%3Bqo=investopediaSiteSearch&%3Bqsrc=0&%3Bo=40186

Stanley, J., & Stanley, J. (2004). Improving public transport to meet community needs: A Warrnambool case-study. Report for Bus Association Victoria and Warrnambool Bus Lines, Melbourne, Australia.

Stefanescu, P., Mocan, M., Stefanescu, W., & Neculai, P. V. (2014). Trip Planners Used in Public Transportation. Case Study on the City of Timisoara. *Procedia: Social and Behavioral Sciences*, 124.

Steininger, G. (2008). *Ziele und Vorgaben der österreichischen Verkehrspolitik in den Koalitionsabkommen der Bundesregierung - ein Zeitvergleich 1945-2007. Schriftenreihe des Instituts für Transportwirtschaft und Logistik - Verkehr, 02/2008. Institut für Transportwirtschaft und Logistik*. Vienna: WU Vienna University of Economics and Business.

STEPS. (2004-2006). *Scenarios for the Transport system and the Energy supply and their Potential effectS*. Retrieved from EU-Project, 6th FP: http://www.steps-eu.com/

Sterman, J. (2000). *Business Dynamics - Systems Thinking and Modelling for a Complex World*. Boston: McGraw-Hill Higher Education.

Sterman, J. D. (2000). *Business Dynamics - Systems Thinking and Modeling for a Complex World*. McGraw-Hill Higher Education.

Stift, W. (2002). *Stop der Stadtvernichtung*. Wien, Österreich: Bundesgremium Textilhandel.

Stradling, S., Carreno, M., Rye, T., & Noble, A. (2007). Passenger perceptions and the ideal urban bus journey experience. *Transport Policy*, *14*(4), 283–292. doi:10.1016/j.tranpol.2007.02.003

Straßenverkehrsunfälle im Jahr 1970 – Straßenverkehrssicherheit im Jahre. (1989). Beiträge zur Österreichischen Statistik. Herausgegeben vom Österreichischen Statistischen Zentralamt, Wien.

Sun, G., Zacharias, J., Ma, B., & Oreskovic, N. M. (2016). How do metro stations integrate with walking environments? Results from walking access within three types of built environment in Beijing. *Cities (London, England)*, *56*, 91–98. doi:10.1016/j.cities.2016.03.001

Swoveland, C., Uyeno, D., Vertinsky, I., & Vickson, R. (1973). Ambulance location: A probabilistic enumeration approach. *Management Science*, *20*(44-II), 686–698. doi:10.1287/mnsc.20.4.686

Talavera-Garcia, R., & Soria-Lara, J. A. (2015). Q-PLOS, developing an alternative walking index. A method based on urban design quality. *Cities (London, England)*, *45*, 7–17. doi:10.1016/j.cities.2015.03.003

Tan, P., Steinbach, M., & Kumar, V. (2006). Introduction to data mining. Addison-Wesley.

Tan, D., Wang, W., Lu, J., & Bian, Y. (2007). Research on Methods of Assessing Pedestrian Level of Service for Sidewalk. *Journal of Transportation Systems Engineering and Information Technology*, *7*(5), 74–79. doi:10.1016/S1570-6672(07)60041-5

Taniguchi, E. (2001). City Logistics. Network Modelling and Intelligent Transport Systems. Pergamon. doi:10.1108/9780585473840

Taniguchi, E., & Thomson, R. G. (Eds.). (2001). *City Logistic II*. The Second International Conference on City Logistics, 27th – 29th 2001, Okinawa, Japan.

Teodorović, D., & Janić, M. (2017). *Public Transportation Systems*. Transportation Engineering.

Teufel, D., Bauer, P, Beker, G., & Wagner, T. (1989). *Gesellschaftliche Kosten des Straßengüterverkehrs. Kasten-Deckungsgrad im Jahr 1987 und Vorschläge zur Realisierung des Verursacherprinzips*. UPI Umwelt- und Prognose-Institut Heidelberg, UPI-Bericht Nr. 14.

Theodoridis, S., & Koutroubas, K. (1999). *Pattern Recognition*. London: Academic Press.

Tirol. (2015). *Stellplatzhöchstzahlenverordnung 2015*. Innsbruck, Österreich: Tiroler Landesregierung.

TOD Handbook. (2004, January). *The City of Calgary Land Use Policy*. Retrieved from http://www.reconnectingamerica.org/assets/Uploads/bestpractice031.pdf

Topp, H. H. (1998). Getrennte Märkte für Wohnung und Stellplatz. *Internationales Verkehrswesen*, *50*(7-8), 322-324.

Townsend, A. M. (2014). Re-Programming Mobility: how the Tech-Industry Is Driving Us Towards A Crisis In Transportation Planning. In Cities on the Move. New Cities Foundation Geneva.

TRANSPLUS. (2003). *D4 Barriers, Solutions and Transferability*. Retrieved from http://www.ivv.tuwien.ac.at/forschung/projekte/international-projects/transplus-2000.html

TRB. (1985, 1994). Highway Capacity Manual, Third Edition 1985, Updated 1994. Transport Research Board.

TRB. (2001). *Highway Capacity Manual* (4th ed.). Transport Research Board.

Tsompanoglou, S., & Photis, Y. N. (2013). Measuring urban concentration: A spatial cluster typology based on public and private sector service patterns. *World Review of Science. Technology and Sustainable Development*, *10*(4), 185–202.

TÜİK. (2011b). *2000 Bina sayımı veri seti: Ankara kenti*. Ankara: Türkiye İstatistik Kurumu.

Türk-İş. (2015). Retrieved from http://www.webcitation.org/query?url=http%3A%2F%2Fwww.turkis.org.tr%2Fdosya%2F8BPrv9CducUP.pdf+&date=2016-06-30

TurkStat. (2016). *Road Motor Vehicle Statistics*. Author.

Tyrinopoulos, Y., & Antoniou, C. (2008). Public transit user satisfaction: Variability and policy implications. *Transport Policy*, *15*(4), 260–272. doi:10.1016/j.tranpol.2008.06.002

Uber. (2016). Retrieved from https://www.uber.com/

United Nations. (1998). *Kyoto Protocol -Report of the conference of the partners on its third session, held at Kyoto from 1 to 11 December 1997, Addendum, Part two: Action taken by the conference of the parties at its third session, Framework Convention on Climate Change, FCCC/CP/1997/Add.1*. UN.

United Nations. (2015). *Framework Convention on Climate Change - Paris*. UN.

Van Dyck, D., Cardon, G., Deforche, B., Owen, N., & De Bourdeaudhuij, I. (2011). Relationships between neighborhood walkability and adults physical activity: How important is residential self-selection? *Health & Place*, *17*(4), 1011–1014. doi:10.1016/j.healthplace.2011.05.005 PMID:21596613

Vandenberghe, L., & Boyd, S. (1996). Semidefinite programming. *SIAM Review*, *38*(1), 49–95. doi:10.1137/1038003

VCÖ. (2006). *Gute Verkehrslösungen in Gemeinden und Städten*. Wien, Österreich: Verkehrsclub Österreich. VCÖ-Magazin.

VCÖ. (2014). *Multimodale Lebensstile prägen Mobilität der Zukunft*. Factsheet.

Verkehrsplanung & Verkehrstechnik. (2016). *MARS (Metropolitan Activity Relocation Simulator)*. Retrieved from http://www.ivv.tuwien.ac.at/forschung/mars-metropolitan-activity-relocation-simulator/overview/

Vienna City Administration. (2014a). Urban Mobility Plan Vienna, STEP 2025 Thematic Concept. Municipal Department 18 - Urban Development and Planning.

Vienna City Administration. (2014b). Vienna's Urban Development Plan-STEP 2025. Municipal Department 18 - Urban Development and Planning.

Viera, A. J., & Garrett, J. M. (2005). Understanding Interobserver Agreement: The Kappa Statistic. *Family Medicine*, *37*(5), 360–363. PMID:15883903

Vigar, G. (2000). Local Barriers to Environmentally Sustainable Transport Planning. *Local Environment*, *5*(1), 19–32. doi:10.1080/135498300113246

Vliet, J. V., White, R., & Dragicevic, S. (2009). Modeling urban growth using a variable grid cellular automaton. *Computers, Environment and Urban Systems*, *33*(1), 35–43. doi:10.1016/j.compenvurbsys.2008.06.006

Vuchic, V. R. (1999). *Transportation for livable cities*. Rutgers Center for Urban Policy Research.

Wald, A. (1939). Contributions to the theory of statistical estimation and testing hypotheses. *The Annals of Mathematics*, *10*(4), 299–326. doi:10.1214/aoms/1177732144

Wald, A. (1945). Statistical decision functions which minimize the maximum risk. *The Annals of Mathematics*, *46*(2), 265–280. doi:10.2307/1969022

Wald, A. (1950). *Statistical Decision Functions*. John Wiley.

Walkscore ®. (2011). Retrieved from https://www.walkscore.com/

Walther, K. (1973). Die Fußweglänge zur Haltestelle als Attraktivitätskriterium im öffentlichen Personennahverkehr. *Verkehr und Technik, 26*(11), 480-484.

Walther, K. (1973). *Nachfrageorientierte Bewertung der Streckenführung im ÖPNV*. Frankfurt/Oder: Fakultät für Bauwesen, TH-Aachen.

Walther, K. (1975). Die fahrzeugäquivalente Reisezeit im öffentlichen Personennahverkehr. *Verkehrs und Technik*, 7.

Walther, K. (1991). Maßnahmenreagibler Modal-Split für den städtischen Personenverkehr - Theoretische Grundlagen und praktische Anwendung (Vol. 45). Aachen: Academic Press.

Walther, K., Oetting, A., & Vallée, D. (1997). *Simultane Modellstruktur für die Personenverkehrsplanung auf der Basis eines neuen Verkehrswiderstands* (Vol. 52). Aachen: Academic Press.

Wang, F., & Sun, J. (2012). *Distance metric learning in data mining.SDM Conference (Tutorial).*

Weber-Fechner Law. (1975). In The Encyclopedia Americana. Americana Corporation.

Weinberger, R. (2012). Death by a thousand curb-cuts: Evidence on the effect of minimum parking requirements on the choice to drive. *Transport Policy*, *20*, 93–102. doi:10.1016/j.tranpol.2011.08.002

Westhauser, C., & Hofecker, C. (2011). *Empfehlung über Fahrradabstellplätze bei verschiedenen Bauten*. Unpublished manuscript, St. Pölten.

Westminster City Council. (2010). *Unitary Development Plan*. Retrieved from http://www.westminster.gov.uk/services/environment/planning/unitarydevelopmentplan/

Wheat, P., & Batley, R. (2015). Quantifying and decomposing the uncertainty in appraisal value of travel time savings. *Transport Policy*, *44*, 134–142. doi:10.1016/j.tranpol.2015.06.010

Wiedmann, K.-P., Hennigs, N., Pankalla, L., Kassubek, M., & Seegebarth, B. (2011). Adoption barriers and resistance to sustainable solutions in the automotive sector. *Journal of Business Research*, *64*(11), 1201–1206. doi:10.1016/j.jbusres.2011.06.023

Wien. (2016). *Parking regulation*. Retrieved from https://www.wien.gv.at/verkehr/parken/

Wikipedia. (2016). *Smart City*. Retrieved April 05, 2016, from https://de.wikipedia.org/wiki/Smart_City

Willson, R. W. (1995). Suburban Parking Requirements: A Tacit Policy for Automobile Use and Sprawl. *Journal of the American Planning Association*, *61*(1), 29–42. doi:10.1080/01944369508975617

Wilson, S. S. (1973). Bicycle Technology. *Scientific American*, *228*(March), 81–90. doi:10.1038/scientificamerican0373-81

Winder, A. D., May, A., & El-Araby, K. (2005). Urban transport performance and policy: results of a European public perception survey. *Association for European Transport and contributors 2005*. Retrieved from http://abstracts.aetransport.org/paper/index/id/2083/confid/11

Witter, R. (2010). Public urban transport, travel behaviour and social exclusion–the case of Santiago de Chile.*XII World Conference on Transportation Research, Lisbon*, 1-23.

World Bank. (2016). Retrieved from http://lpi.worldbank.org/international/global

Yazicioglu Halu, Z., & Yürekli, F. (2011). Yürünebilirlik Kavramı ve Kentsel Mekanlarda Yürüme. *İTÜ Journal Series A: Architecture, Planning, Design, 10*(2), 29–38.

Yildiz, E. V. (2003). *The Analysis of Spatial Structure and Mobility in Ankara* (Unpublished Ph.D. Thesis). TU Wien Institut für Verkehrsplanung und Verkehrstechnik.

Yılmaz, Ö., & Temurlenk, M. S. (2005). Türkiye'deki Istatistik Bölgelerin Kişi Başina Düşen Gelir Açisindan Hiyerarşik ve Hiyerarşik Olmayan Kümeleme Analizi Ile Değerlendirilmesi: 1965-2001. *Atatürk Üniversitesi İktisadi ve İdari Bilimler Dergisi, 19*(2), 75–92.

Yukubousky, R. (1994). *Level of Service Standards* (Vol. 7926). Academic Press.

Zhang, Z.-H., & Jiang, H. (2014). A robust counterpart approach to the bi-objective emergency medical service design problem. *Applied Mathematical Modelling, 38*(3), 1033–1040. doi:10.1016/j.apm.2013.07.028

Zhao, P., Lu, B., & de Roo, G. (2011). The impact of urban growth on commuting patterns in a restructuring city: Evidence from Beijing. *Papers in Regional Science, 90*(4), 735-754.

About the Contributors

Hermann Knoflacher was trained in civil engineering, mathematics, and geodesy. He became full Professor in Transport Planning and Traffic Engineering in the year 1975 at the Vienna University of Technology. Since then he worked in transport and city planning, carried out more than 100 transport master plans for cities. He was Guest Professor in Universities in Europe, China, Japan and USA. He was also advisor to the Minister of Transport for over eight years. He was member of the advisory groups for the EU Commission for Telematics in the 4.FP and for Intermodality and Sustainable Mobility in the 5. FP. Member and Chairman of Road Research working Groups of the OECD and PIARC, and expert for the WHO. He published 8 books, more than 500 scientific publications, 500 expert opinions, and more than 1,000 lectures in the field of transport planning and traffic engineering worldwide (USA, Europe, Asia, Australia). His key areas of expertise are transport planning, city planning, civil engineering, transport policy.

Ebru Ocalir-Akunal is associate professor at the City and Regional Planning Department of the Gazi University, Ankara, Turkey. She graduated from the City and Regional Planning Department of Gazi University in 1993. She completed her master's studies in the Regional Planning Department of the METU in 1996. She got her Ph. D. degree from the Institute for Transport Planning and Traffic Engineering of Vienna University of Technology in 2003. She carried out post-doctoral studies in the Research Centre of Transport Planning and Traffic Engineering of Vienna University of Technology in 2010. Her main topics of research are the development of spatial structures and mobility and decision support systems.

* * *

Darcin Akin has undergraduate and master's degrees in Civil Engineering from Dokuz Eylul University, Turkey in 1987 and 1992, respectively. He received his doctoral degree from Michigan State University (MSU), USA in 2000 in Transportation Engineering & Planning program. After graduation from MSU, he joined the Department of City and Regional Planning at Gebze Institute of Technology (GIT), Kocaeli. He taught graduate courses at GIT in transportation engineering and planning and supervised graduate theses in several aspect of urban planning. He also served as a technical advisor during the preparation of Istanbul Transport Master Plan from 2005 to 2010. He led the project of Trip Generation Handbook for the metropolitan city of Istanbul. His research interests include transportation engineering and planning. He has published and presented papers at national and international journals and conferences, including ASCE journals, TRB Meetings and the TRB Journal. During his position as an associate professor in Department of Urban & Regional Planning at Yildiz Technical University,

Istanbul from 2012 to 2014, he was involved in Sakarya's 2023 and Manisa's 2034 Transport Master Plans. Currently, he is a full professor and the chairman in Civil Engineering Department at Gediz University, Izmir, Turkey.

Aykan Akıncılar is a research assistant in department of industrial engineering, Gazi University as well as Celal Bayar University. He is currently a PhD candidate and his research interests include robust optimization, large-scale optimization, MCDM in terms of methodological aspect while include operations planning and scheduling in transportation and robust planning in disaster management in terms of application areas.

Ece Akıncılar is a production planning engineer in Turkish Aerospace Industries Inc, which is one of the most leading corporation serving for Turkish Defense Industry. She is as well as a M.Sc. candidate from department of industrial engineering, Gazi University. Her main research fields are production planning and multi-criteria decision making.

Leyla Alkan received her undergraduate degree in City and Regional Planning and postgraduate degree in Economics at Middle East Technical University. Earned her PhD. degree in City and Regional Planning at Middle East Technical University with a thesis entitled Tenure Choice and Demand for Homeownership in Ankara. Currently works as a Research Assistant Dr. at Gazi University. Major research interests include urban economics, real estate economics, and urban property development.

Hakan Aplak is currently an assistant professor in Industrial and System Engineering Department of Turkish Military Academy.

Tadej Brezina studied civil engineering at Vienna University of Technology and is researcher and lecturer at its Institute of Transportation, Research Center of Transport Planning and Traffic Engineering. His recent research interests comprise of public transport, cycling, policies of urban stationary traffic and the impact appraisal of electric-driven commercial mobility.

Ziya Cakici received his BSc degree in Civil Engineering from the Celal Bayar Univesity, Manisa, Turkey, in 2011. He received the MSc degree in the Civil Engineering from the Pamukkale University, Denizli, Turkey in 2015. Mr. Cakici joined to Pamukkale University in 2012. He has been working on different subjects of transportation egineering since 2011.

Alberto Castro Fernandez is a PhD independent researcher, holding a MSc. in Civil Engineering from the Madrid Polythechnic University. His research work mainly focuses on topics such as sustainability, bicycle mobility, specially bike-sharing, accessibility and behaviour change theories.

Guenter Emberger, Dr., born in Austria 1967, presently holds the position of a a.o.Univ.Prof. at the Vienna University of Technology, Institute for Transportation, Research Center for Transport Planning and Traffic Engineering. He was Guest Research Fellow at the Institute for Transport Studies (ITS), University of Leeds, UK, (7/2002 to 6/2004). G. Emberger 0studied Computer Science and Business Administration at the University of Vienna, Austria. G. Emberger is working in the field of transport research since 1990. He cumulated experience in travel demand behaviour analyses, design and implementation of 4 stage transport models and strategic transport models on urban, regional and European level. He carried out research in traffic safety and environmental impact analysis of car traffic, and worked in the field of public transport. He was involved in more than 30 international research projects like OPTIMA, PROSPECTS, SPECTRUM, etc. and in more than 35 national research projects and published more 100 scientific articles and book chapters.

Ozge Ercoskun is an associate professor in the City and Regional Planning Department of the Gazi University, Ankara, Turkey. She graduated from the City and Regional Planning Department of the Istanbul Technical University in 1998. She completed her master's studies in the Geodetic and Geographic Information Technologies Department of the METU in 2002. She got her Ph.D. degree from the City and Regional Planning Department of the Gazi University in 2007. She has attended several national and international congresses; summer schools and workshops related to ecological urban planning and geographic information systems. She has written 3 books and more than 50 papers on sustainable urban design and ecological and smart urban planning, geographic information systems and information technologies. She worked as a researcher in many national and institutional projects. She has awards about sustainability and urban growth, sustainable tourism. She has written a chapter named 'Green Urban Planning and Design for Smarter Communities' in 2010 published by Information Science Publishing. She has published an edited book named 'Green and Ecological Technologies for Urban Planning: Creating Smart Cities' from the same publisher in 2012.She has written chapters named 'Creating Resilient Communities: Local Currencies and Time Banks in Green Economy' in 2015 and 'Smart Technologies for Sustainable Mobility' in 2016 published by Information Science Publishing.

Serpil Erol is currently a professor in Industrial Engineering Department of Gazi University.

Pinar Karatas received her BS degree in Civil Engineering from the Çukurova University, Adana and her MS degree in transportation from the Civil Engineering Department at the Middle East Technical University (METU) in Ankara, where she is also currently a PhD candidate. In her MS study, she evaluated Pedestrian Level of Service (PLOS) of METU Campus walkways using three different methods. She also conducted surveys and route-based data collection studies on pedestrian mobility in the campus.

Markus Mailer is holding the chair for Transport Planning at the University of Innsbruck since December 2010 and is head of the unit of Intelligent Transport Systems. He studied Civil Engineering with specialisation in transportation at TU Wien (Vienna University of Technology). After his graduation he started to work as university researcher and lecturer (Universitätsassistent) at the Institute of Transport Planning and Traffic Engineering at TU Wien in 1996. In 2002 he obtained his doctorate. In 2003 he started to work at the BMW Group in Munich in the unit "Transport Concepts Munich". In 2008 he became head of the "Traffic Management" team at BMW focusing on Intelligent Transport Systems, Cooperative Highway Vehicle Systems, Future of Mobility and Mobility Services.

Yetis Sazi Murat received his BS degree in Civil Engineering from the Dokuz Eylul University, Izmir, Turkey, in 1992 and the MSc degree in the Civil Engineering from Pamukkale University, Turkey in 1996 and the PhD degree in the Transportation Engineering from Istanbul Technical University, Turkey in 2001, respectively. Dr. Murat is a full time Professor at Department of Civil Engineering, Pamukkale

University, Denizli, Turkey. He has been working on various topics of transportation science (traffic engineering, transportation planning, public transportation etc.) since 1994. Use of Artificial Intelligence Methods in Transportation Science is one of the main research interests of him. He has many research papers on these subjects.

Nihan Özdemir Sönmez earned B.CP and M.CP. degrees at Middle East Technical University (Ankara), Department of City and Regional Planning, in 1987 and 1990 respectively. She completed her PhD (The Transformation of Squatter settlements into Authorised Apartment Blocks: A Case Study of Ankara, Turkey) at University of Kent at Canterbury, (UK) in 1998. Between 1989-2016 she worked at the Gazi University, Department of City and Regional Planning as a Research Assistant, Lecturer, Assistant Professor, and Associate Professor. Currently she teaches at Ankara University, Department of Real Estate Development and Management. Her research interests are the social and political implications of urban planning and real estate development and management decisions.

Seher Özkazanç was born in Ankara, Turkey, in 1981. She received her bachelor's degree from the department of City and Regional Planning at Gazi University, Turkey, in 2004. She took her Master and Ph.D. degrees from Gazi University, Graduate School of Natural and Applied Sciences, Department of City and Regional Planning in 2008 and 2016, respectively. In January 2006, she joined the Department of City and Regional Planning at University of Yüzüncü Yıl, as a Lecturer, later in June 2006 she was appointed at Gazi University to continue her research. She worked as a research assistant at Gazi University till 2016, and since June 2016, she has been working at Yüzüncü Yıl University, Department of City and Regional Planning. Her current research interests include transportation, urban sociology, disadvantaged groups and women's studies and urban policies.

Paul Pfaffenbichler has twenty years experience in the research fields interdisciplinary transport planning, land-use and transport interaction models, computer simulation, alternative propulsion technologies and system dynamics. Involvement in numerous national and international research projects dealing with sustainability and the development and assessment of mobility and land use plans.

Josef Michael Schopf studied mechanical engineering at Vienna University of Technology (VUT) with a specialization in transport technology and transport means. He became a docent in 1993 with his post-doctoral thesis "Velocity in road traffic". After becoming professor in 2003 he became head of the Research center of Transport Planning and Traffic Engineering at VUT's Institute of Transportation in 2008. His research interest are among others: urban parking policies, active transportation modes and mobility management of businesses.

Hediye Tuydes-Yaman received her MS and PhD degrees from Northwestern University, IL, USA based on her works on developing airport access mode choice models for the San Francisco Region and "Network Traffic Management under Disaster Conditions", respectively. Joining Middle East TEchnical University, (METU) Civil Engineering Department, as a full-time faculty member, she conducted research on both network traffic modeling and traffic safety. More recently, she has conducted projects and surveys assessing sustainability, accessibility, pedestrian level of service (PLOS) and walkability of METU Campus for more than 5 years. She and her research group have presented their findings on walking and Pedestrian Level of Service (PLOS) at the Transportation Research Board Annual Meetings for the last three years, consecutively.

Zafer Yilmaz graduated from Turkish Military Academy in 1996. He got his master degree from Modelling and Simulation Department of Middle East Technical University. He got his PhD. degree in 2015 from Supply and Logistics Department of Defence Sciences Institute of Turkish Military Academy. He currently is an officer in Turkish Army.

Index

H

I

L

M

O

P

Q

R

S

T

U

V

W

Y

Z

www.ingramcontent.com/pod-product-compliance
Lightning Source LLC
LaVergne TN
LVHW060408300926
849162LV00006B/1371